ELEMENTS OF MATHEMATICAL LOGIC AND SET THEORY

BY

J. SŁUPECKI and L. BORKOWSKI

Translated by

O. WOJTASIEWICZ

PERGAMON PRESS

OXFORD · LONDON · EDINBURGH · NEW YORK

TORONTO · SYDNEY · PARIS · BRAUNSCHWEIG

PWN—POLISH SCIENTIFIC PUBLISHERS

WARSZAWA

Pergamon Press Ltd., Headington Hill Hall, Oxford
4 & 5 Fitzroy Square, London W. 1
Pergamon Press (Scotland) Ltd., 2 & 3 Teviot Place, Edinburgh 1
Pergamon Press Inc., 44–01 21st Street, Long Island City, New York 11101
Pergamon of Canada, Ltd., 6 Adelaide Street East, Toronto, Ontario
Pergamon Press (Aust.) Pty. Ltd., 20–22 Margaret Street, Sydney, N.S.W.
Pergamon Press S.A.R.L., 24 rue des Écoles, Paris 5e
Vieweg & Sohn GmbH, Burgplatz 1, Braunschweig

First English edition 1967

This is a translation from the original Polish
Elementy logiki matematycznej i teorii mnogości,
published in 1963 by Polish Scientific Publishers,
Warszawa

Library of Congress Catalog Card No. 66–17815

Printed in Poland (DRP)
2167/67

ERRATA

Page and line	For	Read
ix before l. 19 from bottom		add: "$\phi_1, ..., \phi_{n-1} \vdash \phi_n$" means "the formula ϕ_n is derivable from the formulae $\phi_1, ..., \phi_{n-1}$" 80
xi_7	s	is
5_{12}	Section 5	Section 6
31_{14}	ϕ	ψ
38_9	T18	T28
57_1	next Section	Section 6
81_9	χ	ψ
101_2	term constants	term variables
131_3	$\iota(x \ x) \atop x$	$\iota(x \neq x) \atop x$
132_{11}	J1	D1
136_{15}	a_2	b_1
$136_{17,20}$	$a \| b$	$a_1 \| b_1$
145_1	$\varphi(\alpha_1, ..., \alpha_k)$	$\psi(\alpha_1, ..., \alpha_k)$
173_{11}	$X \nsubseteq Z$	$X \subsetneq Z$
173_{15}	$X \doteq \overline{Y}$	$X \doteq Y$
188_1	$D(R)$	$Ɑ(R)$
197_9	Chap. III, Sec. 3 1.2	Chap. 1. Sec. 3, 1.2
209_2	$\overline{X.Y} = \overline{Y.X}$	$\overline{\overline{X}}.\overline{\overline{Y}} = \overline{\overline{Y}}.\overline{\overline{X}}$
211_{11}	T6.1	T6.1a
213_1	173	173, 7
214_{17}	4	8, 4
215_{17}	$\aleph_0 < c$	$\aleph_0 \leqslant c$
229_{13}	T10.1	T10.3
230_7	T9.5b	T9.5a, b
237_4	$\mathfrak{M}_Z \ \emptyset$	$\mathfrak{M}_Z \neq \emptyset$
238_1	T5.1e	T5.1e. Ch. 1
258_5	T4.2b	T4.2a, b
260_{16}	$Z^0 = \bigcup_{Y \in X} Y$	$Z^0 = \bigcup_{Y \in X} Y^0$
268_9	denoted	is denoted
285_8	rules.	rules and T8.3.
289_6	$v \prec_{W_3} u$	$v \prec_{W_3} w$
304_4	$x =_1 x$	$x =_0 x$
308_{12}	$y =_1 z$	$y =_0 z$
329_8	$D_l(R) \cap D_r(R)$	$D_l(R) \cup D_r(R)$
329 after l. 6 from top		add: D14.1 $\alpha \in \mathfrak{M}_Z = \sum_{M \in Z} (\alpha = \{M\} \times M)$

J. Słupecki and L. Borkowski, *Elements of Mathematical Logic and Set Theory*

CONTENTS

[v]

INTRODUCTION

This textbook consists of two parts.

The material on logic in Part I is essentially confined to the simplest and most widely applicable logical calculi: the classical sentential calculus and the functional calculus which is based upon it. The calculus of relations is partially dealt with in Part II and in the Supplement. Methodological problems in the deductive sciences are not the subject matter of the book; hence they are given only an elementary treatment.

The logical calculi presented here are based on a system of rules which define the methods used in proofs from assumptions. The authors have found this approach valuable in their own didactic work. It is their hope that this textbook, which is the first to set forth this system so comprehensively, will extend its use in the teaching of mathematical logic.

Part II presents set theory in a fairly comprehensive but elementary formulation. The scope it covers lies midway between the two extremes found in Polish textbooks on this branch of mathematics. On the one hand, it is more comprehensive than Sierpiński's *Wstęp do teorii mnogości i topologii* (Introduction to Set Theory and Topology) and Kuratowski's textbook having the same title. On the other hand, it is much narrower than the treatment in Sierpiński's *Zarys teorii mnogości* (An Outline of Set Theory) and in *Teoria mnogości* (Set Theory) by Kuratowski and Mostowski.

The treatment of set theory presented in Part II is so formulated as to employ to a large extent the logical apparatus introduced in Part I, which should make for easier reading. Definitions and theorems are usually set out in terms of logical symbolism, and whenever possible proofs are carried out in a formalized (though often abbreviated) form by the methods discussed in Part I.

All definitions and those theorems which are more frequently referred to are listed at the end of the book to make it easier for the reader to check proofs.

For didactic reasons the material on set theory presented in Part II is neither formulated axiomatically nor based on the theory of types but these two approaches are discussed in considerable detail in the Supplement, where the connections and differences between them are pointed out.

This textbook is intended for students of philosophy. It appears that a knowledge of set theory, as given in this textbook, is necessary for them, in particular for those who specialize in logic. Moreover, there are very close links between logic and set theory, and the more comprehensive textbooks of mathematical logic usually include chapters on set theory.

The authors also hope that this textbook may prove useful to mathematical students seeking a fairly easy approach to the elements of mathematical logic and set theory.

After mastering the material contained in this book the student will be ready for specialized works and comprehensive textbooks. A comprehensive exposition of mathematical logic can be found in Mostowski's *Logika matematyczna* (Mathematical Logic) and of set theory in the books by Sierpiński, and Kuratowski and Mostowski mentioned above.

The authors wish to thank Dr Witold Pogorzelski for his many valuable suggestions while the book was being prepared. They also express thanks to Professor Klemens Szaniawski for certain additions which he suggested in his review of the book.

JERZY SŁUPECKI
LUDWIK BORKOWSKI

SYMBOLS

(*Mathematical symbols used in the sense known from the secondary school curriculum are not included. Numbers refer to the pages where a given symbol is explained.*)

¬	symbol of negation 1
∧	symbol of conjunction 1
∨	symbol of disjunction 1
→	symbol of implication 1
≡	symbol of equivalence 1
RD	rule of detachment 8
JC	rule of joining a conjunction 8
OC	rule of omitting a conjunction 9
JD	rule of joining a disjunction 9
OD	rule of omitting a disjunction 10, 21, 25
JE	rule of joining an equivalence 10
OE	rule of omitting an equivalence 10
a.	assumption(s) 18
a.i.p.	assumption of indirect proof 20
contr.	contradiction 20
⊢	"⊢ ϕ" means "ϕ is a thesis" 21
ON	rule of omitting double negation 25
JN	rule of joining double negation 25
Toll.	rule *modus tollendo tollens* 27
RD$_E$	rule of detachment for equivalence 29
ad.a.	additional assumption of a proof 31
ND	rule of negating a disjunction 34
//	symbol of replacement 43
1	truth of a sentence 47, the universal set 158
0	falsehood of a sentence 47
⊻	symbol of exclusive disjunction 50
/	symbol of alternative negation 57, symbol of substitution 96
↓	symbol of joint negation 57
$\overset{=}{df}$	symbol of definitional equality 58, 75
≺	symbol of strict implication 64
◇	functor of possibility 64
∏	the universal quantifier 88
∑	the existential quantifier 88

$\bigcap$	symbol for the product of a family of sets 183
$\langle x, y \rangle$	the ordered pair x, y 184
$X \times Y$	the Cartesian product of the sets X and Y 185
$\underset{a}{E}\Phi(a)$	the set of a's such that $\Phi(a)$ 186
funct	function 188
D_l	the left domain of a relation 188
D_r	the right domain of a relation 188
$R(x)$	the value of the function R for the argument x 188
$R(X)$	the R image of the set X 189
R^{-1}	the converse of the relation R 191
$1-1$	one-one (perfect) relations 191
R_X	the relation R limited to the set X 193
$R; S$	the relative product of the relations R and S 194
$\sim$	the relation of equinumerosity of sets 199
$\overline{\overline{X}}$	the power of the set X 202
Nc	the set of cardinal numbers 203
$\aleph_0$	the power of the set of natural numbers 204
$\mathfrak{c}$	the power of the set of real numbers 204
$+$	symbol of: sum of cardinal numbers 206, sum of order types 254
$\cdot$	symbol of: product of cardinal numbers 206 product of order types 254
X^Y	the mapping of the set Y onto the set X 206
$\mathfrak{m}^n$	a power of cardinal numbers 207
$<$	the relation "less than" between: cardinal numbers 212, segments of a well-ordered set 274, ordinal numbers 280
2^X	the set of all the subsets of the set X 216
$Z(R)$	the set of those values of the argument of the function R which are not elements of the function values assigned to them by R 216
φ_A	the characteristic function of the set A 217
$C(R)$	the field of the relation R 240
is_R	an isomorphism established by a relation R 240
s	the relation of isomorphism 241
con(A)	relations connected in a set A 246
trans (A)	relations transitive in a set A 246
asym(A)	relations asymmetric in a set A 246
ord(A)	relations ordering a set A 246
A^0	an ordered set with the range A (by analogy, B^0, C^0, etc.) 248
$x \prec_A y$	"x precedes y in a set A^0" 248

$\simeq_R$	similarity of ordered sets established by a relation R 248
$\simeq$	the relation of similarity of ordered sets 249
$\overline{A^0}$	the order type of an ordered set A^0 250
Ot	the set of order types 250
ω	the order type of the set of natural numbers 264
η	the order type of the set of rational numbers 266
λ	the order type of the set of real numbers 268
α^*	an order type converse to a given order type 270
On	the set of ordinal numbers 272
A^+	a well-ordered set with the range A (by analogy, A^+, B^+, etc.) 273
$A(a)$	the segment of a well-ordered set A^+ determined by the element a 274
if(A^+)	increasing functions defined in a set A^+ 275
Ln	the set of limit numbers 282
α^β	a power of ordinal numbers 286
$\overline{\alpha}$	the cardinal number of an ordinal number α 293
$Z(\mathfrak{m})$	the set of all ordinal numbers whose cardinal number equals $\mathfrak{m}$ 293
ω_ξ	the initial number such that the set of all smaller initial numbers is of type ξ 294
$\aleph_\xi$	the power of the set of all ordinals $< \omega_\xi$ 294
Ω	the initial number ω_1 294

Fundamental Logical Calculi

CHAPTER 1

THE SENTENTIAL CALCULUS

1. Symbols and Formulae

In the sentential calculus formulae are constructed from symbols of the following kinds: (1) sentential variables, (2) constants, (3) parentheses.

As sentential variables we shall use the letters

$$p, q, r, s, p_1, q_1, r_1, s_1, \ldots$$

In a concrete piece of reasoning using the formulae of the sentential calculus these variables may be replaced by any sentences. Note in this connection that the term "sentence" as used in logic denotes only expressions which are either *true* or *false*. Hence neither the interrogative nor the imperative sentences referred to in grammar are sentences in this sense of the term; it applies only to *declarative* sentences.

The constants of the sentential calculus are connectives with which we form compound sentences from component sentences. The most frequently used connectives of this kind are

not; and; or; if (... then); if and only if.

Instead of these words the following symbols will be used as the constants of the sentential calculus:

(1) $\neg, \wedge, \vee, \rightarrow, \equiv$.

The symbol $\neg$ is called the *symbol of negation*; the expression "$\neg p$" is read: not p (or: it is not true that p). The symbols $\wedge, \vee, \rightarrow, \equiv$ are called respectively the *symbol of conjunction (logical product)*, *disjunction (logical sum)*, *implication*, and *equivalence*.

The expression "$p \wedge q$" is read: p and q,

    ,,   ,,   "$p \vee q$" ,, ,, p or q,

    ,,   ,,   "$p \rightarrow q$" ,, ,,   if p then q,

    ,,   ,,   "$p \equiv q$" ,, ,, p if and only if q.

The compound sentences formed from a given sentence (or sentences) by means of the symbols (1) are called respectively the *negation, conjunction, disjunction, implication,* and *equivalence of the given sentence or sentences.*

The sentences of which a compound sentence is formed are called its *elements.* The elements of a conjunction are often called *factors,* those of a disjunction, *components.* The first element of an implication is called its *antecedent,* the second, its *consequent.*

The constants of the sentential calculus belong to the class of words which are useful and necessary in formulating theorems in all fields of science. In view of this and of the fact that any sentences may be substituted for sentential variables, the laws and rules of the sentential calculus are applicable in all disciplines, and in everyday reasoning as well.

The constants of the sentential calculus are included among the expressions called *functors.* Functors are expressions which, together with certain component expressions, called their *arguments,* yield expressions that are built correctly, and are usually referred to as well-formed formulae. The constants of the sentential calculus are included in the class of sentence-forming functors of sentential arguments, since together with the component sentences they form new sentences. The negation functor is a functor of one argument, since combined with one sentence it forms a compound sentence; the functors $\land$, $\lor$, $\to$, $\equiv$ are functors of two arguments, since they connect any two sentences to form a new compound sentence. Only sentence-forming functors of sentential arguments may appear in the sentential calculus. The functors specified above are the most important, but not the only ones, in that calculus.

Sentential formulae in the sentential calculus are the sentential variables and the compound formulae formed from them by means of functors of the sentential calculus. The concept of the sentential formula (in the sentential calculus) can be defined more precisely by the following definition consisting of three conditions[1]:

(1) Sentential variables are sentential formulae.

(2) If ϕ and ψ are sentential formulae, then $\neg(\phi)$, $(\phi) \land (\psi)$, $(\phi) \lor (\psi)$, $(\phi) \to (\psi)$, $(\phi) \equiv (\psi)$ are sentential formulae.

---

[1] See below, p 140.

(3) Every sentential formula in the sentential calculus either is a sentential variable or is formed from sentential variables by a single or repeated application of rule (2).

The Greek letters "ϕ", "ψ", "χ" are used here as variables for which *names* of any formulae of the sentential calculus may be substituted, while for the sentential variables "p", "q", etc., we may substitute, in the sentential calculus, any sentential formulae belonging to that calculus. The strings of symbols: $\neg \phi$, $\phi \wedge \psi$, etc., stand respectively for a compound formula consisting of the symbol of negation, followed by the formula ϕ, a formula consisting of formula ϕ, the symbol of conjunction and the formula ψ (in that order), etc. The variables "ϕ", "ψ", "χ", and the compound formulae, such as "$\neg\phi$", "$\phi \wedge \psi$", do not belong to the symbols and formulae of the sentential calculus; they belong to the system in which we discuss the sentential calculus and its formulae. Such a system is called a *metasystem of the sentential calculus*. The metasystem also includes the names of definite symbols and formulae belonging to the system. As can be seen, we form these names by placing the formulae being named in quotation marks. These quotation marks will often be dropped in cases when the name of a symbol of a formula is preceded by one of the following words: symbol, sign, functor, variable, formula.

Thus, for instance,

instead of: the symbol (sign, functor) "$\neg$", we may write: the symbol (sign, functor) $\neg$;

instead of: the variable "p", we may write: the variable p;

instead of: the formula "$p \rightarrow q$", we may write: the formula $p \rightarrow q$.

By condition (2), compound formulae are formed by means of functors belonging to the sentential calculus, the arguments being placed in parentheses. The use of parentheses ensures a unique interpretation of compound formulae. In order to avoid an excessive number of parentheses we assume that in compound formulae we may omit parentheses around variables.

Thus instead of: $\neg(p)$ we write: $\neg p$;

instead of: $(p) \wedge (q)$ we write $p \wedge q$, and so on.

We also assume that instead of the formulae $(\neg\phi) \wedge \psi$ and $\psi \wedge (\neg\phi)$ we may write, respectively, $\neg\phi \wedge \psi$ and $\psi \wedge \neg\phi$.

An analogous convention is adopted for the formulae containing the symbol $\neg$ and one of the symbols $\vee$, $\rightarrow$, $\equiv$.

It is further assumed that the symbol of conjunction binds more strongly than the symbol of disjunction, so that in the formulae $(\phi \wedge \psi) \vee \chi$ and $\chi \vee (\phi \wedge \psi)$ we may omit the parentheses. This is analogous to the convention used in arithmetic stating that the parentheses are superfluous in the formula $a+(b.c)$. It is also assumed that the symbol $\vee$ binds more strongly than the symbol $\rightarrow$, and the latter more strongly than the symbol $\equiv$.

Thus, instead of the formula $(\phi \vee \psi) \rightarrow \chi$ we write the formula $\phi \vee \psi \rightarrow \chi$;

instead of the formula $(\phi \rightarrow \psi) \equiv \chi$ we write the formula $\phi \rightarrow \psi \equiv \chi$;

instead of the formula $\phi \equiv (\psi \wedge \chi)$ we write the formula $\phi \equiv \psi \wedge \chi$; and so on.

It follows from the definition given above that in the sentential calculus every formula consists of a finite number of symbols. This is because the property is shared by the simple formulae of the sentential calculus, i.e. sentential variables, and it is also shared by the compound formulae formed by rule (2) from formulae sharing that property.

Exercises

1. Rewrite the following formulae without the superfluous parentheses:

$$[(p \vee q) \wedge r] \rightarrow (p \wedge r),$$
$$[(p \wedge q) \wedge (r \wedge s)] \equiv (p \equiv s),$$
$$\{[(p \equiv q) \rightarrow r] \wedge s\} \equiv (p \wedge \neg s).$$

2. The Polish logician Jan Łukasiewicz introduced a parenthesis-free notation (in literature on logic, it is called *Łukasiewicz's notation* or the *Polish notation*). In this notation we first write the functors of the sentential calculus and then their arguments in the order in which they appear, the letters: N, K, A, C, E being used respectively as the symbols of negation, conjunction, disjunction, implication, and equivalence. Thus, e.g., instead of "$p \wedge q$" we write "Kpq", instead of "$p \wedge q \rightarrow r$" we write "$CKpqr$", etc.

Write out the formulae given in Ex. 1. using Łukasiewicz's notation.

3. Write down in the notation used in this textbook the following formulae set out in Łukasiewicz's notation:

$$NCKpqArs,$$
$$ECpqCqp,$$
$$NNCpCqp.$$

4. Use the notation introduced in this textbook to write down the formulae

if p and q, then r;

if p then if q then r;

if p if and only if q, then if p then q.

5. Use the notation introduced in this textbook to write down the various interpretations of the formulae:

if p and q or r then s,

if p and q or r and s then q or s.

2. Primitive Rules

There are two methods of constructing the sentential calculus: the *axiomatic method* and the *method from assumptions*. The axiomatic method will be discussed under section 5. But the system developed in detail will be based on assumptions. We do this because such systems come closer to common intuition in that the proofs carried out in them differ hardly at all from mathematical proofs or from reasoning in other disciplines. The systems based on assumptions owe the name of *natural deduction,* by which they are sometimes called, to this property.

The first systems based on assumptions were developed in 1934 and 1935 by Jaśkowski and Gentzen. The system presented below differs from them in some details. It had been presented in the paper: L. Borkowski, J. Słupecki, *A Logical System Based on Rules and Its Application in Teaching Mathematical Logic*, Studia Logica 7 (1958).

To help the reader understand the concepts to be introduced below we shall first analyse two proofs by way of example.

EXAMPLE 1. In this example it is assumed that the variables range over the set of all integers.

The theorem to be proved is:

If a_1 and a_2 are both divisible by b, then $a_1 . c_1 + a_2 . c_2$ is divisible by b.

The proof is as follows:

The assumption is that:

(1) $\qquad\qquad a_1$ and a_2 are both divisible by b.

By the definition of divisibility [1] it follows from (1) that there exist integers m_1 and m_2 such that

(2)
$$a_1 = m_1.b,$$
$$a_2 = m_2.b.$$

It follows from (2) that

(3)
$$a_1.c_1 = m_1.c_1.b,$$
$$a_2.c_2 = m_2.c_2.b.$$

It follows from (3) that

(4)
$$a_1.c_1+a_2.c_2 = (m_1.c_1+m_2.c_2).b.$$

Hence if follows, by the definition of divisibility, that

$$a_1.c_1+a_2.c_2 \text{ is divisible by } b.$$

EXAMPLE 2. In this example, too, we assume that the variables range over the set of all integers.

The theorem to be proved is:

If n is prime and divides a.a then n divides a.

Proof. The assumption is that

(1) n is prime and divides $a.a$.

For an indirect proof we assume that

(2) n does not divide a.

Then from the law

(3) if n is prime and divides $a.b$ and n does not divide a, then n divides b

we have

(4)   if n divides $a.a$ and n does not divide a, then n divides a.

It follows from (1) and (2) that

(5) n is prime and divides $a.a$ and n does not divide a.

It follows from (4) and (5) that

(6) n divides a.

But the last statement contradicts (2).

---

[1] We use the definition: a is divisible by b if and only if there exists an integer m such that $a = m.b$.

This contradiction concludes the indirect proof of the theorem.

If we consider these proofs as typical examples we can formulate the rules that determine their structure. Both theorems in question are implications (conditional sentences). The proofs are sequences of sentences (formulae) which we call *proof lines*. But the two proofs differ as to their structure. In the first example, the antecedent of the theorem to be proved is adopted as the assumption of the theorem; from that assumption we draw conclusions by resorting to previously known theorems; the proof is considered complete when we obtain the consequent of the theorem to be proved. In the second example, the antecedent of the theorem to be proved is adopted as the assumption of the theorem; by adopting the negation of the consequent of that theorem as the assumption of the indirect proof we draw further conclusions from both these assumptions; the proof is considered complete when we obtain two contradictory statements.

The rules determining the structure of each such proof will be called *rules of proof construction*. In analysing the proofs quoted above we may, in addition to these rules, distinguish rules of the second kind, according to which we proceed in carrying out the proofs. For instance, when passing, in the second proof, from lines (4) and (5) to line (6) we follow the rule which permits us to deduce from an implication and its antecedent its consequent as the conclusion. Line (5) of the same proof we obtain from lines (1) and (2) on the strength of the rule which permits us to make a conjunction of two sentences. Rules of this kind will be called *rules of joining new lines to the proof*. Such rules lay down the conditions under which formulae of a definite type and resulting from formulae already appearing in the proof may be joined to the proof.

Both among the rules of proof construction and the rules of joining new lines to the proof we distinguish primitive and secondary rules. The primitive rules are adopted without proof, though we try to choose them so as to make them as intuitive as possible. The secondary rules are proved by means of primitive ones and of those theorems which can be proved by primitive rules. The secondary rules serve to make proofs shorter, but theoretically they are superfluous: every thesis proved by means of secondary rules can be proved exclusively by means of primitive rules.

In the sentential calculus we adopt the following primitive rules of joining new lines to the proof on the strength of those lines which already appear in the proof.

(1) *The rule of detachment*:

The rule states that if both an implication and its antecedent are contained in the proof, then its consequent may also be joined to the proof.

The rule of detachment, hereafter to be denoted by RD, will be written in the form of the following schema, consisting of two parts, upper and lower, separated by a horizontal line:

$$\text{RD} \qquad \frac{\begin{array}{c} \phi \rightarrow \psi \\ \phi \end{array}}{\psi.}$$

Written in this way, the rule states that the formula ψ may be joined to the proof as a new proof line, if the proof already includes the formulae $\phi \rightarrow \psi$ and ϕ (if the formulae $\phi \rightarrow \psi$ and ϕ already form some of the proof lines), regardless of the order in which these formulae appear in the proof.

The remaining rules for joining new lines to the proof will be written in the form of analogous schemata.

The rule of detachment is applied, for instance, in the following inference:

$$\frac{\begin{array}{c} \text{If the sum of the digits of the number 30612 is divisible by 3,} \\ \text{then the number 30612 is divisible by 3.} \\ \text{The sum of the digits of the number 30612 is divisible by 3.} \end{array}}{\text{The number 30612 is divisible by 3.}}$$

(2) *The rule of joining a conjunction*:

$$\text{JC} \qquad \frac{\begin{array}{c} \phi \\ \psi \end{array}}{\phi \wedge \psi.}$$

The rule of joining a conjunction states that a conjunction may be joined to the proof if both elements of the conjunction are already contained in the proof.

Here is an example of inference by the rule of joining a conjuction:

$$a < x$$
$$\frac{x < b}{a < x \land x < b}$$

(or in an abbreviated form: $a < x < b$).

(3) *The rule of omitting a conjunction:*

$$\text{OC} \qquad \frac{\phi \land \psi}{\phi,} \qquad \frac{\phi \land \psi}{\psi.}$$

The rule of omitting a conjunction states that if a conjunction is contained in the proof, any element of the conjunction may be joined to the proof.

The rule OC may also be written in the form of a single schema:

$$\frac{\phi \land \psi}{\phi}$$
$$\psi.$$

Generally, the rule of the form

$$\phi_1$$
$$\vdots$$
$$\frac{\phi_n}{\psi_1}$$
$$\vdots$$
$$\psi_k$$

states that if the formulae $\phi_1, ..., \phi_n$ are contained in the proof, any of the formulae $\psi_1, ..., \psi_k$ may be joined to the proof.

Here is an example of inference by the rule of omitting a conjunction:

$$\frac{a < x \land x < b}{a < x} \quad (\text{or: } a < x < b) \quad \frac{a < x \land x < b}{x < b.}$$

(4) *The rule of joining a disjunction:*

$$\text{JD} \qquad \frac{\phi}{\phi \lor \psi,} \qquad \frac{\psi}{\phi \lor \psi.}$$

The rule of joining a disjunction states that a disjunction may be joined to the proof if any of its elements is already contained in the proof.

Here is an example of inference by the rule of joining a disjunction:

$$\frac{a > 0}{a > 0 \lor a = 0} \quad (\text{or: } a \geqslant 0) \quad \frac{a = 0}{a > 0 \lor a = 0.}$$

(5) *The rule of omitting a disjunction*:

OD
$$\frac{\phi \lor \psi}{\psi.}$$
$$\neg \phi$$

The rule of omitting a disjunction states that if a disjunction and the negation of one of its elements are contained in the proof, the other element of the disjunction may be joined to the proof.

Here is an example of inference by the rule of omitting a disjunction:

$$a > 0 \lor a = 0$$
$$\frac{\neg a > 0}{a = 0.}$$

(6) *The rule of joining an equivalence*:

JE
$$\phi \to \psi$$
$$\frac{\psi \to \phi}{\phi \equiv \psi.}$$

The rule of joining an equivalence states that the equivalence $\phi \equiv \psi$ may be joined to a proof if that proof already includes the implication $\phi \to \psi$ and the inverse implication $\psi \to \phi$ (obtained by interchanging the antecedent and the consequent).

Here is an example of inference by the rule of joining an equivalence:

If the sides a and b of a triangle are equal, then the angles opposite the sides a and b are equal.

If the angles opposite the sides a and b are equal, then the sides a and b of a triangle are equal.

The sides a and b of a triangle are equal if and only if the angles opposite the sides a and b are equal.

(7) *The rule of omitting an equivalence*:

OE
$$\frac{\phi \equiv \psi}{\phi \to \psi,} \quad \frac{\phi \equiv \psi}{\psi \to \phi.}$$

The rule of omitting an equivalence states that if an equivalence belongs to a proof then we may join to the proof the implication having as antecedent the first element of the equivalence and as consequent the second element of the equivalence, and the implication which is the inverse of the former. The former implication will be called a simple implication.

Here are examples of inference by the rule of omitting an equivalence:

$$\frac{\neg x \geqslant y \equiv y > x}{\neg x \geqslant y \to y > x,} \qquad \frac{\neg x \geqslant y \equiv y > x}{y > x \to \neg x \geqslant y.}$$

To help the reader memorize these rules we draw attention to the fact that for implication we have only one primitive rule, namely the rule of detachment, and for each of the remaining functors we have two rules: of joining and of omitting; moreover, in the rules for joining the constant appears in the lower part of the schema but not in the upper, and in the rules for omitting it appears in the upper part of the schema but not in the lower.

The primitive rules of joining new proof lines to a proof, as specified above, are quite obvious in the light of the meaning of the functors $\neg$, $\wedge$, $\vee$, $\to$, $\equiv$, as explained in Paragraph 1. Formulae joined in accordance with these rules are consequences of the formulae on the strength of which they are joined. For instance, every element of a conjunction is a consequence of that conjunction; the consequent of an implication is a consequence of that implication and its antecedent, etc.

These rules have been known to logicians for a long time. For instance, the rule of detachment was formulated by the Stoics in the 3rd century B.C. as one of the primitive ("unprovable") rules of a system of sentential calculus which they built for the first time in history.[1] That rule was known as *modus ponens*, or more precisely, *modus ponendo ponens*. The Stoics also used the rule of omitting

[1] The historical information given in the present textbook has been drawn mainly from: I. M. Bocheński, *Formale Logik*, 1956; I. M. Bocheński, *Ancient Formal Logic*, 1951; T. Kotarbiński, *Wykłady z dziejów logiki* (Lectures on the History of Logic), 1957; J. Łukasiewicz, *Z historii logiki zdań* (From the History of the Logic of Propositions), 1934; E. A. Moody, *Truth and Consequence in Mediaeval Logic*, 1953.

a disjunction, but formulated for an exclusive disjunction,[1] for which the rule of joining a disjunction is not valid. The rule of omitting a disjunction was known as *modus tollendo ponens*. The rule of joining a disjunction appeared in the works of 13th and 14th century logicians (Robert Kildwarby, Albert of Saxony, John Buridan), who also formulated the rule of omitting a non-exclusive disjunction. For instance, Robert Kildwarby (13th century) formulated the rule of joining a disjunction as the rule stating that a disjunction is a consequence of each of its elements. He also gave the following example of inference by that rule: you are sitting, hence you are sitting or you are not sitting.

Albert of Saxony formulated the rule of omitting a conjunction ("any element of a conjunction is a consequence of that conjunction of which it is an element"). The joining of a conjunction was known to the Stoics, who stated that a conjunction is true if and only if both its elements are true.

Although certain laws for equivalence were formulated by Boethius, who lived at the turn of the 5th century, it seems that the rules for joining and omitting an equivalence, which lay down the connection between equivalence and implication, date from a much later time.

Among the rules for constructing a proof from assumptions we differentiate the rule for constructing a direct proof from assumptions and the rule for constructing an indirect proof from assumptions.

The way of constructing a direct proof from assumptions will be discussed first by means of the example of the proof of the law of the hypothetical (conditional) syllogism:

$$(p \rightarrow q) \rightarrow [(q \rightarrow r) \rightarrow (p \rightarrow r)].$$

The proof begins with writing down the assumptions of the theorem. The first assumption is the antecedent of the entire implication, i.e., the formula $p \rightarrow q$. The second assumption is the antecedent of the remaining part of the formula (obtained from the entire formula by cancelling out the first assumption and the symbol of implication by which it is followed), i.e. the formula $q \rightarrow r$. The third assumption is the antecedent of the remaining part of the formula (obtained

---

[1] See Sec. 4 below. The disjunction for which the rules JD and OD have been formulated, is also called *inclusive disjunction*.

by cancelling out the previous assumptions and the symbols of implication by which they are followed), i.e., the formula p. This concludes
the writing down of the assumptions, since the remaining part of the
formula is no longer an implication. After writing down the assumptions of the theorem we continue the proof by trying to obtain, as
the last proof line, the expression obtained by cancelling out the
assumptions and the symbols of implication by which they are followed, in our case, the formula r. During this procedure we may join
new lines to the proof on the strength of existing proof lines, following the primitive rules given above. To the right of each proof
line thus obtained we indicate on the strength of which rules and
proof lines it has been obtained. This yields the following proof of
the theorem:

$$(p \rightarrow q) \rightarrow [(q \rightarrow r) \rightarrow (p \rightarrow r)].$$

Proof. (1) $p \rightarrow q$ ⎫
　　　 (2) $q \rightarrow r$ ⎬　　　　　　　　　{assumptions}
　　　 (3) p ⎭
　　　 (4) q 　　　　　　　　　{RD: 1; 3}
　　　　 r 　　　　　　　　　{RD: 2; 4}

Generally, the direct proof from assumptions of the expression

(I) 　　　　　$\phi_1 \rightarrow \{\phi_2 \rightarrow [\phi_3 \rightarrow ... \rightarrow (\phi_{n-1} \rightarrow \phi_n) ...]\}$

is constructed in the following way.

1. In the first $n-1$ lines we write down, consecutively, the formulae
$\phi_1, \phi_2, ..., \phi_{n-1}$ as assumptions of the theorem.
2. To the proof we may join:
   (a) theorems proved previously, as new proof lines;
   (b) new proof lines on the strength of existing proof lines in accordance with the rules RD, JC, OC, JD, OD, JE, OE.
3. The proof is complete if in its last line the formula ϕ_n appears.
The last proof line is not numbered, to indicate that the proof
is complete.

If the theorem to be proved is not an implication then no assumptions appear in the proof. In this case the proof begins with writing
down one or more theorems previously proved, as authorized by
(2a) of the description of direct proof. Every proof line is then a *theorem*

and the proof is called an *ordinary direct proof*. An example is provided by the proof of T2b below. Thus it is not excluded that formula (I) may have the form ϕ_1, so that n equals 1.

The method of constructing an indirect proof from assumptions will first be discussed by means of the example of the proof of one of the laws of transposition:

$$(\neg p \to q) \to (\neg q \to p).$$

As in the case of a direct proof, we begin the indirect proof from assumptions by writing down the assumptions of the theorem — in the present case the expressions $\neg p \to q$ and $\neg q$. Next we write down the *negation* of the remaining part of the original formula (obtained by cancelling out the assumptions of the theorem and the corresponding implication symbols) as the assumption of the indirect proof. In the present case, the assumption of the indirect proof is the formula $\neg p$. We continue the proof by joining new proof lines on the strength of the rules adopted, so as to obtain two contradictory proof lines, that is proof lines of the form ψ and $\neg \psi$, respectively. In this way we obtain the following proof of the theorem:

$$(\neg p \to q) \to (\neg q \to p).$$

Proof. (1) $\neg p \to q$ ⎫
   (2) $\neg q$ ⎬ {assumptions}
   (3) $\neg p$ {assumption of indirect proof}
   (4) q {RD: 1; 3}
   a contradiction.   {2; 4}

Generally, the indirect proof from assumptions of formula (I) is built in the following way.

1. (a) In the first $n-1$ lines we write out, consecutively, the formulae $\phi_1, \phi_2, \ldots, \phi_{n-1}$ as the assumptions of the theorem;
 (b) in the nth line we write out the formula $\neg \phi_n$ as the assumption of the indirect proof.

2. To the proof we may join:
 (a) theorems previously proved, as new proof lines;
 (b) new proof lines on the strength of existing proof lines in accordance with the rules RD, JC, OC, JD, OD, JE, OE.

3. The proof is complete if two contradictory lines appear in it. The completion of the proof is indicated by writing "contradiction" (or "contr.") in the last (non-numbered) proof line and writing out on the right the numbers of the two contradictory proof lines.

The assumptions of the theorem and the assumption of the indirect proof will be called assumptions of the proof.

If the theorem to be proved is not an implication then the proof will begin with the assumption of the indirect proof, which in this case is the negation of the theorem to be proved. Such a proof is called an *ordinary indirect proof.* An example is provided by the proof of T22 below. As in the comment on the direct proof, we do not exclude that formula (I) may have the form ϕ_1.

It can easily be seen that every direct proof from assumptions of formula (I) can be transformed into an indirect proof from the assumptions of that formula. To do so it suffices to insert the formula $\neg \phi_n$ after the $(n-1)$th proof line. Then this inserted line and the last proof line of the direct proof of formula (I), i.e. the formula ϕ_n, give two contradictory proof lines, which complete the indirect proof from assumptions. Hence it follows that it suffices to adopt the rule of constructing an indirect proof from assumptions as the only primitive rule of building proofs, for the rule of building a direct proof from assumptions is derived from that rule. In general, we say that the rule R_1 of constructing proofs is derived from the rule R_2 and theorems T or rules R if and only if, whenever there exists a proof of a theorem constructed in accordance with rule R_1, there exists a proof of that theorem built in accordance with rule R_2, in which theorems T or rules R are used. In proving theorems and rules of the sentential calculus (and also in the case of the functional calculus, discussed in Chapter II) we shall not confine ourselves to primitive rules alone, but we shall also make use of certain derived rules which make it possible to abbreviate some proofs.

Proofs from assumptions have long been used in logic and in mathematics. Even in the works of Aristotle (384–322 B.C.), the founder of logic, we find examples of direct and indirect proofs from assumptions. We shall here analyse the following proofs, taken from Aristotle's works.

EXAMPLE 1[1]. "If M belongs to no N,[2] but to some X, then it is necessary that N should not belong to some X. For since the negative premiss is convertible, N will belong to no M; but M was admitted to belong to some X; therefore N will not belong to some X. The conclusion is reached by means of the first figure."

Here Aristotle proves a syllogistic mood of figure 2 (the mood called *Festino*):

If no N are M and some X are M, then some X are not N.

The proof is carried out by Aristotle in the following manner. He adopts as assumptions (premisses) both elements of the conjunction that forms the antecedent of that mood. From the first premiss, on the strength of the law of conversion of a universal affirmative sentence (i.e., the law stating that: If no N are M, then no M are N), he deduces the sentence: no M are N. And from that sentence and the second premiss he deduces, as a conclusion, the consequent of the theorem to be proved: some X are not N—on the strength of a mood in figure 1 (the mood called *Ferio*):

If no M are N and some X are M, then some X are not N.

We thus have here an example of a direct proof from assumptions. Likewise, other proofs of moods in figures 2 and 3 on the strength of the moods in figure 1, called proofs by conversion, are direct proofs from assumptions.

EXAMPLE 2[3]. "If M belongs to all N, but not to some X, it is necessary that N should not belong to some X; for if N belongs to all X, and M is predicated also of all N, M must belong to all X; but it was assumed that M does not belong to some X."

Here Aristotle proves a certain mood in figure 2 (the mood called *Baroco*):

---

[1] J. Łukasiewicz, *Aristotle's Syllogistic from the Standpoint of Modern Formal Logic*. Oxford 1951 (the second enlarged edition of this book appeared in 1957), p. 51.

[2] In formulating syllogistic moods Aristotle used the following expressions: "M belongs to no N" for "no N is M"; "M belongs to some X" for "some X is M"; "N does not belong to some X" for "some X is not N". Instead of "every N is M" Aristotle also says "M belongs to all N" or "M is predicated of all N".

[3] J. Łukasiewicz, *Aristotle's Syllogistic* ..., p. 54.

If all *N* are *M* and some *X* are not *M*, then some *X* are not *N*.

The proof is carried out as follows: both premisses of that mood are adopted as assumptions of the theorem, and the sentence contradictory to the consequent of that mood (i.e. the proposition "all *X* are *N*") is adopted as the assumption of indirect proof. From that sentence and from the first premiss Aristotle deduces the sentence "all *X* are *M*"—on the strength of a mood in figure 1 (the mood called *Barbara*):

If all *X* are *N* and all *N* are *M*, then all *X* are *M*.

But the sentence obtained in this way contradicts the second premiss. The proofs ends with stating the contradiction. It is thus a typical example of an indirect proof from assumptions.

Likewise, other Aristotelian proofs of moods in figures 2 and 3 by means of moods in figure 1 by reduction to the impossible (*reductio ad absurdum*) are indirect proofs from assumptions. Aristotle adopts as assumptions the premisses of the mood to be proved and the sentence which is in contradiction to the consequent of that mood. The proofs end with the deduction from these assumptions of a proposition that contradicts one of the premisses.

In the works of the Stoics we also find proofs of rules of the sentential calculus which are examples of proofs from assumptions.

The indirect proof from assumptions of the theorem

If n is prime and divides a . a then n divides a,

given above (p. 6), is found in the works of Euclid, a Greek mathematician who lived after Aristotle.

Although proofs from assumptions have long been employed in practice both in mathematics and in logic, the rules for these proofs have only recently been formulated in contemporary logic. As far as can be ascertained, the rule for constructing indirect proofs from assumptions, adopted in this book as a primitive rule, has not been formulated in a previously constructed system based on assumptions.

Among the theorems of the sentential calculus which, together with theorems of other formal systems, we call theses, we distinguish theses of order 1, of order 2, etc.; in general, theses of order *n*, where *n* is a natural number $\geqslant 1$. The definition of these concepts is as follows:[1]

---

[1] This is an inductive definition. Cf. Chap. 2, Sec. 5.

ϕ is a *thesis of order* 1 if and only if there is a (direct or indirect) proof from assumptions of the formula ϕ, in which we resort only to the rules specified under 2b in the description of the rules for constructing a (direct or indirect) proof from assumptions, in joining new proof lines to the proof.

ϕ is a *thesis of order n* if and only if there exists a proof from the assumptions of the formula ϕ in which—in accordance with the condition 2a of the description of such a proof—we use theses of order at most $n-1$ and if ϕ is not a thesis of order less than n.

ϕ is a *thesis* if and only if there is a natural number n such that ϕ is a thesis of order n.

3. Theses and Derived Rules

On the strength of the rule OC we shall use, as a derived rule, the rule for constructing a (direct or indirect) proof from assumptions, which authorizes us, when we write out the assumptions of the theorem, to break up every formula which is a conjunction into its elements. For if an assumption of a theorem, written out in accordance with the primitive rule, is a conjunction of certain formulae, then on applying the rule OC we obtain the various elements of that conjunction in the proof.

It is in this way that we prove, for instance, the second law of hypothetical syllogism (the first, T1a: $(p \rightarrow q) \rightarrow [(q \rightarrow r) \rightarrow (p \rightarrow r)]$, was given above):

T1. $(p \rightarrow q) \wedge (q \rightarrow r) \rightarrow (p \rightarrow r)$.

Proof. (1) $p \rightarrow q$ ⎫
     (2) $q \rightarrow r$ ⎬   {a.}[1]
     (3) p ⎭
     (4) q {RD: 1; 3}
         r {RD: 2; 4}

In antiquity, a distinction was made between hypothetical (conditional) syllogisms, formulated for conditional sentences, and Aristotle's categorical syllogisms, formulated for categorical sentences. Hence the name of Thesis T1 — the *law of the hypothetical syllogism.*

T2. $(p \wedge q \rightarrow r) \rightarrow [p \rightarrow (q \rightarrow r)]$ the *law of exportation.*

---

[1] "a." stands for "assumptions".

Proof. (1) $p \wedge q \rightarrow r$

     (2) p {a.}

     (3) q

     (4) $p \wedge q$ {JC: 2; 3}

          r {RD: 1; 4}

T2a. $[p \rightarrow (q \rightarrow r)] \rightarrow (p \wedge q \rightarrow r)$ *the law of importation.*

Proof. (1) $p \rightarrow (q \rightarrow r)$

     (2) p {a.}

     (3) q

     (4) $q \rightarrow r$ {RD: 1; 2}

          r {RD: 4; 3}

The theses given so far have been of order 1, since in their proofs no recourse was made to earlier theses. The following one is of order 2.

T2b. $p \wedge q \rightarrow r \equiv p \rightarrow (q \rightarrow r)$.

Proof. (1) $(p \wedge q \rightarrow r) \rightarrow [p \rightarrow (q \rightarrow r)]$ {T2}

     (2) $[p \rightarrow (q \rightarrow r)] \rightarrow (p \wedge q \rightarrow r)$ {T2a}

         $p \wedge q \rightarrow r \equiv p \rightarrow (q \rightarrow r)$ {JE: 1; 2}

The proof of Thesis T2b is an ordinary direct proof. Each line of such a proof is a thesis. But in proofs from assumptions the proof lines are in most cases not theses.

Proofs of nearly all theses that have the form of equivalence are carried out as in the case of T2b: we first prove the corresponding simple and converse implications and next apply the rule of joining the equivalence. Further proofs of theses that are equivalences will be written down so that the proof consists of two parts, (a) and (b), the former being the proof of simple implication, and the latter being the proof of the converse implication.

T3. $p \rightarrow (q \rightarrow r) \equiv q \rightarrow (p \rightarrow r)$ *the law of commutation.*

This law makes it possible to interchange the antecedent of an implication with the antecedent of its consequent. The proof of this law is left to the reader.

The proof of the following thesis is an indirect proof from assumptions.

T4. $p \vee q \rightarrow (\neg q \rightarrow p)$.

Proof. (1) $p \lor q$ ⎱   {a.}
       (2) $\neg q$ ⎰
       (3) $\neg p$ {a.i.p.}[1]
       (4) q {OD: 1; 3}
           contr.[2]   {2; 4}

It can easily be observed that in the same way as T4 we can prove any formula of the form $\phi \lor \psi \to (\neg \psi \to \phi)$, where ϕ and ψ are any formulae of the sentential calculus. The proof of such a formula is so constructed that as assumptions (1) and (2) we adopt the formulae $\phi \lor \psi$ and $\neg \psi$, and as the assumption of the indirect proof, the formula $\neg \phi$. By applying the rule OD to (1) and (3) we obtain in line (4) the formula ψ, which contradicts the formula $\neg \psi$ that appears in line (2). Hence the schema of the proof is as follows:

      (1) $\phi \lor \psi$ ⎱   {a.}
      (2) $\neg \psi$ ⎰
      (3) $\neg \phi$ {a.i.p.}
      (4) ψ {OD: 1; 3}
         contr.   {2; 4}

For instance, the formula

(a) $(p \equiv q) \lor p \land r \to [\neg(p \land r) \to (p \equiv q)]$

is of the form $\phi \lor \psi \to (\neg \psi \to \phi)$ (where $\phi = $ "$p \equiv q$" and $\psi = $ "$p \land r$"). The following proof of (a) falls under the schema given above:

      (1) $(p \equiv q) \lor p \land r$ ⎱   {a.}
      (2) $\neg(p \land r)$ ⎰
      (3) $\neg(p \equiv q)$ {a.i.p.}
      (4) $p \land r$ {OD: 1; 3}
         contr.   {2; 4}

Thus, by means of the primitive rules of the system we can prove the following theorem:

M4. *Any formula of the sentential calculus which has the form*

$$\phi \lor \psi \to (\neg \psi \to \phi)$$

     is a thesis.

---

[1] "a.i.p." stands for "assumption of indirect proof".
[2] "contr." stands for "contradiction".

By introducing the notation "$\vdash \phi$" as an abbreviation of the statement "the formula ϕ is a thesis", we may symbolize M4 as follows:

M4. $\qquad\qquad \vdash \phi \vee \psi \to (\neg \psi \to \phi).$

The statement M4 is not a thesis of the sentential calculus, but a thesis of the metasystem of that calculus. The statement M4 will be called the *metathesis corresponding to thesis* T4. In general, a metathesis corresponding to a given thesis T (of the sentential calculus) is the theorem M of the metasystem, stating that any formula whose schema we obtain by replacing the sentential variables in T by the variables of the metasystem ("ϕ", "ψ", etc.) is a thesis of the sentential calculus (the same sentential variable being replaced everywhere by the same variable, and different sentential variables being replaced by different variables).

We shall now prove the following theorem:

Th. 1. *For every thesis of the sentential calculus its corresponding metathesis can be proved.*

The proof is by induction on the order of the thesis.

(1) If T is a thesis of order 1, and if M is the metathesis corresponding to T, then the proof schema of every formula of the sentential calculus which is a thesis by M is obtained by replacing the sentential variables in the proof of thesis T by the variables "ϕ", "ψ", etc., the same sentential variable being replaced everywhere by the same variable and different sentential variables being replaced by different variables, in the same way in which the proof of T4 was transformed into the proof of M4.

(2) It is assumed that Th. 1 is true for all theses of order lower than n. Let T be a thesis of order n, and M its corresponding metathesis. The proof schema of every formula of the sentential calculus, which is a thesis by M, is obtained in the same way as in the case (1). The induction hypothesis guarantees that metatheses corresponding to theses of order less than n can be proved.

The derived rule for omitting the disjunction having the form

$$OD \qquad \frac{\begin{array}{c} \phi \vee \psi \\ \neg \psi \end{array}}{\phi}$$

can be based on Metathesis M4. For if the formulae $\phi \lor \psi$ and $\neg \psi$ belong to the proof, then by joining to the proof the thesis $\phi \lor \psi \to$ $\to (\neg \psi \to \phi)$ and by applying twice the rule of detachment we obtain the formula ϕ. This rule authorizes us to join to the proof the first element of a disjunction on the strength of that disjunction and the negation of its second element.

Instead of speaking of a derived rule based on the metathesis corresponding to a given thesis, we shall speak of a rule *derived* from that thesis. (In general, we say that the rule R_1 of joining new lines to the proof is *derived* from the rules R or theses T if and only if, whenever the rule R_1 permits a formula to be joined to the proof, it may be joined on the strength of rules R or theses T.)

A certain derived rule can be based on each thesis that has the form of an implication. We shall now prove this assertion.

Let T be a thesis of the sentential calculus in the form of an implication, that is the form

(I) $$\phi_1 \to \{\phi_2 \to \ ... \ \to (\phi_{n-1} \to \phi_n) \ ...\},$$

where $n \geqslant 2$. The rule whose schema is

$$\begin{array}{c} \phi_1 \\ \phi_2 \\ \vdots \\ \phi_{n-1} \\ \hline \phi_n \end{array}$$

is then a derived rule. For if in a proof each of the formulae ϕ_1, $\phi_2, ..., \phi_{n-1}$ appear, then on the strength of the metathesis corresponding to thesis T, by $n-1$-fold application of the rule of detachment we obtain the formula ϕ_n in the proof.

It can be seen that form (I) is not uniquely determined for every thesis of the sentential calculus. For instance, thesis T4 may be treated as a thesis of the form $\phi_1 \to (\phi_2 \to \phi_3)$, by putting $\phi_1 = $ "$p \lor q$", $\phi_2 = $ "$\neg q$", $\phi_3 = $ "p". But the same thesis may also be treated as one of the form $\phi_1 \to \phi_2$, if we put $\phi_1 = $ "$p \lor q$", $\phi_2 = $ "$\neg q \to p$". Consequently, rules with different schemata may be based on the same thesis. For instance, apart from the rule OD, the derived rule which has the schema

$$\frac{\phi \lor \psi}{\lnot \psi \to \phi}$$

is also based on Thesis T4.

Analogously, it can be demonstrated that the rule with the schema

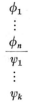

$$\phi_1$$
$$\vdots$$
$$\frac{\phi_n}{\psi_1}$$
$$\vdots$$
$$\psi_k$$

is derived from a thesis of the form

$$\phi_1 \land \phi_2 \land \ \cdots \ \land \phi_n \to \psi_1 \land \psi_2 \land \ \cdots \ \land \psi_k$$

and the rules JC and OC.

One of the derived rules based on a certain thesis will be denoted by the same name by which the thesis is denoted. For instance, the rule whose schema is

$$\phi \to \psi$$
$$\frac{\psi \to \chi}{\phi \to \chi}$$

is called the rule of hypothetical syllogism, derived from T1.

A substitution of the formula ϕ, belonging to the sentential calculus, is a formula obtained from ϕ by replacing a variable, for instance the variable p, which occurs in it, by an arbitrary formula, provided that if the variable p occurs more than once in ϕ, it is replaced by the same formula everywhere it occurs in ϕ. If more than one variable occurs in the formula ϕ then more than one of them may be simultaneously replaced by arbitrary formulae, provided that the same variables are replaced by the same formulae.

Thus, e.g., formula (a) on p. 20 above, is a substitution of thesis T4, obtained by the substitution of $p \equiv q$ for the variable p, and $p \land r$ for the variable q.

Formula (a), like any other substitution of Thesis T4, is a formula of the form $\phi \lor \psi \to (\lnot \psi \to \phi)$, and hence, by M4, it is a thesis.

In general, *every substitution of a thesis of the sentential calculus is a formula of the form of the metathesis corresponding to the thesis, and on the strength of that metathesis it is also a thesis.*

Thus the rule of substitution, stating that the substitution of any thesis is a thesis, is derived.

The proof of the thesis

T4a. $\qquad\qquad p \lor q \to (\neg p \to q)$

is analogous to the proof of T4.

By means of the derived rule OD we prove

T4b. $\qquad\qquad \neg p \lor q \to (p \to q).$

Proof. (1) $\neg p \lor q$ $\}$ $\qquad\qquad$ {a.}
(2) p
(3) $\neg q$ $\qquad\qquad\qquad\qquad$ {a.i.p.}
(4) $\neg p$ $\qquad\qquad\qquad\qquad$ {OD: 1; 3}
$\qquad$ contr. $\qquad\qquad\qquad\qquad$ {2; 4}

Theses T4, T4a and T4b may be called the *laws of relationship between disjunction and implication*. It is worth while noting that intuitive examples of inference by rules derived from these theses can be given in this connection. For instance,

I'll go to London by train today or I'll go to London by car today.
If I do not go to London by train today then I'll go to London
by car today.

In the example below the letters a and b stand for sides of a triangle and the letters α and β, for the angles opposite those sides.

$$\frac{a \neq b \lor \alpha = \beta}{a = b \to \alpha = \beta,}$$

It is not daylight now or it is not night now.
If it is daylight now then it is not night now.

The intuitive character of these rules is emphasized here, since one schema, namely the rule derived from T4b, will be used in Sec. 4 to construct the truth table for implication.

T5. $\qquad\qquad\qquad \neg\neg p \to p.$

Proof. (1) $\neg\neg p$ $\qquad\qquad\qquad\qquad$ {a.}
(2) $\neg p$ $\qquad\qquad\qquad\qquad$ {a.i.p.}
$\qquad$ contr. $\qquad\qquad\qquad\qquad$ {1; 2}

The *rule of omitting double negation*, derived from T5, will be denoted by ON.

T5a. $p \to \neg \neg p.$

Proof. (1) p {a.}
     (2) $\neg\neg\neg p$ {a.i.p.}
     (3) $\neg p$ {ON: 2}
       contr.   {1; 3}

The *rule of joining double negation*, derived from T5a, will be denoted by JN.

T5b. $\neg\neg p \equiv p.$ {JE: T5; T5a}

Theses T5, T5a and T5b are called the *laws of double negation*. The law T5b, stating that the double negation of a sentence is equivalent to that sentence, was known even to the Stoics.

Derived from rule JN is the rule of omitting the disjunction in the form:

$$\neg \phi \lor \psi \qquad\qquad \phi \lor \neg \psi$$

OD $\dfrac{\phi}{\psi}$ $\dfrac{\psi}{\phi}$

Proof. (1) $\neg \phi \lor \psi$ ⎱ {a.}   *Proof.* (1) $\phi \lor \neg \psi$ ⎱ {a.}
     (2) ϕ ⎰   (2) ψ ⎰
     (3) $\neg\neg \phi$ {JN: 2}   (3) $\neg\neg \psi$ {JN: 2}
       ψ {OD: 1; 3} ϕ {OD: 1; 3}

The intuitive meaning of this rule is the same as the intuitive meaning of the rule OD in its previous form, for these rules state that an element of a disjunction is a consequence of that disjunction and a formula contradictory to the other element.

On the strength of T5, or the rule ON, it is possible to deduce, as a derived rule, the rule for constructing an indirect proof from assumptions, a rule which differs from that given before only in that if ϕ_n in formula (I) has the form $\neg \psi$ it permits us to adopt the formula ψ as the assumption of indirect proof. For when we adopt the formula $\neg\neg \psi$, as an assumption of indirect proof and apply to it rule ON, we obtain a proof in accordance with the primitive rule. This derived rule will be employed in the proof of T6.

T6. $p \to q \equiv \neg q \to \neg p$ *the law of transposition.*

Proof. (a) (1) $p \to q$ ⎫
            (2) $\neg q$ ⎭   {a.}
            (3) p {a.i.p.}
            (4) q {RD: 1; 3}
               contr.   {2; 4}

Part (b) of the proof is left to the reader.

The transposition of an implication is the implication obtained from the original one by interchanging the antecedent with the consequent while simultaneously negating both these elements. The *rule of transposition* permits us to join to a proof the transposition of an implication which belongs to the proof.

Here is an example of inference based on the rule of transposition:

$$\frac{a = b \to \alpha = \beta}{\alpha \neq \beta \to a \neq b}$$

(here a and b stand for sides of a triangle, and α and β for the angles opposite to those sides).

The rule of transposition (in the form of an implication) is one of the few laws of the propositional calculus which were known even to Aristotle.

T7. $p \wedge q \to r \equiv p \wedge \neg r \to \neg q$
                      the law of compound transposition.

Proof. (a) (1) $p \wedge q \to r$ ⎫
            (2) p ⎬   {a.}
            (3) $\neg r$ ⎭
            (4) q {a.i.p.}
            (5) $p \wedge q$ {JC: 2; 4}
            (6) r {RD: 1; 5}
               contr.   {3; 6}

The proof of the converse implication is left to the reader.

Here is an example of inference by the rule of compound transposition:

If the sequence a_n is monotonic and the sequence a_n is bounded, then the sequence a_n is convergent.

If the sequence a_n is monotonic and the sequence a_n is not convergent, then the sequence a_n is not bounded.

Among the rules proved by the Stoics by means of their primitive rules we find the following one which is derived from the law of compound transposition:

$$\frac{\phi \wedge \psi \to \chi}{\neg \chi \wedge \phi \to \neg \psi.}$$

T8. $(p \to q) \wedge \neg q \to \neg p$ *modus tollendo tollens* (the mood which refutes by refuting).

The proof of this law is left to the reader.

From this we have the derived rule:

Toll.[1]
$$\frac{\begin{array}{c}\phi \to \psi \\ \neg \psi\end{array}}{\neg \phi.}$$

This rule, of which we shall often make use, permits us to obtain the negation of the antecedent of any implication on the strength of the implication and the negation of its consequent. For instance,

$$\frac{\begin{array}{c}\text{If the number 306121 is divisible by 3, then the sum of the} \\ \text{digits of the number 306121 is divisible by 3.} \\ \text{The sum of the digits of the number 306121 is not divisible by 3.}\end{array}}{\text{The number 306121 is not divisible by 3.}}$$

The rules with the schemata:

Toll.
$$\frac{\begin{array}{c}\phi \to \neg \psi \\ \psi\end{array}}{\neg \phi,} \qquad \frac{\begin{array}{c}\neg \phi \to \psi \\ \neg \psi\end{array}}{\phi,} \qquad \frac{\begin{array}{c}\neg \phi \to \neg \psi \\ \psi\end{array}}{\phi}$$

can be derived from the rule formulated above and the rules ON or JN.

To put this in a generalized form, *the rule* Toll. *permits us to obtain, on the strength of an implication and a formula contradicting its consequent, a formula contradicting its antecedent.*

We shall now prove, by way of example, the last form of this rule. In the proof, the rule Toll. will be used in its first formulation.

Proof. (1) $\neg \phi \to \neg \psi$ }
      (2) ψ }   {a.}
      (3) $\neg \neg \psi$ {DN: 2}
      (4) $\neg \neg \phi$ {Toll:1;3}
          ϕ {ON: 4}

----

[1] "Toll." stands for "tollens".

The rule *modus tollendo tollens* (also called briefly *modus tollens*), in its first formulation, was one of the primitive ("unprovable") rules of the sentential calculus adopted by the Stoics. All four formulations of the rule are found in the works of Boethius.

T9. $$p \to (\neg p \to q).$$

Proof. (1) p ⎱
　　　　(2) $\neg p$ ⎰ $\qquad\qquad\qquad$ {a.}
　　　　(3) $\neg q$ $\qquad\qquad\qquad\qquad$ {a.i.p.}
　　　　　contr. $\qquad\qquad\qquad\qquad$ {1; 2}

This law shows why contradiction is dangerous: on the strength of the law two contradictory sentences yield any sentence whatever. It is worth while pointing out that an arbitrary sentence ψ can be derived from two contradictory sentences, ϕ and $\neg \phi$, if we resort merely to the rules JD and OD:

　　　　(1) ϕ ⎱
　　　　(2) $\neg \phi$ ⎰ $\qquad\qquad\qquad$ {a.}
　　　　(3) $\phi \lor \psi$ $\qquad\qquad\qquad$ {JD: 1}
　　　　　ψ $\qquad\qquad\qquad\qquad$ {OD. 3; 2}

Thesis T9 has been called by Łukasiewicz the *law of Duns Scotus*, after one of the greatest mediaeval philosophers who flourished at the turn of the 13th century, because Thesis T9 is derivable, by the law of exportation (T2), from the thesis: $p \land \neg p \to q$, which corresponds to the theorem formulated by Duns Scotus stating that a contradictory sentence has any sentence as its consequence. Duns Scotus proved this theorem by means of the following example, resorting to the explicitly formulated rules OC, JD, OD:

Th. *Socrates is running and Socrates is not running, hence you are in Rome.*

Proof. (1) Socrates is running and Socrates is not running {a.}
　　　　(2) Socrates is running ⎱
　　　　(3) Socrates is not running ⎰ $\qquad\qquad$ {OC: 1}
　　　　(4) Socrates is running or you are in Rome {JD: 2}
　　　　　You are in Rome. $\qquad\qquad\qquad$ {OD: 4; 3}

T10. $\qquad q \to (p \to q)$ $\qquad$ *the law of simplification.*

Proof. (1) q ⎫
      (2) p ⎭ $\{a.\}$

      (3) $\neg q$ $\{a.i.p.\}$

          contr. $\{1; 3\}$

T11. $p \rightarrow p$ *the law of identity for implication.*

Proof. (1) p $\{a.\}$

      (2) $\neg p$ $\{a.i.p.\}$

          contr. $\{1; 2\}$

T11a. $p \equiv p$ *the law of identity for equivalence.*

Proof. (1) $p \rightarrow p$ $\{T11\}$

      (2) $p \equiv p$ $\{JE: 1; 1\}$

The laws of identity are among the logical theses which are sometimes used as premisses in mathematical proofs (as will be seen in Part II).

T12. $(p \equiv q) \rightarrow (q \equiv p)$.

Proof. (1) $p \equiv q$ $\{a.\}$

      (2) $p \rightarrow q$ $\{OE: 1\}$

      (3) $q \rightarrow p$ $\{OE: 1\}$

          $q \equiv p$ $\{JE: 3; 2\}$

T13. $(p \equiv q) \wedge (q \equiv r) \rightarrow (p \equiv r)$.

The proof of T13 is left to the reader.

Theses T11a, T12 and T13 state that *equivalence is reflexive, symmetric and transitive.*[1]

On the other hand, when it comes to implication only the laws of reflexivity (T11) and transitivity (T1) are valid, and the law of symmetry is not, since implication is not always reversible.

The connection between implication and equivalence is laid down by the following thesis:

T14. $(p \equiv q) \equiv (p \rightarrow q) \wedge (q \rightarrow p)$.

The easy proof of this thesis is left to the reader.

The following rule of detachment is valid for equivalence:

$$\text{RD}_\text{E} \qquad \frac{\begin{array}{c} \phi \equiv \psi \\ \phi \end{array}}{\psi,} \qquad \frac{\begin{array}{c} \phi \equiv \psi \\ \psi \end{array}}{\phi.}$$

[1] See the footnote on p. 125.

These rules are derived from the rules OE and RD:

$$(1) \quad \phi \equiv \psi \left.\right\} \quad \{a.\}$$
$$(2) \quad \phi$$
$$(3) \quad \phi \to \psi \quad \{OE: 1\}$$
$$\qquad \psi \qquad \{RD: 3; 2\}$$

$$(1) \quad \phi \equiv \psi \left.\right\} \quad \{a.\}$$
$$(2) \quad \psi$$
$$(3) \quad \psi \to \phi \quad \{OE: 1\}$$
$$\qquad \phi \qquad \{RD: 3; 2\}$$

T15. $(p \to \neg p) \to \neg p.$

T15a. $(\neg p \to p) \to p.$

T15b. $(p \to q \wedge \neg q) \to \neg p.$

T15c. $(\neg p \to q \wedge \neg q) \to p.$

T15d. $(p \to q) \wedge (p \to \neg q) \to \neg p.$

The proofs of these theses, sometimes called the laws of *reductio ad absurdum*, are left to the reader.

The rule which is derived from T15d and has the schema

$$\frac{\begin{array}{c} \phi \to \psi \\ \phi \to \neg \psi \end{array}}{\neg \phi}$$

was formulated by the Stoics, who were also the authors of the following example of inference by this rule:

> If you know that you are dead then you are dead.
> If you know that you are dead then you are not dead.
> _____
> You do not know that you are dead.

T16. $(p \to q) \wedge (p \to r) \to (p \to q \wedge r)$ *the law of multiplication of consequents.*

The proof of the law is left to the reader.

In the proof of the theorem converse to T16 we shall resort to the derived rule of joining an implication to the proof. By this rule, in a direct or indirect proof from assumptions of formula (I) it is permissible to introduce an arbitrary formula ψ as an additional assumption of the proof. If this additional assumption and the assumptions of the proof yield the formula χ, then the implication $\psi \to \chi$ may be joined to the proof.

In fact, if on the strength of the assumptions of the proof $\phi_1, \ldots, \phi_{n-1}, (\neg \phi_n)$ and the additional assumption ψ we obtain the formula χ, we thereby prove the thesis: $\phi_1 \to \{\phi_2 \to \ldots \to [\phi_{n-1} \to (\psi \to \chi)] \ldots\}$ (or the thesis: $\phi_1 \to \{\phi_2 \to \ldots \to [\phi_{n-1} \to (\neg \phi_n \to (\psi \to \chi))] \ldots\}$).

On joining this thesis to the proof of formula (I) and on detaching the assumptions $\phi_1, \ldots, \phi_{n-1}$ ($\sqcap \phi_n$), we obtain the formula $\psi \to \chi$. Of course, the additional assumption ψ and the formulae obtained on the strength of this formula, except for the conclusion $\psi \to \chi$, may not be used later in the proof. It is also obvious that more than one additional assumption may appear in the proof.

Additional assumptions of the proof will be preceded by double numbers, e.g., the first additional assumption will have the number "1.1", the second, "2.1", etc. Proof lines obtained on the strength of lines with double numeration will also have double numbers, the first digit indicating from which assumption the line has been obtained, the second digits being consecutive. The lines obtained by the rule of joining an implication to the proof are preceded by single numbers. If on the strength of an additional assumption ϕ preceded, e.g., by the number 1.1, we obtain in the proof the formula χ preceded, e.g., by the number 1.3, and on the strength of the rule of joining an implication we join to the proof the line $\psi \to \chi$, then in the parentheses on the right of that line we write: $1.1 \to 1.3$ to indicate that the rule of joining an implication has been used in the proof.

T16a. $\qquad (p \to q \wedge r) \to (p \to q) \wedge (p \to r).$

Proof.
(1)	$p \to q \wedge r$	{a.}
(1.1)	p	{ad. a.}[1]
(1.2)	$q \wedge r$	{RD: 1.1}
(1.3)	q	{OC: 1.2}
(1.4)	r	{OC: 1.2}
(2)	$p \to q$	{1.1 → 1.3}
(3)	$p \to r$	{1.1 → 1.4}
	$(p \to q) \wedge (p \to r)$	{JC: 2; 3}

To make it easier for the reader to understand the derived rule introduced above we give the proof of T16a in a formulation which does not resort to the rule. We must then first prove the following two auxiliary theses:

(α) $\qquad\qquad (p \to q \wedge r) \to (p \to q),$

(β) $\qquad\qquad (p \to q \wedge r) \to (p \to r).$

---

[1] "ad. a." stands for "additional assumption of the proof".

The easy proofs of these theses are left to the reader.
The proof of T16a is now as follows:

(1)	$p \to q \wedge r$	{a.}
(2)	$(p \to q \wedge r) \to (p \to q)$	{α}
(3)	$(p \to q \wedge r) \to (p \to r)$	{β}
(4)	$p \to q$	{RD: 2; 1}
(5)	$p \to r$	{RD: 3; 1}
	$(p \to q) \wedge (p \to r)$	{JC: 4; 5}

The rule of joining an implication to the proof makes it unnecessary to introduce superfluous auxiliary theses.

T16b. $\qquad p \to q \wedge r \equiv (p \to q) \wedge (p \to r).$ {JE: T16; T16a}

T17. $\qquad (p \to r) \wedge (q \to r) \equiv p \vee q \to r.$

Proof. (a)

(1)	$p \to r$	
(2)	$q \to r$	{a.}
(3)	$p \vee q$	
(4)	$\neg r$	{a.i.p.}
(5)	$\neg p$	{Toll.: 1; 4}
(6)	$\neg q$	{Toll.: 2; 4}
(7)	q	{OD: 3; 5}
	contr.	{6; 7}

(b)

(1)	$p \vee q \to r$	{a.}
(1.1)	p	{ad. a.}
(1.2)	$p \vee q$	{JD: 1.1}
(1.3)	r	{RD: 1; 1.2}
(2)	$p \to r$	{1.1 → 1.3}
(2.1)	q	{ad. a.}
(2.2)	$p \vee q$	{JD: 2.1}
(2.3)	r	{RD: 1; 2.2}
(3)	$q \to r$	{2.1 → 2.3}
	$(p \to r) \wedge (q \to r)$	{JC: 2; 3}

By T17, the condition $x \leqslant -2 \to f(x) > 0$, or $x < -2 \vee x = -2 \to f(x) > 0$, is equivalent to the conjunction of the conditions: $x < -2 \to f(x) > 0$, $x = -2 \to f(x) > 0$.

The simple implication contained in T17 will be called the *law*

THE SENTENTIAL CALCULUS   33

of addition of antecedents. The rule of adding antecedents, with the
schema

$$\phi \to \chi$$
$$\psi \to \chi$$
$$\overline{\phi \lor \psi \to \chi}$$

is derived from this law.

The law of addition of antecedents yields, by the law of importation
(T2a), the *law of simple constructive dilemma*:

T18. $(p \to r) \land (q \to r) \land (p \lor q) \to r.$

The rule of simple constructive dilemma, derived from T18, permits
us to deduce from two implications with the same consequent and
the disjunction of their antecedents, the consequent of the implications.

Here is an example of inference by the rule of simple constructive
dilemma:

$$n = 1 \to (n+1)^2 > n^2$$
$$n > 1 \to (n+1)^2 > n^2$$
$$\underline{n = 1 \lor n > 1}$$
$$(n+1)^2 > n^2.$$

Derivable from the rule of joining an implication to a proof is
the rule stating that if from some additional assumption two
contradictory formulae are obtained in a proof, then the formula
which contradicts the additional assumption may be joined to the
proof. For if from the additional assumption ψ we obtain in the proof
two contradictory lines, χ and $\neg \chi$, then by the rule JC and the rule
of joining an implication to the proof we may join to the proof the
implication $\psi \to \chi \land \neg \chi$. Hence by T15b we obtain the formula
$\neg \psi$. Likewise, if from the additional assumption $\neg \psi$ we obtain in
the proof two contradictory lines, χ and $\neg \chi$, then we may join to
the proof the implication $\neg \psi \to \chi \land \neg \chi$. Hence by T15c we obtain ψ.

This rule will be used in the proof of the next thesis.

T19. $\neg (p \lor q) \equiv \neg p \land \neg q$ *the law of negating a disjunction.*

Proof. (a) (1) $\neg (p \lor q)$ {a.}
            (1.1) p {ad. a.}

$$(1.2) \quad p \lor q \qquad \{\text{JD: } 1.1\}$$
$$(2) \quad \neg p \qquad \{1.1 \to \text{contr. } (1; 1.2)\}$$
$$(2.1) \quad q \qquad \{\text{ad. a.}\}$$
$$(2.2) \quad p \lor q \qquad \{\text{JD: } 2.1\}$$
$$(3) \quad \neg q \qquad \{2.1 \to \text{contr. } (1; 2.2)\}$$
$$\neg p \land \neg q \qquad \{\text{JC: } 2; 3\}$$

Part (b) of the proof is left to the reader.

By T19, *the negation of a disjunction is equivalent to the conjunction of its negated elements.* For instance, the statement $\neg a \geqslant b$, that is the statement $\neg (a > b \lor a = b)$, is equivalent to the statement $\neg a > b \land \neg a = b$.

Derivable from T19 is the rule of negating a disjunction:

$$\text{ND} \qquad \frac{\neg(\phi \lor \psi)}{\neg \phi} \qquad \frac{\neg(\phi \lor \psi)}{\neg \phi \land \neg \psi.}$$
$$\neg \psi,$$

T20. $\neg (p \land q) \equiv \neg p \lor \neg q$ *the law of negating a conjunction.*

Proof. (a)

$$(1) \quad \neg(p \land q) \qquad \{\text{a.}\}$$
$$(2) \quad \neg(\neg p \lor \neg q) \qquad \{\text{a.i.p.}\}$$
$$(3) \quad \neg\neg p \qquad \{\text{ND: } 2\}$$
$$(4) \quad \neg\neg q \qquad \{\text{ND: } 2\}$$
$$(5) \quad p \qquad \{\text{ON: } 3\}$$
$$(6) \quad q \qquad \{\text{ON: } 4\}$$
$$(7) \quad p \land q \qquad \{\text{JC: } 5; 6\}$$
$$\text{contr.} \qquad \{1; 7\}$$

Part (b) of the proof is left to the reader.

By T20, *the negation of a conjunction is equivalent to the disjunction of its negated elements.* E.g., the sentence $\neg (a < x < b)$, that is the sentence $\neg (a < x \land x < b)$, is equivalent to the sentence $\neg a < x \lor \neg x < b$.

Theses T19 and T20 are analogous in structure. They are called *De Morgan's laws*, after the 19th-century English logician. However, they were already known in the Middle Ages. We find them, for instance, in the works of Occam (14th c.), of Burleigh (14th c.) and of other 14th and 15th century logicians.

T21. $\neg (p \to q) \equiv p \land \neg q$ *the law of negating an implication.*

The proof is left to the reader.

T22. $\quad \neg(p \wedge \neg p)$ *the law of non-contradiction.*

Proof. (1) $p \wedge \neg p$ $\qquad\qquad\qquad$ {a.i.p.}

(2) p $\quad\big\}$

(3) $\neg p$ $\qquad\qquad\qquad\qquad$ {OC: 1}

contr. $\qquad\qquad\qquad\qquad\qquad$ {2; 3}

The proof of this thesis is a simple indirect proof; it does not include any assumptions of the theorem, but only an assumption of indirect proof, which is a sentence contradictory to the thesis to be proved.

The law of non-contradiction was formulated by Aristotle, who also gave its metalogical formulation, stating that two contradictory sentences cannot both be true (or, in an equivalent formulation: of two contradictory sentences one must be false).

T23. $\quad p \vee \neg p$ *the law of the excluded middle.*

Proof. (1) $\neg(p \vee \neg p)$ $\qquad\qquad\qquad$ {a.i.p.}

(2) $\neg p$ $\quad\big\}$

(3) $\neg\neg p$ $\qquad\qquad\qquad\qquad$ {ND: 1}

contr. $\qquad\qquad\qquad\qquad\qquad$ {2; 3}

The law of the excluded middle is a logical thesis which is often used as a premiss in mathematical proofs, especially in ramified proofs to which additional assumptions are joined (this type of proof will be discussed below).

The law of the excluded middle was formulated by Aristotle, too. He also gave its metalogical formulation, according to which one of two contradictory sentences must be true.

In his works Aristotle assumed the validity of the law, and even devoted a paragraph of his *Metaphysics* to the defence of the law of the excluded middle. Yet in one of his texts he questioned the validity of the law with reference to future undetermined events. He stated in this connection that the adoption of this law for such events would lead to the conclusion that everything which will happen is necessary. Referring to this idea of Aristotle, in 1918–1920 the Polish logician Jan Łukasiewicz constructed a system called the three-valued sentential calculus, in which the law of the excluded middle does not hold.

T24. $\quad (p \to q) \to (p \wedge r \to q \wedge r)$ *the law of a new factor.*

The proof is left to the reader.

An example of inference by the law of a new factor:

$$\frac{a > 2 \to a > 0}{a > 2 \wedge a < 9 \to a > 0 \wedge a < 9}$$

or

$$2 < a < 9 \to 0 < a < 9.$$

T25. $(p \to q) \wedge (r \to s) \to (p \wedge r \to q \wedge s)$ *the law of multiplying implications by sides.*

The proof is left to the reader.

An example of inference by the law of multiplying an implication by sides is:

$$a < x \to c < f(x)$$

$$\frac{x < b \to f(x) < d}{a < x \wedge x < b \to c < f(x) \wedge f(x) < d}$$

or

$$a < x < b \to c < f(x) < d.$$

We shall now justify the derivative rule of constructing ramified proofs with joined additional assumptions. By this rule:

(a) a direct proof from assumptions of formula (I) [1] is completed if we obtain in it the formula ϕ_n on the strength of each of the additional assumptions $\psi_1, \ldots, \psi_k$, a disjunction of which is one of the proof lines;

(b) an indirect proof from assumptions of formula (I) is completed if we obtain in it a contradiction on the strength of each of the additional assumptions $\psi_1, \ldots, \psi_k$, a disjunction of which is one of the proof lines.

To prove (a) it suffices to note that if in the direct proof from assumptions of formula (I) we obtain formula ϕ_n on the strength of each of the additional assumptions $\psi_1, \ldots, \psi_k$, we may, by the rule of joining an implication to the proof, join the formulae $\psi_1 \to \phi_n$, $\psi_2 \to \phi_n, \ldots, \psi_k \to \phi_n$ as new proof lines. By applying to these formulae the rule of adding the antecedents (generalized so as to hold for k implications) we obtain the formula $\psi_1 \vee \psi_2 \vee \ldots \vee \psi_k \to \phi_n$. By applying to it and to the disjunction $\psi_1 \vee \psi_2 \vee \ldots \vee \psi_k$ (which

---

[1] This formula is given on p. 13.

we assume to be one of the proof lines) the rule of detachment we obtain the formula ϕ_n as the last proof line.

To prove (b) it suffices to note that if in the indirect proof from assumptions of formula (I) we obtain a contradiction on the strength of each of the additional assumptions $\psi_1, ..., \psi_k$, we may—by the derived rule given on p. 33—join to the proof the negation of each of these additional assumptions, that is the formulae: $\neg \psi_1, ..., \neg \psi_k$. By applying the rule OD to the disjunction $\psi_1 \lor \psi_2 \lor ... \lor \psi_k$ (which we assume to be one of the proof lines) and to the formulae $\neg \psi_1, ..., \neg \psi_{k-1}$ we obtain ψ_k, which contradicts the formula $\neg \psi_k$, previously obtained. The appearance of this contradiction terminates the indirect proof from assumptions of formula (I).

The proofs of T26 and T27 are examples of proofs in which the first of these derivative rules is employed. The second rule, which is frequently used in mathematical reasoning, will be illustrated by the following example:

Thesis: $(p \to q) \land (r \to s) \land \neg(q \lor s) \to \neg(p \lor r)$.

Proof.

(1)	$p \to q$	
(2)	$r \to s$	{a.}
(3)	$\neg(q \lor s)$	
(4)	$p \lor r$	{a.i.p.}
(1.1)	p	{ad. a.}
(1.2)	q	{RD: 1; 1.1}
(1.3)	$q \lor s$	{JD: 1.2}
(2.1)	r	{ad. a.}
(2.2)	s	{RD: 2; 2.1}
(2.3)	$q \lor s$	{JD: 2.2}
	contr.	{1.1 $\to$ contr. (1.3; 3); 2.1 $\to$ contr. (2.3; 3); 4}

T26. $(p \to q) \to (p \lor r \to q \lor r)$ *the law of a new element.*

Proof.

(1)	$p \to q$	{a.}
(2)	$p \lor r$	
(1.1)	p	{ad. a.}
(1.2)	q	{RD: 1; 1.1}
(1.3)	$q \lor r$	{JD: 1.2}

(2.1) r {ad. a.}

(2.2) $q \lor r$ {JD: 2.1}

       $q \lor r$ {1.1 → 1.3;

                                           2.1 → 2.2; 2}

T27. $(p \to q) \land (r \to s) \to (p \lor r \to q \lor s)$ *the law of adding implications by sides.*

Proof. (1) $p \to q$ ⎫

       (2) $r \to s$ ⎬   {a.}

       (3) $p \lor r$ ⎭

    (1.1) p {ad. a.}

    (1.2) q {RD: 1; 1.1}

    (1.3) $q \lor s$ {JD: 1.2}

    (2.1) r {ad. a.}

    (2.2) s {RD: 2; 2.1}

    (2.3) $q \lor s$ {JD: 2.2}

          $q \lor s$ {1.1 → 1.3; 2.1 → 2.3; 3}

Here is an example of inference by the rule of adding implications by sides:

$$a > b \to a^2 > b^2$$
$$a = b \to a^2 = b^2.$$
$$\overline{a > b \lor a = b \to a^2 > b^2 \lor a^2 = b^2}$$

or:

$$a \geqslant b \to a^2 \geqslant b^2.$$

T27 yields by the law of importation the law of compound constructive dilemma:

T28. $(p \to q) \land (r \to s) \land (p \lor r) \to (q \lor s).$

The rule of compound constructive dilemma, derived from T18, permits us to deduce, from two implications and the disjunction of their antecedents, the disjunction of their consequents.

An example of inference by the rule of compound constructive dilemma is:

$$n = 1 \to 2n = 2$$
$$n > 1 \to 2n > 2$$
$$\underline{n = 1 \lor n > 1}$$
$$2n = 2 \lor 2n > 2.$$

The forms of inference called *dilemmas* were already known in antiquity. Besides the simple and compound constructive dilemmas, simple and compound destructive dilemmas are also known. The rule of simple destructive dilemma permits us to deduce from two implications with the same antecedent and the disjunction of the negations of their consequents, the negation of the antecedent of these implications. The rule of compound destructive dilemma permits us to deduce from two implications and the disjunction of the negations of their consequents, the disjunction of the negations of their antecedents. The formulation of the laws on which these rules are based, as well as their proofs, are left to the reader.

T29. $$\neg p \lor q \equiv (p \to q).$$

Thesis T4b is the simple implication for this equivalence. Hence only part (b) of the proof is given below.

Proof. (b)

(1)	$p \to q$	{a.}
(1.1)	p	{ad. a.}
(1.2)	q	{RD: 1; 1.1}
(1.3)	$\neg p \lor q$	{JD: 1.2}
(2.1)	$\neg p$	{ad. a.}
(2.2)	$\neg p \lor q$	{JD: 2.1}
	$\neg p \lor q$	{1.1 → 1.3; 2.1 → 2.2}

The proof of the converse implication contained in Thesis T29 is a ramified proof in which the disjunction of the additional assumptions is a logical thesis: the law of the excluded middle. Mathematical proofs are often carried out in such a way that the disjunction of the additional assumptions is a special case of the law of the excluded middle, e.g., if a thesis is proved first with the assumption $x = 0$, and next with the assumption $x \neq 0$. In other cases the disjunction of the additional assumptions is a thesis or a special case of a thesis from a given field, e.g., when the proof is carried out for the assumptions: $x > 0$, $x = 0$, $x < 0$, where the law of trichotomy: $x > 0 \lor x = 0 \lor x < 0$ is implicitly assumed. In ramified proofs from assumptions in which the disjunction of additional assumptions is the law of the excluded middle or a substitution of it, we shall not write out the law in the proof, nor shall we, in the last proof line, indicate the fact that we refer to that law.

T29a. $\qquad p \lor q \equiv (\neg p \to q).$

T4a is the simple implication. The proof of the converse implication is analogous to part (b) of the proof of T29.

The rule corresponding to the converse implication contained in T29 is found in the works of Paul of Venice (Paulus Nicolettus Venetus), who lived at the turn of the 15th century (died 1429) and was the author of a comprehensive textbook on logic. His formulation was that an implication has as its consequence the disjunction formed of a sentence contradictory to its antecedent and its consequent; he also gave an example of inference by this rule: If you are a man, then you are a living being, and hence you are not a man or you are a living being.

T30. $\qquad p \land (q \lor r) \equiv p \land q \lor p \land r.$

Only part (b) of the proof is given below.

Proof. (b)

(1)	$p \land q \lor q \land r$	{a.}
(1.1)	$p \land q$	{ad. a.}
(1.2)	$p\rbrace$	
(1.3)	$q\rbrace$	{OC: 1.1}
(1.4)	$q \lor r$	{JD: 1.3}
(1.5)	$p \land (q \lor r)$	{JC: 1.2; 1.4}
(2.1)	$p \land r$	{ad. a.}
(2.2)	$p\rbrace$	
(2.3)	$r\rbrace$	{OC: 2.1}
(2.4)	$q \lor r$	{JD: 2.3}
(2.5)	$p \land (q \lor r)$	{JC: 2.2; 2.4}
	$p \land (q \lor r)$	{1.1 → 1.5; 2.1 → 2.5; 1}

By T30 the following conditions are equivalent:

x is a rational number $\land$ $(x < -2 \lor x > 5)$,

x is a rational number $\land$ $x < -2 \lor x$ is a rational number
$\land$ $x > 5$.

Thesis T30 is called the *distributive law for conjunction into disjunction*, analogous to the distributive law in arithmetic for multiplication into addition: $a.(b+c) = ab+ac$.

T31. $p \lor q \land r \equiv (p \lor q) \land (p \lor r)$.

Thesis T31, the proof of which is left to the reader, is the *distributive law for disjunction into conjunction*. By this thesis, and the symmetry of equivalence (T12), the conditions

$$(a|b.c \lor \neg a|b) \land (a|b.c \lor \neg a|c),$$
$$a|b.c \lor \neg a|b \land \neg a|c \ ^{(1)}$$

are equivalent.

T32. $(p \to q) \land (r \to s) \land (p \lor r) \land \neg (q \land s) \to (q \to p) \land (s \to r)$.

Proof. (1)	$p \to q$	
(2)	$r \to s$	{a.}
(3)	$p \lor r$	
(4)	$\neg (q \land s)$	
(5)	$\neg q \lor \neg s$	{RD$_E$: T20; 4}
(1.1)	q	{ad. a.}
(1.2)	$\neg s$	{OD: 5; 1.1}
(1.3)	$\neg r$	{Toll.: 2; 1.2}
(1.4)	p	{OD: 3; 1.3}
(6)	$q \to p$	{1.1 → 1.4}
(2.1)	s	{ad. a.}
(2.2)	$\neg q$	{OD: 5; 2.1}
(2.3)	$\neg p$	{Toll.: 1; 2.2}
(2.4)	r	{OD: 3; 2.3}
(7)	$s \to r$	{2.1 → 2.4}
	$(q \to p) \land (s \to r)$	{JC: 6; 7}

Thesis T32 is called the *law of a closed system of theorems* or *Hauber's law*. It might also be called the *law of conversion of implications*. The following rule of converting implications is based on T32:

$$\phi_1 \to \psi_1$$
$$\phi_2 \to \psi_2$$
$$\phi_1 \lor \phi_2$$
$$\underline{\neg (\psi_1 \land \psi_2)}$$
$$\psi_1 \to \phi_1$$
$$\psi_2 \to \phi_2.$$

Thesis T32 and its corresponding rule may be generalized to n implications. The rule of converting implications then has the form

---

(1) "$a|b$" means: a divides b.

$$\phi_1 \to \psi_1$$
$$\phi_2 \to \psi_2$$
$$\cdots\cdots$$
$$\phi_n \to \psi_n$$
$$\phi_1 \lor \phi_2 \lor \ldots \lor \phi_n$$
$$\underline{\neg(\psi_i \land \psi_j) \quad \text{for} \quad 1 \leqslant i \neq j \leqslant n}$$
$$\psi_1 \to \phi_1$$
$$\psi_2 \to \phi_2$$
$$\cdots\cdots$$
$$\psi_n \to \phi_n.$$

Thus if we have n confirmed implications and the disjunction of their antecedents, and if their consequents mutually exclude one another, then we may convert each of these implications.

Here is an example of the application of the rule of converting implications (where "a" and "b" stand for sides of a triangle, and "α" and "β" for their opposite angles):

$$a = b \to \alpha = \beta$$
$$a > b \to \alpha > \beta$$
$$a < b \to \alpha < \beta$$
$$a = b \lor a > b \lor a < b$$
$$\neg(\alpha = \beta \land \alpha > \beta)$$
$$\neg(\alpha = \beta \land \alpha < \beta)$$
$$\underline{\neg(\alpha > \beta \land \alpha < \beta)}$$
$$\alpha = \beta \to a = b$$
$$\alpha > \beta \to a > b$$
$$\alpha < \beta \to a < b.$$

T33a. $(p \equiv q) \to (\neg p \equiv \neg q)$

   b. $(p \equiv q) \to (p \land r \equiv q \land r)$

   c. $(p \equiv q) \to (r \land p \equiv r \land q)$

   d. $(p \equiv q) \to (p \lor r \equiv q \lor r)$ *the laws of exten-*

   e. $(p \equiv q) \to (r \lor p \equiv r \lor q)$ *sionality of equiv-*

   f. $(p \equiv q) \to (p \to r \equiv q \to r)$ *alence*

   g. $(p \equiv q) \to (r \to p \equiv r \to q)$

   h. $(p \equiv q) \to [(p \equiv r) \equiv (q \equiv r)]$

   i. $(p \equiv q) \to [(r \equiv p) \equiv (r \equiv q)]$

j. $(p \equiv q) \wedge (r \equiv s) \rightarrow (p \wedge r \equiv q \wedge s)$ *the laws of exten-*

k. $(p \equiv q) \wedge (r \equiv s) \rightarrow (p \vee r \equiv q \vee s)$ *sionality of equiv-*

l. $(p \equiv q) \wedge (r \equiv s) \rightarrow (p \rightarrow r \equiv q \rightarrow s)$ *alence (cont.)*

m. $(p \equiv q) \wedge (r \equiv s) \rightarrow [(p \equiv r) \equiv (q \equiv s)]$

The proofs of Theses T33a–g are in a certain respect analogous: they can be proved if use is made of the rule OE, the corresponding laws for implication, and the rule JE. In the proofs of T33h and T33i we use the laws stating that equivalence is symmetric and transitive. The proof of T33f is given by way of example:

Proof. (1) $p \equiv q$ {a.}

       (2) $p \rightarrow q$

       (3) $q \rightarrow p$ {OE: 1}

       (4) $(p \rightarrow q) \rightarrow [(q \rightarrow r) \rightarrow (p \rightarrow r)]$

       (5) $(q \rightarrow p) \rightarrow [(p \rightarrow r) \rightarrow (q \rightarrow r)]$ {T1a}

       (6) $(q \rightarrow r) \rightarrow (p \rightarrow r)$ {RD: 4; 2}

       (7) $(p \rightarrow r) \rightarrow (q \rightarrow r)$ {RD: 5; 3}

           $p \rightarrow r \equiv q \rightarrow r.$ {JE: 6; 7}

The proofs of Theses T33j–T33m are based on theses T33b–T33i and T13.

The derived rule of extensionality of the sentential calculus is based on Theses T33; the scheme of that rule is

$$\frac{\phi \equiv \psi}{\chi \equiv \chi\,(\phi \,//\, \psi)}.$$

The symbol $\chi \equiv \chi\,(\phi \,//\, \psi)$ stands for a formula obtained from χ by the replacement of its parts ψ by the formula ϕ. If the formulae

$$(p \rightarrow q) \rightarrow [(q \rightarrow r) \rightarrow (p \rightarrow r)],$$
$$q \rightarrow r,$$
$$\neg r \rightarrow \neg q,$$

correspond, respectively, to χ, ψ, and ϕ then the formula

$$(p \rightarrow q) \rightarrow [(\neg r \rightarrow \neg q) \rightarrow (p \rightarrow r)]$$

corresponds to the formula $\chi(\phi//\psi)$.

Note that if ψ occurs more than once in χ as a part, then the symbol $\chi(\phi//\psi)$ stands for any of the formulae obtained from χ by the replacement of ψ by ϕ in an arbitrary number of, but not necessarily all, places. In such a case the symbol $\chi(\phi \,//\, \psi)$ is thus not unique.

The proof of the rule of extensionality is by induction:

1) If χ has one of the following forms:

$$\neg \psi, \; \psi \rightarrow \vartheta, \; \vartheta \rightarrow \psi, \; \psi \vee \vartheta, \; \vartheta \vee \psi, \; \psi \wedge \vartheta, \; \vartheta \wedge \psi, \; \psi \equiv \vartheta,$$
$$\vartheta \equiv \psi,$$

then the equivalence $\chi \equiv \chi(\phi//\psi)$ is a consequence of the assumption $\phi \equiv \psi$ and the laws of extensionality T33a–T33i.

2) Suppose that the following equivalences are consequences of the assumption that $\phi \equiv \psi$:

$$\chi_1 \equiv \chi_1(\phi//\psi), \quad \chi_2 \equiv \chi_2(\phi//\psi).^{(1)}$$

These equivalences and the laws of extensionality T33a, T33j–T33m yield the equivalence $\chi \equiv \chi(\phi//\psi)$, both when χ is the negation of χ_i and when χ is an implication, disjunction, conjunction, or equivalence constructed from χ_1 and χ_2. The rule of extensionality is thus a derived rule.

On the strength of the rule of extensionality and $\mathrm{RD_E}$ we state that the rule with the schema

$$\phi \equiv \psi$$
$$\frac{\chi}{\chi(\phi//\psi)}$$

is a derived rule. This rule is also called the rule of extensionality.

T20 yields, by T33a and T5b, the following thesis:

T34. $p \wedge q \equiv \neg(\neg p \vee \neg q).$

We shall give some more theses, the proofs of which are omitted:

T35. $(p \vee q) \vee r \equiv p \vee (q \vee r).$

T36. $(p \wedge q) \wedge r \equiv p \wedge (q \wedge r).$

These theses are the associative laws for disjunction and conjuction; they permit us to drop parentheses in formulae that are disjunctions or conjunctions of a number of component parts.

T37. $p \vee q \equiv q \vee p.$

T38. $p \wedge q \equiv q \wedge p.$

---

(1) It is not excluded that formula ψ is replaced by formula ϕ in only one of the formulae χ_1 and χ_2.

These theses are the laws of commutativity for disjunction and conjunction; on the strength of these laws, in any disjunction or conjunction of many elements any two elements may be transposed.

T31 and T32 may be used to prove:

T39. $(p \lor q) \land (r \lor s) \equiv p \land r \lor p \land s \lor q \land r \lor q \land s.$

T40. $p \land q \lor r \land s \equiv (p \lor r) \land (p \lor s) \land (q \lor r) \land (q \lor s).$

T39 is analogous to the law of arithmetic:

$$(a+b) \times (c+d) = ac+ad+bc+bd.$$

The Theses T39 and T40 can be generalized so as to be applicable to disjunctions and conjunctions of many elements. On the strength of these theses a conjunction of disjunctions may be transformed into an equivalent disjunction of conjunctions, and vice versa.

We conclude by listing the derived rules of proof construction and the more important rules of joining new proof lines to a proof, as introduced in this section.

We have introduced the following derived rules for constructing proofs from assumptions:

(1) The rule which permits us to break into their elements assumptions that are structurally conjunctions, and to write out these elements as assumptions (p. 18).

(2) The rule by which we prove those theses which have the form of an equivalence by proving the corresponding simple and converse implications (p. 19).

(3) The rule which permits us to write the assumption of an indirect proof by omitting the negation symbol (p. 25).

(4) The rule of introducing an additional assumption and of joining an implication to the proof (pp. 30–31).

(5) The rule of joining to the proof a formula which contradicts any additional assumption from which two contradictory proof lines are derived (p. 33).

(6) The rule of constructing ramified proofs from assumptions (p. 36).

In the proofs of theses and rules we have employed the following derived rules of joining new proof lines:

1. The derived rule of omitting a disjunction (pp. 21, 25).

2. The rules of omitting and joining double negation (ON and JN, p. 25).
3. The rule *modus tollens* (Toll., p. 27).
4. The rule of negating a disjunction (ND, p. 34).
5. The rule of detachment for equivalence (RD$_E$, p. 29).
6. The rule of adding antecedents (pp. 32–33).

We have proved the rule of substitution (p. 33) and the rule of extensionality (pp. 43–44) to be derived rules.

Exercises

1. Prove the following laws of the sentential calculus:
   a) $[(p \rightarrow q) \rightarrow p] \rightarrow p$,
   b) $[(p \rightarrow q) \rightarrow r] \rightarrow [(r \rightarrow p) \rightarrow (s \rightarrow p)]$,
   c) $(p \equiv q) \land \neg p \rightarrow \neg q$,
   d) $(p \equiv q) \land \neg q \rightarrow \neg p$,
   e) $(p \equiv q) \equiv p \land q \lor \neg p \land \neg q$,
   f) $p \land (q \rightarrow r) \rightarrow (p \land q \rightarrow r)$,
   g) $p \land (q \rightarrow r) \rightarrow (q \rightarrow p \land r)$,
   h) $\neg(p \equiv q) \equiv p \land \neg q \lor q \land \neg p$,
   i) $p \rightarrow q \lor r \equiv (p \rightarrow q) \lor (p \rightarrow r)$,
   j) $p \land q \rightarrow r \equiv (p \rightarrow r) \lor (q \rightarrow r)$.

2. By making use of the rule of extensionality prove:
   a) the laws of symmetry and transitivity for equivalence,
   b) the thesis $(p \lor q) \rightarrow r \equiv (\neg p \rightarrow q) \rightarrow r$ from the theses
   $$p \lor q \equiv \neg p \rightarrow q \text{ and } p \equiv p,$$
   c) the thesis $(p \rightarrow \neg p) \rightarrow \neg p$ from the theses $\neg\neg p \equiv p$ and $(\neg p \rightarrow p) \rightarrow r$,
   d) the thesis $p \rightarrow (\neg p \rightarrow q)$ from the theses $p \lor q \equiv \neg p \rightarrow q$ and $p \rightarrow (p \lor q)$.

3. a) Formulate and prove the logical law by which, from the thesis
   (I) Given any two points there exists at most one straight line passing through them,
   you can prove the thesis:
   (II) Two different straight lines have at most one point in common,
   and conversely, prove (I) from (II).
   Hint: Write down (I) in the form:
   $$\neg(A = B) \land a \text{ and } b \text{ pass through } A \text{ and } B \rightarrow a = b,$$
   and by analogy write down (II).

   b) Formulate and prove the logical law by which the statement $c > 0 \land |a| > |b| \rightarrow ca^2 > cb^2$ is a consequence of the statements $|a| > |b| \rightarrow a^2 > b^2$ and $c > 0 \land a^2 > b^2 \rightarrow ca^2 > cb^2$.

4. Give formalized proofs of
   a) the thesis $\neg(x > x)$ from the thesis $x > y \rightarrow \neg(y > x)$,

b) the thesis $a > b \wedge \neg (c = 0) \to \neg (ac = bc)$ from the theses
$a > b \wedge c > 0 \to ac > bc, a > b \wedge c < 0 \to ac < bc, \neg (a = b) \equiv a > b \vee a < b$.

5. From the theses: $a > b \vee a = b \vee a < b, a^3 > b^3 \to \neg (a^3 = b^3 \vee a^3 < b^3)$,
$a^3 = b^3 \to \neg (a^3 < b^3 \vee a^3 > b^3), a^3 < b^3 \to \neg (a^3 = b^3 \vee a^3 > b^3), a > b \to$
$a^3 > b^3, a = b \to a^3 = b^3, a < b \to a^3 < b^3$, prove the theses $a^3 > b^3 \to a > b$,
$a^3 = b^3 \to a = b, a^3 < b^3 \to a < b$. Why cannot the proof be given for the
second powers of a and b?

4. Zero-One Verification. Truth Functors

The zero-one method of verifying the formulae of the sentential
calculus makes it possible to determine in a finite number of definite
steps whether any given formula of the sentential calculus is true,
that is, whether all the sentences obtainable from that formula through
the substitution of any sentences for the variables are true. The name
of this method comes from the fact that we usually symbolize the
truth of a sentence as "1", and its falsehood as "0". Truth and falsehood are called *truth values*. The concept of truth has been defined
precisely in the methodology of the deductive sciences, but we shall
not analyse that definition here. We shall confine ourselves to the
statement that, in accordance with the intuitive sense of the terms
"truth" and "falsehood", we adopt the principle of bivalence by
which only two truth values exists, i.e., every sentence is either
true or false. It is also assumed as obvious that these two truth
values are mutually exclusive, i.e., that no sentence is at the same
time both true and false.

The zero-one method is also called the *truth table method*, because
we use in it what are called *truth tables*, which show how the truth
value of a compound sentence, formed by means of a given functor
of the sentential calculus, is determined by the truth values of the
component sentences.

The properties of negation are shown in the following table:

ϕ	$\neg \phi$
1	0
0	1

The first row under the line states that the sentence $\neg \phi$ is false when
ϕ is true, and the second, that $\neg \phi$ is true when ϕ is false. These
properties of negation fully agree with its common usage.

The properties of negation can be presented in a simple table because the truth value of negation depends only on the truth value of the negated sentence, and not on its meaning. Conjunction, disjunction, implication, and equivalence share the same property: the truth values of these compound sentences are determined by the truth values of their component sentences.

The functors which, together with their arguments, form compound sentences with the property described above are called *truth functors*. Thus the functors of the sentential calculus are truth functors. Not every functor of sentential arguments, however, is a truth functor. For instance, the functor "John knows that" in the formula

$$\text{John knows that } p$$

is not a truth functor, unless John is omniscient.

Taking the table for negation as an example we may deduce analogous tables for conjunction, disjunction, implication, and equivalence, if we assume that the primitive rules of joining new proof lines to proofs always lead from true sentences to true sentences, and that the formulae provable by the rule of constructing indirect proofs from assumptions are true. Hence it follows that the derived rules also always lead from true sentences to true sentences. Thus if all the formulae which occur in the schema of a given rule above the horizontal line are true, then the formula below the line must also be true, and if that formula is false, then at least one of the formulae above the horizontal line must be false. Thus, for instance, from the schema of the rule JC

$$\frac{\begin{array}{c}\phi\\ \psi\end{array}}{\phi \wedge \psi}$$

we conclude that a conjunction of two true sentences is true, and from the schemata of the rule OC

$$\frac{\phi \wedge \psi}{\phi} \qquad \frac{\phi \wedge \psi}{\psi}$$

we conclude that a conjunction of which the first or the second element is false, is false. We have thus justified the truth table for conjunction:

ϕ	ψ	$\phi \wedge \psi$
1	1	1
0	1	0
1	0	0
0	0	0

By this table, a conjunction is true if and only if both its elements are true; it is false if at least one element is false.

We shall now justify the truth table for disjunction. It follows from the schemata of the rule JD

$$\frac{\phi}{\phi \vee \psi} \qquad \frac{\psi}{\phi \vee \psi}$$

that a disjunction, the first or the second element of which is true, is true. Suppose that the formula ψ which occurs below the line in the schema

$$\frac{\phi \vee \psi}{\neg \phi}$$
$$\psi$$

of the rule OD is false. If the formula ϕ is also false, so that the formula $\neg \phi$ is true, then the disjunction $\phi \vee \psi$ must be false, for if a false sentence occurs below the line, then the sentences above the line cannot both be true. Consequently, the truth table for disjunction has the following form:

ϕ	ψ	$\phi \vee \psi$
1	1	1
0	1	1
1	0	1
0	0	0

In conformity with this table, a disjunction is true if and only if at least one of its elements is true, and is false if both its elements are false.

We distinguish between a disjunction formed by means of the functor $\vee$ and an exclusive disjunction, which is characterized by the table:

ϕ	ψ	$\phi \veebar \psi$
1	1	0
0	1	1
1	0	1
0	0	0

Thus, an exclusive disjunction is true if and only if one and only one of its elements is true, and is false if both its elements are true or if both are false. It is called exclusive because the truth of one element excludes the truth of the other element when the whole disjunction is true. In contradistinction to exclusive disjunction, disjunction formed by means of the functor $\vee$ is sometimes called non-exclusive, since the truth of one element does not exclude the truth of the other element when the whole disjunction is true.

In current English there are no strict rules governing the use of "either ... or ..." and "... or ...", which are both used to mean non-exclusive disjunction. In the present book we adopt the convention that "$p \vee q$" will be read "p or q", and "$p \veebar q$" will be read "either p or q".

In justifying the truth table for implication we resort to the rule RD and to the rule, derived from T4b, of the connection between disjunction and implication, the schema of which is

$$\frac{\neg \phi \vee \psi}{\phi \to \psi.}$$

By the latter rule and the truth tables for negation and disjunction we obtain the following auxiliary table:

ϕ	ψ	$\neg \phi$	$\neg \phi \vee \psi$	$\phi \to \psi$
1	1	0	1	1
0	1	1	1	1
1	0	0	0	
0	0	1	1	1

Thus, in the first, second, and fourth cases the implication is true. In the third it must be false in view of the rule RD:

$$\frac{\phi \to \psi}{\phi}$$
$$\overline{\psi.}$$

For if ψ is false and ϕ is true, then $\phi \to \psi$ cannot be true, and hence must be false.

We thus obtain the following truth table for implication:

ϕ	ψ	$\phi \to \psi$
1	1	1
0	1	1
1	0	0
0	0	1

In accordance with this truth table, an implication is false if and only if its antecedent is true and its consequent is false; in all the remaining cases the implication is true.

The truth table for equivalence is obtained from the truth table for implication and the rules JE and OE:

ϕ	ψ	$\phi \to \psi$	$\psi \to \phi$	$\phi \equiv \psi$
1	1	1	1	1
0	1	1	0	0
1	0	0	1	0
0	0	1	1	1

In the first and the fourth cases the equivalence is true in view of the rule JE, which states that if the simple and converse implications are both true then the equivalence is also true. In the second case the converse implication $\psi \to \phi$ is false and hence it follows by the second schema of the rule OE

$$\frac{\phi \equiv \psi}{\psi \to \phi}$$

that the equivalence $\phi \equiv \psi$ is also false. In the third case the simple implication $\phi \to \psi$ is false and hence by the first schema of the rule OE

$$\frac{\phi \equiv \psi}{\phi \to \psi}$$

it follows that the equivalence $\phi \equiv \psi$ is also false.

In accordance with this truth table, an equivalence is true if and only if both its elements have the same truth value, and it is false if and only if its elements have different truth values.

The truth tables given above will be used in the zero-one verification of the formulae of the sentential calculus. Verification can be performed in a full and in an abbreviated form.

The full form of verification will be explained by means of the following example:

(a) $$\neg(p \wedge q) \to (\neg p \vee \neg q).$$

In this formula two different variables, p and q, appear. Let us first assume that the value of the variable p is 1, and that of the variable q is 0. To indicate this, we rewrite formula (a) writing the symbol 1 under the letter p whenever it occurs in (a) and, in an analogous way, the symbol 0 under the letter q:

$$\neg(p \wedge q) \to (\neg p \vee \neg q)$$
$$\quad 1 \quad 0 \qquad\ 1 \qquad 0$$

We can now compute the values of the formulae $p \wedge q$, $\neg p$ and $\neg q$. In accordance with the truth tables for conjunction and negation, the first two have in this case the value 0, while the third has the value 1. We rewrite formula (a) and under the functors of conjunction and negation write the values of compound formulae formed by them:

$$\neg(p \wedge q) \to (\neg p \vee \neg q)$$
$$\qquad 0 \qquad\quad 0 \qquad 1$$

The values of the formulae $\neg(p \wedge q)$ and $\neg p \vee \neg q$ are computed in turn. Each of them has the value 1 which we indicate by writing the symbol 1 under the symbol of negation at the beginning of the entire formula, and under the symbol for disjunction:

$$\neg(p \wedge q) \to (\neg p \vee \neg q)$$
$$1 \qquad\qquad\qquad 1$$

Thus, when the values of the variables p and q are, respectively, 1 and 0, the antecedent and the consequent of formula (a) both have the value 1, and hence, in accordance with the truth table for implication, the value of formula (a) in this case is 1, which we might indicate by writing the symbol "1" under the implication sign.

The procedure carried out above has several "levels". In practice, however, we write the 1 and 0 under the constants and variables of

the formula being verified, in the same line. In the case of (a), this
has the form

$$\neg (p \wedge q) \rightarrow (\neg p \vee \neg q)$$
$$1 \quad 1 \ 0 \ 0 \ \ 1 \quad 0 \ 1 \ 1 \quad 1 \ 0$$

The symbols underlined the same number of times have been obtained at the same "level" of verification; the greater the number
of lines the "higher" the "level" at which a given value has been
obtained. Whether a given set of values of the variables verifies
a given formula, can be seen from the value which is underlined the
greatest number of times. If a given formula is verified, that value must
be 1.

We have verified formula (a) for the case in which the values of the
variables p and q are respectively 1 and 0. This formula must still be
verified for the remaining three combinations of values of these
variables. We now give the notation of the verification of formula
(a) in all the four possible cases:

$$\neg (p \wedge q) \rightarrow (\neg p \vee \neg q)$$
$$0 \quad 1 \ 1 \ 1 \quad 1 \quad 0 \ 1 \ 0 \ 0 \ \ 1$$

$$1 \ 0 \ 0 \ 1 \ \ 1 \ \ 1 \ 0 \ 1 \ 0 \ 1$$
$$1 \ 1 \ 0 \ 0 \ \ 1 \ \ 0 \ 1 \ 1 \ 1 \ 0$$
$$1 \ 0 \ 0 \ 0 \ \ 1 \ \ 1 \ 0 \ 1 \ 1 \ 0$$

As we can see, the formula considered always proves to be true.
As such it will be classed as a true formula. If a formula cannot
be verified for all combinations of values of its variables, we class it
as a false formula. For instance, $(p \rightarrow q) \rightarrow (q \rightarrow p)$ is false, for it
cannot be verified when the variable p has the value 0, and when the
variable q has the value 1:

$$(p \rightarrow q) \rightarrow (q \rightarrow p)$$
$$0 \ 1 \ 1 \ \ 0 \ \ 1 \ 0 \ 0$$

Expressions which prove true for every combination of values
of their respective variables are also true in the sense that every

sentence obtained from any of them by the substitution of arbitrary sentences for all the variables, is a true sentence regardless from what discipline the substituted sentences are taken and what is their content. This follows from the fact that every sentence has one of the two truth values, truth or falsehood, and from the fact that the truth values of negation, conjunction, disjunction, implication, or equivalence depend not on the content of the component sentences, but only on their truth values.

Here is a non-abbreviated verification of two more formulae. The first contains only one variable, p, and the second three variables, p, q, and r.

$$p \lor \lnot p$$

```
1 1 0 1

    -- --

  0 1 1 0
```

$$(p \to q) \to [(q \to r) \to (p \to r)]$$

```
1 1 1  1  1 1 1  1  1 1 1

  --    --   --    --    --
        --

0 1 1  1  1 1 1  1  0 1 1
1 0 0  1  0 1 1  1  1 1 1
0 1 0  1  0 1 1  1  0 1 1
1 1 1  1  1 0 0  1  1 0 0
0 1 1  1  1 0 0  1  0 1 0
1 0 0  1  0 1 0  0  1 0 0
0 1 0  1  0 1 0  1  0 1 0
```

Note that when we write under the variables their truth values then under the first variable we write alternately 1 and 0, under the second we write alternately 1 twice and 0 twice, under the third variable alternately 1 four times and 0 four times, and so on. It can easily be seen that this exhausts all the possible combinations of values of the variables. Note also that if in the formulae being verified the number of different variables increases by one, then the number of combinations of values of these variables is doubled.

Abbreviated verification is most often used for formulae that have the form of an implication. In this case we assume that the antecedent of the whole implication is true and examine whether its consequent can be false, or else we assume that the consequent of the whole implication is false and examine whether its antecedent can be true. If it is excluded that the antecedent is true and the consequent is false, then, in accordance with the truth table for implication, the entire formula must be true. If, however, such a possibility is admitted in even one case, then the entire formula is false.

There follow a few examples of abbreviated verification:

$$(1) \qquad (p \rightarrow q) \rightarrow (\neg q \rightarrow \neg p)$$
$$\qquad\qquad 1 \ 0 \ 0 \quad 1 \quad 1 \ 0 \ 0 \ 0 \ 1$$

Explanation: We assume that the consequent of the entire implication, i.e. the formula $\neg q \rightarrow \neg p$, has the value 0. Then "$\neg q$" must have the value 1, and "$\neg p$", the value 0. But then "p" has the value 1, and "q", the value 0, and hence the antecedent "$p \rightarrow q$" has the value 0. Thus, the case in which the antecedent has the value 1 while its consequent has the value 0 is excluded and hence the whole implication is true.

$$(2) \qquad (p \rightarrow q) \rightarrow (q \rightarrow p)$$
$$\qquad\qquad 0 \ 1 \ 1 \quad 0 \quad 1 \ 0 \ 0$$

We assume that the consequent has the value 0. Then "q" has the value 1, and "p" has the value 0. But then the antecedent "$p \rightarrow q$" has the value 1, and thus the whole formula has the value 0.

$$(3) \qquad \neg(p \lor q) \rightarrow \neg p \land \neg q$$
$$\qquad\qquad 1 \ 0 \ 0 \ 0 \quad 1 \quad 1 \ 0 \ 1 \quad 1 \ 0$$

We assume that the antecedent, i.e. the formula "$\neg(p \lor q)$", has the value 1. Then "$p \lor q$" has the value 0, and hence both "p" and "q" have the value 0. But in this case both "$\neg p$" and "$\neg q$" have the value 1, and hence the formula "$\neg p \land \neg q$" has the value 1. When the antecedent is true the consequent is also true, and therefore the whole formula is true.

Not only formulae of the sentential calculus, but also schemata of rules of the sentential calculus can be verified by the zero-one method. A rule of the sentential calculus is correct if it never leads from true to false formulae. The correctness of rules can also be examined by abbreviated and non-abbreviated methods.

There follow two examples of an abbreviated verification of

schemata of rules of the sentential calculus:

(1)
$$\phi \rightarrow \psi$$
$$1\ 1\ 1$$
$$\neg\,\psi$$
$$0\ 1$$
$$\overline{\neg\,\phi}$$
$$0\ 1$$

If the conclusion, i.e. $\neg\,\phi$, is false, then ϕ has the value 1. If the first premiss is to be true, then ψ must have the value 1, but then $\neg\,\psi$, i.e. the second premiss, has the value 0. Thus it is excluded that both premisses are true when the conclusion is false, and hence the rule is correct.

(2)
$$\neg\,\phi \wedge \neg\,\psi$$
$$1\ 0\ \ 1\ 1\ 0$$
$$\overline{\neg\,(\phi \vee \psi)}$$
$$1\ \ \ 0\ 0\ 0\ .$$

If the premiss is true, then both ϕ and ψ have the value 0, but then $\phi \vee \psi$ has the value 0, and the conclusion, i.e. $\neg\,(\phi \vee \psi)$, has the value 1. Thus, if the premiss is true, the conclusion is also true, and therefore the rule is correct.

The functors whose truth tables have been discussed above are not the only truth functors. For instance, there are four truth functors of one argument, as is shown by the table below:

ϕ				
0	0	1	0	1
1	1	0	0	1

Of these, however, only the functor of negation is interesting.

There are sixteen truth functors of two arguments, because that is the number of ways in which the last column of the table below can be filled:

ϕ	ψ	
1	1	
0	1	
1	0	
0	0	

We shall also give here the truth tables of two truth functors of two arguments, namely the symbol of alternative negation "/" and the symbol of joint negation " ↓ " [the former is also called "Sheffer's stroke", "non-conjunction" (Church), "alternative denial" (Quine), and the latter, "non-disjunction" (Church), "joint denial" (Quine)]. The formula "p/q" is read "not both p and q", and the formula "$p \downarrow q$", "neither p nor q". The formula "p/q" is the negation of the formula "$p \wedge q$", and the formula "$p \downarrow q$" is the negation of the formula "$p \vee q$". These functors are defined by the following table:

ϕ	ψ	ϕ/ψ	$\phi \downarrow \psi$
1	1	0	0
0	1	1	0
1	0	1	0
0	0	1	1

These functors can also be introduced into the system of the sentential calculus by means of the following rules of joining and omitting:

JA
$$\frac{\neg\phi}{\phi/\psi} \qquad \frac{\neg\psi}{\phi/\psi.}$$

OA
$$\frac{\phi/\psi}{\begin{array}{c}\phi\\\hline \neg\psi.\end{array}}$$

JJ
$$\frac{\neg\phi}{\begin{array}{c}\neg\psi\\\hline\end{array}}{\phi \downarrow \psi.}$$

OJ
$$\frac{\phi \downarrow \psi}{\neg\phi} \qquad \frac{\phi \downarrow \psi}{\neg\psi.}$$

The symbols of alternative negation and joint negation differ from the other truth functors of two arguments in that either suffices to define all the remaining truth functors.[1] This will be explained by means of the example of the symbols of alternative negation and conjunction. "The symbol of conjunction can be defined

[1] Detailed remarks on definitions in the sentential calculus will be given in the next Section.

by means of the symbol of alternative negation" means the same as "there exists a formula, written down exclusively by means of the symbol of alternative negation and the variables p and q, which, for every combination of the values of these variables takes on the same values as the formula $p \wedge q$ does for this combination of values of the variables."

Given below, as examples, are definitions of negation, conjunction and disjunction by means of the symbols of alternative negation and joint negation:

$$\neg p \underset{\mathrm{df}}{=} p/p,$$

$$p \wedge q \underset{\mathrm{df}}{=} (p/q)/(p/q),$$

$$p \vee q \underset{\mathrm{df}}{=} (p/p)/(q/q),$$

$$\neg p \underset{\mathrm{df}}{=} p \downarrow p,$$

$$p \wedge q \underset{\mathrm{df}}{=} (p \downarrow p) \downarrow (q \downarrow q),$$

$$p \vee q \underset{\mathrm{df}}{=} (p \downarrow q) \downarrow (p \downarrow q).$$

These definitions can be verified by the zero-one method, showing in every case that the formulae on both sides of these definitions have the same truth value for the same combinations of values of the variables. We verify all the definitions of the sentential calculus in the same way.

Functors of three, four and more arguments can also be introduced into the sentential calculus, but they are uninteresting from the formal and from the intuitive point of view. Note also that all truth functors of more than two arguments can easily be defined by means of functors of one and two arguments.

As has already been mentioned the symbols $\neg$, $\wedge$, $\vee$, $\rightarrow$, $\equiv$ are truth functors, and hence the truth value of the sentences formed by means of these functors does not depend on the contents of the component sentences. No intuitive objections arise in the case of the first three functors, but things look different in the case of the last two, i.e. the functors of implication and equivalence.

By the table given above, an implication is true if its antecedent is false; it is also true if its consequent is true. Hence the following implications are true:

(a) Paris is the capital of Turkey → London lies in the mountains;

(b) New York lies on the Mediterranean → Mars is a planet.

Such examples, however, give rise to certain intuitive objections. The source of these objections lies in the fact that the formula "$p \to q$" is read: if p then q, but the connective "if... then ..." is often used in meanings for which the sentences obtained from the implications (a) and (b) by the replacement of the symbol → by the connectives "if ... then ..." are not true. We may distinguish the following meanings of the formula "if p then q":

(1) from the fact that p follows the fact that q (or: the sentence q is a logical consequence of the sentence p),

(2) it is not possible that (p and not q),

(3) it is not true that (p and not q),

(4) the fact that p is the cause of the fact that q.

As can be seen from the truth table for implication, the third of the meanings given above for the connective "if ... then ... " corresponds to the implication symbol " → ". If this meaning is adopted, it is obvious, e.g., that the implication (a) is true, since it means the same as the sentence:

It is not true that (Paris is the capital of Turkey and London does not lie in the mountains),

since the conjunction negated in this sentence is false, its first element being false, and thus its negation is true.

It must, however, be admitted that the connective "if ... then ..." is only rarely used in everyday speech in the third meaning specified above; it is much more frequently used in the remaining three meanings, for which it is not true that an implication which has a false antecedent or a true consequent is true.

The first two of the meanings given above as interpretations of the connective "if ... then ..." have been given precision in logic. The second has been given a precise interpretation in what is called the system of strict implication, built by the American logician C.I. Lewis. In his system the analogues of certain theses given in the preceding section, obtained by replacing the symbol " → " by the symbol of strict implication, are not valid; this holds in particular for the analogue of the thesis

$$\neg p \vee q \rightarrow (p \rightarrow q),$$

and the analogues of the theses

$$\neg p \rightarrow (p \rightarrow q), \quad q \rightarrow (p \rightarrow q),$$

on the strength of which the truth table for implication might be justified.

To distinguish it from strict implication, the implication formed by means of the truth functor " $\rightarrow$ " is called *material implication*.

The concept of *consequence*[1] has been defined in the methodology of deductive sciences. One of the definitions of that concept, in which the concept of a law of logic is used, will be given below. In the definition of the concept of a law of logic we resort in turn to the concept of a logical constant.

Logical constants are functors of the sentential calculus, the symbols "*a*", "*i*", "*e*", "*o*" of Aristotle's syllogistic, the word "*is*", the quantifiers, the symbol of identity, etc.[2] They are terms which are used in formulating theorems in various disciplines.

Laws of logic are formulae which have the following two properties:

(1) they are built exclusively of logical constants and variables,

(2) they are true,[3] i.e. they are verifiable for any values of the variables they include.

As examples of laws of logic we may give the theses of the sentential calculus as well as the laws which the reader probably knows from the rudiments of logic as taught in secondary schools:

(a) *If no S are P, then no P are S*; in symbols: $S e P \rightarrow P e S$;

(b) *If all S are P, then it is not true that some S are not P*; in symbols $S a P \rightarrow \neg (S o P)$.

Many such laws of logic will be introduced in the next chapter.

---

[1] The term "consequence" is used in this Section as an alternative for the term "logical consequence".

[2] The quantifiers and the identity relation will be discussed in the next Chapter.

[3] On the concept of truth, cf. Chap. 2, Sec. 3, p. 121.

We now use the concept of a law of logic to define the concept of (logical) *consequence*:[1]

The sentence S_2 is a consequence of the sentence S_1 if and only if the implication which has S_1 as its antecedent and S_2 as its consequent is a substitution of some law of logic.

The sentence S is a consequence of the sentences $S_1, S_2, ..., S_n$ if and only if the implication which has the conjunction of the sentences $S_1, S_2, ..., S_n$ as its antecedent and the sentence S as its consequent is a substitution of some law of logic.

EXAMPLES. The sentence (2) "No fish is a whale" is a consequence of the sentence (1) "No whale is a fish", since the implication which has (1) as its antecedent and (2) as its consequent is a substitution for the law of logic (a).

The sentence (4) "It is not true that there exists a square that is not a rectangle" is a consequence of the sentence (3) "Every square is a rectangle", since the implication which has (3) as its antecedent and (4) as its consequent is a substitution for the law of logic (b).

The sentence (7) "If today is Friday, then the day after tomorrow is Sunday" is a consequence of the sentences (5) "If today is Friday, then tomorrow is Saturday" and (6) "If tomorrow is Saturday, then the day after tomorrow is Sunday", since the implication which has the conjunction of (5) and (6) as its antecedent and (7) as its consequent is a substitution of the law of logic

T1. $$(p \to q) \wedge (q \to r) \to (p \to r).$$

In the light of these explanations it is not true that any sentence is a consequence of a true sentence, or that a true sentence is a consequence of any sentence. Hence it is not correct to read the formula "$p \to q$" as "q is a consequence of p".

The concept of (logical) consequence is an important concept in scientific methodology, to which we resort in the classification of forms of inference and in the classification of the sciences. Deductive inference is defined as inference in which the conclusion is a (logical) consequence of the premises. All the schemata of correct inference given in the present book are schemata of deductive inference.

---

[1] This definition was given by K. Ajdukiewicz in *Logiczne podstawy nauczania* (The Logical Foundations of Teaching), 1934.

Those disciplines in which exclusively deductive methods of inference are used are called deductive sciences.

While we realize that the connective "if ... then ..." is used with other meanings as well, we should nevertheless bear in mind that the truth table for implication has been based on the primitive rules given in Section 2 and on a certain rule that is derived from those rules. Among the primitive rules, the rules of joining new proof lines to the proof are quite intuitive (and remain valid in the system of strict implication as well), and the rules for constructing proofs do not give rise to special intuitive objections, and are moreover commonly used in mathematical proofs.

A distinction analogous to that in the case of implication can also be drawn in the case of equivalence. *Equivalence* is the conjunction of two implications, simple and converse. If these implications are material, then we speak of a material equivalence, and if they are strict, we speak of a strict equivalence. If there are two sentences such that one is a consequence of the other and vice versa, we say that they are logically equivalent.

Truth functions, i.e. formulae formed by means of truth functors, were investigated in antiquity (the Stoics and the school of Megara) and in the Middle Ages (mainly in the 13th to the 15th century).

The Stoics drew attention to the fact that the negation of a sentence is formed by placing before that sentence the symbol of sentential negation, and drew a distinction between the negation of an entire sentence and the negation of some parts of a sentence. They did not, however, formulate laws that would correspond to the truth table for negation; such laws can be found in the works of mediaeval logicians, for instance those of John Buridan (15th century).

Conjunction was defined by the Stoics as a compound sentence formed by means of the connective "and", and they stated that a conjunction is true if and only if both its elements are true. The same conditions for the truth of a conjunction were formulated by mediaeval logicians.

The term "disjunction" was used by the Stoics mainly for exclusive disjunction, but certain texts suggest that they used it for alternative negation as well.

Since the 13th century the trend has been to use that term mainly for non-exclusive disjunction, although followers of the Stoic interpretation were still met in the 14th century. Peter of Spain (Petrus Hispanus, fl.13th c.) stated that for a disjunction to be true it suffices that one of its elements be true, but it is possible that both its elements be true; for a disjunction to be false it is necessary that both its elements be false. Burleigh (14th c.) was decidedly in favour of interpreting "disjunction" as non-exclusive disjunction; he pointed out that a disjunction is a consequence of any of its elements, and hence it follows that if both elements are true, the disjunction as a whole is also true.

The greatest controversies, both in antiquity and in the Middle Ages, and also in the latest period, were connected with the interpretation of implication.

Philo of Megara (ca. 300 B.C.) was the first to define implication (or the conditional sentence) as a sentence which is false only if its antecedent is true and its consequent is false, and which is true in the remaining three cases. His definition inaugurated controversies about the interpretation of implication, which led to the formulation, within the School of Megara and the Stoic School, of three other definitions. Among them there is the ancient form of strict implication, defined as follows: An implication is true if the negation of its consequent disagrees with its antecedent. Controversies on that subject must have been very lively, since Callimachus, an Alexandrian librarian of the 2nd century B.C., referred to it in his epigram: "Ravens are already croaking on the roofs, which implications are correct."

These controversies revived still more strongly in the 14th and 15th centuries and resulted in distinctions between the various meanings of the conditional sentence and in the construction of different sentential calculi, in particular a system which corresponded to the contemporary system of material implication, and a system which corresponded to the contemporary system of strict implication.

The modern sentential calculus originates with G. Frege (1842–1912), the greatest logician in the period between Aristotle and the 20th century. He based his system, published in 1879, on negation and implication so as it had been interpreted by Philo of Megara. Modern truth tables for implication and other truth functors were introduced *ca.* 1885 by the American logician Peirce, who also pointed out that Philo of Megara had been the first to define material implication. About 1880 Peirce also became aware of the fact that all other truth functors can be defined by means of the symbol of alternative negation or the symbol of joint negation. However, this result of his research was not published at that time. In 1921, Sheffer proved the definability of all other truth functors by means of alternative negation, which he symbolized "/", and which is therefore sometimes called "Sheffer's stroke".

In *Principia Mathematica*, an important work in the history of logic, written by B. Russell and A. N. Whitehead (Vol. 1 appeared in 1910), there is the incorrect reading of "$p \to q$" as "q is a consequence of p". This interpretation led to paradoxical consequences, for instance that of two arbitrary sentences either the first is a consequence of the second, or vice versa. This is so because the law: $(p \to q) \lor (q \to p)$, which the reader can easily test by the zero-one method, is valid for material implication. In 1918, as a reaction against this interpretation, Lewis developed his system of strict implication, in which he introduced the functor of possibility "$\lozenge$" (the formula "$\lozenge p$" is read: *it is possible that p*) and defined the symbol of strict implication "$\prec$" in the following manner:

$$p \prec q = \neg \lozenge (p \land \neg q).^{(1)}$$

The formula "$p \prec q$" is read by Lewis "q is a consequence of p". Hence, q is a consequence of p if and only if it is not possible that (p and not q).

However, attention was later drawn to the fact that the concept of consequence is metasystematic and can be defined without a modification of the sentential calculus.

The first endeavours in the history of logic to define consequence were made by 13th and 14th century logicians, who spoke of "formal consequence" (as distinct from material consequence which corresponds to what is today called material implication). They defined this concept by means of the concept of the form of the sentence or consequence (i.e., reasoning).

Thus, for instance, as Buridan put it, the logical form of the sen-

---

(1) In Lewis's system the symbol "$=$" is the symbol of strict equivalence.

tence included what he called *syncategorematic words* (and what we call *logical constants*), and the number and order of words. Consequently, as he wrote, such sentences as "A man is a man" and "A man is an ass" differ in form, because of the different number of terms which occur in them. Further, such sentences as "Omnis homo est ens vivum" and "Ens vivum omnis homo est" also differ in form because of the order of the words. The consequences "All B are A, hence some B are A" and "All B are A, hence some A are B" also differ in form.

These definitions of logical form come very close to the present-day definition which might be formulated as follows: The logical form of a sentence (or of an inference understood as a sequence of sentences) is the formula (or a sequence of formulae) obtained from that sentence (or that sequence of sentences) by replacing non-logical constants by variables, the same constants being replaced by the same variables and different constants by different variables.

Formal consequence was defined by Albert of Saxony as such that every consequence that has the same form is valid (i.e., does not lead from truth to falsehood). A similar definition is also encountered earlier, for instance in the works of Duns Scotus.

In the first half of the 19th century, B. Bolzano gave a definition of consequence similar to the contemporary definition formulated by Tarski in 1935. Tarski's definition is equivalent to that of Ajdukiewicz, in the case where a given sentence is a consequence of a finite class of sentences. It is, however, more general since it can cover those cases in which a given sentence is a consequence of an infinite class of sentences.

Exercises

1. Classify as true or false, using the zero-one method, the following formulae of the sentential calculus:

   a) $(p \rightarrow q) \land \neg p \rightarrow \neg q$,
   b) $\neg(p \land q) \rightarrow \neg p \land \neg q$,
   c) $\neg p \lor \neg q \rightarrow \neg(p \lor q)$,
   d) $(p \land q \rightarrow r) \rightarrow p \land (q \rightarrow r)$,
   e) $[(p \rightarrow q) \rightarrow p] \rightarrow p$,
   f) $[(p \rightarrow q) \rightarrow q] \rightarrow q$,

   g) $(p \rightarrow q \wedge r) \rightarrow (p \vee r \rightarrow q)$,

   h) $[(p \equiv q) \equiv r] \equiv [q \equiv (p \equiv r)]$.

2. Define

   a) the symbols of disjunction, implication and alternative negation by means of the symbols of conjunction and negation;

   b) the symbols of conjunction, implication and alternative negation by means of the symbols of disjunction and negation;

   c) the symbols of conjunction, disjunction and alternative negation by means of the symbols of implication and negation;

   d) the symbol of implication by means of the symbol of alternative negation;

   e) the symbol of implication by means of the symbol of joint negation.

3. Examine by the zero-one method which of the laws valid for non-exclusive disjunction are also valid for exclusive disjunction.

4. a) Prove that negation cannot be defined by means of implication.

   b) Prove that implication cannot be defined by means of negation and equivalence.

5. Prove that the symbols of disjunction, conjunction and negation suffice to define all other truth functors.

   Hint: Draw attention to the fact that on the basis of a truth table for a truth functor of n arguments it is possible, for any formula consisting of that functor and variables, to give an equivalent formula written down by means of those variables and the symbols of negation, conjunction, and disjunction; for instance, for the formula $p \rightarrow q$ such an equivalent formula is $p \wedge q \vee \neg p \wedge q \vee \neg p \wedge \neg q$.

5. The Consistency and Completeness of the Assumptional System of the Sentential Calculus

Any formula of the sentential calculus which, in accordance with the tables given in Section 4, takes on the value 1 for any combination of the values of its variables is called a *true formula* of the sentential calculus or a *tautology* of that calculus. It can easily be seen that all the theses proved in Section 3 are true formulae. The problem arises whether every thesis provable by the rules of the assumptional system of the sentential calculus is a true formula. The answer is provided by

Th. 1. *Every thesis of the assumptional system of the sentential calculus is a true formula.*

The proof is by induction with respect to the order of the thesis. Assume that the formula

(I) $$\phi_1 \rightarrow \{\phi_2 \rightarrow \dots \rightarrow (\phi_{n-1} \rightarrow \phi_n) \dots\}$$

is a thesis of the 1st order. Suppose, as an assumption of an indirect proof, that formula (I) is not true. Hence for a certain combination of values of its variables formula (I) takes on the value 0, which occurs when for that combination of values of its variables the formulae $\phi_1, \ldots, \phi_{n-1}$ have the value 1, and the formula ϕ_n has the value 0. Then the formula $\neg \phi_n$ has, for that combination of the values of the variables, the value 1. The assumption that (I) is a thesis of the 1st order leads to the conclusion that there is an assumptional indirect proof of formula (I), in which the rules RD, JC, OC, JD, OD, JE, and OE, applied to the assumptions $\phi_1, \ldots, \phi_{n-1}, \neg \phi_n$, yield two contradictory formulae. But the said primitive rules, when applied to formulae which for a given combination of values of their variables have the value 1, yield formulae which for that combination of values of their variables also have the value 1. Now since all the assumptions of the proof, i.e. the formulae $\phi_1, \ldots, \phi_{n-1}, \neg \phi_n$, have the value 1 for a certain combination of values of their variables, then all the formulae obtained from the former on the strength of the said rules have the value 1 for that combination of the values of variables. This, however, is in contradiction to the fact that two contradictory formulae occur in the proof of (I), since if one of them has the value 1 for a given combination of the values of variables, then the other has the value 0 for that combination of the values of variables. Thus the assumption that (I) is not a true formula results in a contradiction. Hence any thesis of the 1st order is a true formula.

Suppose now, as an inductive assumption, that any thesis of an order lower than n is a true formula. Let formula (I) be a thesis of the nth order. There is, then, an assumptional indirect proof which yields two contradictory formulae from the assumptions of the proof $\phi_1, \ldots, \phi_{n-1}, \neg \phi_n$, and from theses of orders lower than n. If the assumption of the indirect proof is that formula (I) is not true, contradiction results as in the previous part of the proof, for, on the one hand, we have to conclude that the assumptions of the proof $\phi_1, \ldots, \phi_{n-1}, \neg \phi_n$, which for a certain combination of the values of their variables have the value 1, and theses of orders lower than n, which always have the value 1, yield, in accordance with the said rules, only such formulae which have the value 1 for that combination of the values of variables; on the other hand, one of the two contradictory formulae which occur in that proof ought to have the value 0 for that combination of the values of variables. Thus, if Th.1 is true for the theses of orders lower than n, it is also true for the theses of the nth order, which completes the proof by induction of Th.1.

A system is called *consistent* if and only if its theses do not include two contradictory formulae.

Th.2. *The assumptional system of the sentential calculus is consistent.*

If every thesis of that system is true (by Th.1), then no two contradictory formulae can both be its theses, since if one of them is true, then the other is false.

The problem now arises whether the rules of the assumptional system of the sentential calculus can be used to prove every true formula of that calculus. The answer is provided by

Th.3. *Every true formula of the sentential calculus is a thesis of the assumptional system.*

The proof of this theorem makes use of what are called *conjunctive normal forms.*

The formulae ϕ and ψ are called *equivalent* if and only if the formula $\phi \equiv \psi$ is a thesis of the sentential calculus. Equivalence defined in this way is transitive, since, by T13, if $\vdash \phi_1 \equiv \phi_2$ and $\vdash \phi_2 \equiv \phi_3$, then $\vdash \phi_1 \equiv \phi_3$. Hence it follows that if in the sequence of the formulae $\phi_1, \phi_2, ..., \phi_n$ every formula is equivalent to that which follows next, then ϕ_1 is equivalent to ϕ_n.

A formula is called an *elementary disjunction* if it is a variable, the negation of a variable, or a disjunction consisting of (two or more) variables or negations of variables.

A formula is called a *conjunctive normal form of a formula of the sentential calculus* if it is equivalent to the latter and is either an elementary disjunction or a conjunction of (two or more) elementary disjunctions.

By making use of several theses given in Section 3, and always applying the rule of extensionality, we reduce any formula of the sentential calculus to its conjunctive normal form by transforming it step by step into equivalent formulae. At each step, by the law of double negation (T5b), we cancel every even number of symbols of negation that follow one another. By T14 and T29, we transform a given formula into a formula recorded exclusively by means of variables and the symbols of conjunction, disjunction, and negation. By the laws of negating conjunction and disjunction (T19 and T20), we transform a formula thus obtained into one in which the symbols of negation occur neither before a conjunction nor before a disjunction. Hence, by T5b, we obtain a formula built of variables or their negations by means of the symbols of conjunction and/or disjunction. We shall now prove by induction that any such formula has its equivalent conjunctive normal form. If such a formula is a variable or its negation then, by definition and by T11a (reflexivity of equivalence), it is its own normal form. Assume that the formulae ϕ^n and ψ^k are, respectively, the conjunctive normal forms of the formulae ϕ and ψ. We shall show that the formulae $\phi \wedge \psi$ and $\phi \vee \psi$ have their equivalent conjunctive normal forms. The formula $\phi^n \wedge \psi^k$ is, of course, equivalent to the formula $\phi \wedge \psi$. Let $\phi^n = \phi_1 \wedge \ldots \wedge \phi_n$, and let $\psi^k = \psi_1 \wedge \ldots \wedge \psi_k$, where each of the formulae $\phi_1, \ldots, \phi_n, \psi_1, \ldots, \psi_k$ is an elementary disjunction. The formula $\phi^n \wedge \psi^k$ is a conjunction, of $n+k$ elements, of elementary disjunctions, and hence it is a conjunctive form of the formula $\phi \wedge \psi$. The formula $\phi^n \vee \psi^k$ is equivalent to the formula $\phi \vee \psi$. By the laws of distributivity (T31 and T32), or possibly the theses T39 and T40, the formula $\phi^n \vee \psi^k$, i.e. the formula $\phi_1 \wedge \ldots \wedge \phi_n \vee \psi_1 \wedge \ldots \wedge \psi_k$, may be transformed into the formula:

$$(\phi_1 \vee \psi_1) \wedge (\phi_1 \vee \psi_2) \wedge \ldots \wedge (\phi_1 \vee \psi_k) \wedge$$
$$\wedge (\phi_2 \vee \psi_1) \wedge (\phi_2 \vee \psi_2) \wedge \ldots \wedge (\phi_2 \vee \psi_k) \wedge$$
$$\cdots\cdots\cdots\cdots\cdots\cdots\cdots\cdots\cdots\cdots\cdots$$
$$\wedge (\phi_n \vee \psi_1) \wedge (\phi_n \vee \psi_2) \wedge \ldots \wedge (\phi_n \vee \psi_k).$$

Each element of that conjunction, being a disjunction of elementary disjunctions, is an elementary disjunction, and hence the entire formula is a conjunctive normal form of the formula $\phi \vee \psi$.

In this way, any formula of the sentential calculus can be reduced to a conjunctive normal form.

A very simple structural criterion of the truth of conjunctive normal forms can be formulated, namely:

A conjunctive normal form is a true formula if and only if each disjunction that is an element of that conjunction includes at least one variable which in one case is negated and in another is not.

It is so because if the above condition is satisfied, then each disjunction which is an element of such a conjunction is a true formula, and consequently the entire conjunction is a true formula. If the condition is not satisfied, i.e. if in a disjunction no variable in one case is negated, and in another is not, then such a disjunction takes on the value 0 when the non-negated variables are given the value 0, and the negated variables—the value 1, so that their negations take on the value 0. In such a case the entire conjunction takes on the value 0 too, which shows that it is not a true formula. By the law of excluded middle (T23), the rule of joining a disjunction, and the laws of associativity and commutativity for disjunction (T35 and T37), which permit the transposition of elements of a disjunction, every disjunction in which at least one variable in one case is negated, and in another case is not, can be proved to be a true formula. By the rule of joining a conjunction every conjunctive normal form can in this way be proved to be a true formula.

Now, if a given formula ϕ of the sentential calculus is a true formula, then the equivalence $\phi \equiv \phi^k$, where ϕ^k is a conjunctive normal form of the formula ϕ, is a thesis of the sentential calculus. By Th.1, this equivalence is a true formula. From this fact, and from the fact that the formula ϕ is a true formula, it follows that the formula ϕ^k is a true formula also. But then, as has been demonstrated previously, the formula ϕ^k is a thesis. From the theses $\phi \equiv \phi^k$ and ϕ^k we obtain the thesis ϕ. We have thus proved that every true formula of the sentential calculus is a thesis of the assumptional system of the sentential calculus.

A system is called *semantically complete* if and only if every true formula recorded in the language of that system is a thesis. Th. 3 may, therefore, be reformulated as follows:

Th.3a. *The assumptional system of the sentential calculus is semantically complete.*

The proof of the semantical completeness of the sentential calculus, as given here, also offers a method of verifying the formulae of that calculus. To verify a formula of that calculus it suffices to reduce it to a conjunctive normal form and to check whether it is a true formula.

In addition to the conjunctive normal form there is also the disjunctive normal form. A disjunctive normal form of a given formula of the sentential calculus is an equivalent formula which consists of a variable, or its negation, or a conjunction, of two or more elements, of variables or their negations, or a disjunction, of two or more elements, of such conjunctions. In reducing a given formula to its disjunctive normal form we make use of the same theses and rules as in the case of reducing it to its conjunctive normal form, the difference being that in the last stage we make different use of the laws of distributivity, namely, we proceed so as to obtain a disjunction of conjunctions.

A system is *syntactically complete* if and only if every sentential formula recorded in the language of that system either is a thesis of that system or, when joined to the theses of that system, results in a contradiction.

Th.4. *The assumptional system of the sentential calculus is syntactically complete.*

Proof. Assume that the formula ϕ of the sentential calculus is not a thesis of the assumptional system. From this and from Th.3 it follows that the formula ϕ is not a true formula. Hence for a certain combination of the values of the variables the formula ϕ has the value 0, for instance, when the variables $p_1, \ldots, p_n$ have the value 1, and the variables $q_1, \ldots, q_k$ have the value 0. We now join the formula ϕ to the theses of the assumptional system. The (secondary) rule of substitution in theses is valid in that system. Hence, in the system expanded by the thesis ϕ, the formula ϕ^*, obtained by the substitution in the thesis ϕ of the formula "$p \to p$" for the variables $p_1, \ldots, p_n$, and of the formula "$\neg (p \to p)$" for the variables $q_1, \ldots, q_k$, is a thesis. The formula ϕ^* has the value 0 for every combination of the values of the variables, and hence its negation, $\neg \phi^*$, is a true formula. By Th.3, $\neg \phi^*$ is a thesis, and hence the system of the sentential calculus expanded by the formula ϕ is contradictory, since the formulae ϕ^* and $\neg \phi^*$ are theses. We have thus proved that if a formula which is not a thesis of the assumptional system is joined to the theses of that system, a contradictory system is obtained.

It can also be proved that the system of the primitive rules of the sentential calculus, as given in Section 2, is independent, which means that none of these rules is secondary with respect to the remaining rules.

Here is an example of reducing a formula to its conjunctive normal form:

$$[(p \equiv q) \to p] \to (p \vee q)$$
$$\equiv [(p \to q) \wedge (q \to p) \to p] \to (p \vee q)$$
$$\equiv \neg [\neg ((\neg p \vee q) \wedge (\neg q \vee p)) \vee p] \vee (p \vee q)$$
$$\equiv [\neg \neg ((\neg p \vee q) \wedge (\neg q \vee p)) \wedge \neg p] \vee (p \vee q)$$
$$\equiv [(\neg p \vee q) \wedge (\neg q \vee p) \wedge \neg p] \vee (p \vee q)$$
$$\equiv (\neg p \vee q \vee p \vee q) \wedge (\neg q \vee p \vee p \vee q) \wedge (\neg p \vee p \vee q)$$
$$\equiv (\neg p \vee q \vee p) \wedge (\neg q \vee p \vee q).$$

In the last step we have made use of the theses: $p \vee p \equiv p$, $p \wedge p \equiv p$, and the laws of commutativity for disjunction and conjunction.

Exercises

1. Reduce the following formulae to the conjunctive normal form:
   a) $[(p \to q) \to q] \to q$,
   b) $[(p \equiv q) \to p] \to p$,
   c) $[(p \to q) \to r] \to (q \to r)$,
   d) $[(p \to q) \to r] \to (\neg p \to r)$,
   e) $[(p \to q) \to r] \to [(r \to p) \to (s \to p)]$.

2. Prove that every formula of the sentential calculus is equivalent to its disjunctive normal form.

3. A variable, its negation, and any formula of the form: $\phi_1 \to \{\phi_2 \to [\ldots \to (\phi_{n-1} \to \phi_n) \ldots]\}$, where each of the formulae $\phi_1, \ldots, \phi_n$ is a variable or a negation of a variable, are called elementary implications. Prove that every formula of the sentential calculus is equivalent to an elementary implication or a conjunction of elementary implications.

4. Formulate a structural criterion of the truth of elementary implications.

6. The Axiomatic Approach to the Sentential Calculus

The sentential calculus can also be constructed by the axiomatic method. When constructing a deductive system by the axiomatic method we choose a certain, usually small, number of sentences which we do not prove by means of other sentences. Such sentences are the *axioms* of the system. We also specify the primitive rules of proving (inference), which make it possible to obtain new theorems. The terms which occur in the axioms or are introduced by the primitive rules of inference are called the *primitive terms* of the system. All other terms of the system must be defined by means of primitive terms or terms which have previously been defined. Definitions may be introduced as the rule of (mutual) replacement of formulae in which the terms being defined do not occur, by formulae in which they do. Definitions may also be introduced as formulae of the system, characterized by a specified structure, usually having the form of an equivalence or an equation, and accepted on the strength of the rules of defining, which lay down what definitions may be joined to the system.[1] The primitive rules of inference and the definitions introduced as the rules of replacement are the primitive rules of the system. Axioms (and possibly the definitions if interpreted as formulae of the system) and all the formulae deducible from these by means of the primitive rules of the system in a finite number of steps are *theses* of the system. Finite sequences of formulae, in which the last term is the theorem being proved and in which every term other than an axiom or a definition is obtained from previous terms of a given sequence by means of some primitive rule of the system, are *proofs* of theorems.

Most known axiomatic systems are systems based on other axiomatic systems in the sense that one system makes use of theses or at least rules of that other system on which it is based. In particular, axiomatic mathematical systems are based on (more or less rich) logical systems. In mathematical systems, all the logical rules of the system serving as the base of a given mathematic system are usually adopted as the rules of inference. On the other hand, in the logical systems there is a clear tendency to specify the least possible number of rules of inference adopted as primitive rules. The axiomatic sentential calculus is constructed as a system which is not based on any other system.

Two systems are *equivalent* if and only if every thesis of one system

---

[1] Cf. Chap. II, Sec. 5.

is a thesis of the other system, and every rule of one system is a rule of the other system.

There are many equivalent systems of the sentential calculus with different axioms, primitive terms, and primitive rules. Among such systems there are some in which all the primitive terms of the assumptional system occur as primitive terms, and also some in which a smaller number of primitive terms occurs. As an example of the former category we may mention the system based on the following axioms:

1. $(p \rightarrow q) \rightarrow [(q \rightarrow r) \rightarrow (p \rightarrow r)]$,
2. $p \rightarrow (q \rightarrow p)$,
3. $[p \rightarrow (p \rightarrow q)] \rightarrow (p \rightarrow q)$,
4. $p \rightarrow \neg \neg p$,
5. $\neg \neg p \rightarrow p$,
6. $(p \rightarrow q) \rightarrow (\neg q \rightarrow \neg p)$,
7. $p \wedge q \rightarrow p$,
8. $p \wedge q \rightarrow q$,
9. $(p \rightarrow q) \rightarrow [(p \rightarrow r) \rightarrow (p \rightarrow q \wedge r)]$,
10. $p \rightarrow p \vee q$,
11. $q \rightarrow p \vee q$,
12. $(p \rightarrow r) \rightarrow [(q \rightarrow r) \rightarrow (p \vee q \rightarrow r)]$,
13. $(p \equiv q) \rightarrow (p \rightarrow q)$,
14. $(p \equiv q) \rightarrow (q \rightarrow p)$,
15. $(p \rightarrow q) \rightarrow [(q \rightarrow p) \rightarrow (p \equiv q)]$.

Instead of Axioms 2 and 3 it suffices to adopt the thesis

$$[p \rightarrow (q \rightarrow r)] \rightarrow [(p \rightarrow q) \rightarrow (p \rightarrow r)],$$

and instead of Axioms 4, 5, and 6 it suffices to adopt the thesis

$$(\neg p \rightarrow \neg q) \rightarrow (q \rightarrow p).$$

The primitive rules of the system are those of substitution and detachment.

Among the axiomatic systems of the sentential calculus with a smaller number of primitive terms there are—according to the choice of primitive terms—systems based on implication and negation, on disjunction and negation, on conjunction and negation, on alternative negation and on the symbol of joint negation. The simplest and best systems of this kind have been constructed by Łukasiewicz. One of his best known systems will be discussed below.[1]

The primitive terms of that system are the symbols of implication and negation. These terms are independent, so that neither of them can be defined by means of the other. But if we have these two terms at our disposal, we can define all truth functors. We give here only the definitions of the symbols of alternation, conjunction, and equivalence:

D1. $$\phi \lor \psi \underset{df}{=\!=} \neg \phi \to \psi,$$

D2. $$\phi \land \psi \underset{df}{=\!=} \neg(\neg \phi \lor \neg \psi),$$

D3. $$\phi \equiv \psi \underset{df}{=\!=} (\phi \to \psi) \land (\psi \to \phi).$$

The symbol "$\underset{df}{=\!=}$", called the *symbol of definitional equality*, is not a functor of the sentential calculus, but a metalogical symbol. In Łukasiewicz's interpretation, definitions are not theses of the system, but parts of what is called the *rule of definitional replacement*. The general schema of this rule has the form

$$\frac{\phi \underset{df}{=\!=} \psi}{\dfrac{\chi}{\chi(\phi//\psi).}}$$

Thus the rule of definitional replacement permits us to join to the system a formula obtained from an arbitrary thesis χ by replacing some part of that thesis, identical with the right-hand side of any definition, by the left-hand side of that definition. This rule hardly differs from the rule of extensionality, and therefore no examples of its application are given here. Note only that the rule which permits the replacement of the left side of a definition by its right side is in Łukasiewicz's system a derived rule (with respect to the thesis $p \equiv p$ and the rule of extensionality, which is derivable in his system). That is why the rule of the mutual replacement of elements of definitions is not required as a primitive rule.

---

[1] That system is discussed in detail in Jan Łukasiewicz's, *Elements of Mathematical Logic*, Warsaw 1963.

Łukasiewicz's system has three axioms:

A1. $$(p \rightarrow q) \rightarrow [(q \rightarrow r) \rightarrow (p \rightarrow r)],$$
A2. $$(\neg p \rightarrow p) \rightarrow p,$$
A3. $$p \rightarrow (\neg p \rightarrow q).$$

In the assumptional system, as described in the preceding Sections, these axioms occurred as theses (T1a, T15a, and T9).

In Łukasiewicz's system the primitive rules—in addition to the rule of definitional replacement—are those of substitution and detachment.

As an example of a proof in Łukasiewicz's system we give the proof of the thesis

(a) $$p \rightarrow p.$$

By substituting the formula $\neg p \rightarrow q$ for q in A1 we obtain

$$[p \rightarrow (\neg p \rightarrow q)] \rightarrow \{[(\neg p \rightarrow q) \rightarrow r] \rightarrow (p \rightarrow r)\}.$$

By applying the rule of detachment to the above and A3 we obtain

$$[(\neg p \rightarrow q) \rightarrow r] \rightarrow (p \rightarrow r).$$

In this thesis we substitute the variable p for the variables q and r:

$$[(\neg p \rightarrow p) \rightarrow p] \rightarrow (p \rightarrow p).$$

By applying the rule of detachment to the above and A2 we obtain formula (a).

Łukasiewicz used a simple and clear method of writing out proofs. But proofs in an axiomatic system are much more difficult and much more complicated than in an assumptional system. Note also that in an axiomatic system every premiss of the proof is a thesis. Hence, proofs in such systems are ordinary direct proofs (cf. p. 14 above).

Methodological problems concerned with the consistency, completeness, independence, etc., of a given system, arise with reference to a given axiomatic system in ways analogous to assumptional systems.

The most important of these properties is the consistency of a system. As shown in Section 3, two contradictory sentences can, by the rules JD and OD, yield any sentence. Hence, if these rules hold in a system, or if a system includes the thesis $p \rightarrow (\neg p \rightarrow q)$ and the rule RD, or the thesis $p \land \neg p \to q$ and the rules JC and RD, then if two contradictory formulae are theses of such a system, every sentential formula recorded in terms of that system is a thesis. A contradictory system is thus devoid of any cognitive value.

There are two methods of proving the consistency of a system: (1) the interpretation method, (2) the method of hereditary property. The concept of interpretation will be discussed later (cf. Part II, Chap. 1, Sec. 3). Let it only be noted here that the interpretation method can be used to prove the consistency of an axiomatic system by reference to the theorem on the consistency of another system, namely that in which a given system is interpreted. The interpretation method is a special case of the method of hereditary property.

The theoretical foundation of consistency proofs by the method of hereditary property is the theorem resulting from the definition of the concept of a thesis of a system. A property is called *hereditary* with respect to a given rule of inference if and only if, whenever the premisses have that property, the conclusion obtained from them by that rule also has that property. The consistency of a given axiomatic system is proved by the method of hereditary property by proving that there is a property which is an attribute of all the axioms and is inherited according to each of the primitive rules of the system, and is not an attribute of any two contradictory formulae. For if all the axioms have a certain property, which is inherited according to each primitive rule of the system, then that property is transferred from the axioms to all the formulae obtained from the axioms in a finite number of steps in accordance with the primitive rules of the system, and hence that property is an attribute of every thesis of the system. It follows that if a formula does not have that property, it is not a thesis of the system. And since such a property is not an attribute of any two contradictory formulae, two contradictory formulae cannot both be theses of the system, and hence such a system is consistent.

This method is used to prove the consistency of the axiomatic systems of the sentential calculus given in the present book; it is shown that every axiom is a true formula and that this property is inherited according to the rules of substitution and detachment, and is not an attribute of any two contradictory formulae.

It also follows, from the fact that the axioms of these systems are true formulae and that that property is inherited according to the primitive rules of these systems, that every thesis of these systems is a true formula.

It can also be proved that both axiom systems are independent. A given axiom is independent of the remaining axioms of a system if and only if it cannot be proved on the strength of the remaining axioms and in accordance with the primitive rules of the system concerned, or, to put it briefly, if it is not a consequence of the remaining axioms of the system. The method of hereditary property is used to prove that a given axiom is independent of the remaining axioms of the system; to do so it is demonstrated that there is a property which is an attribute of the remaining axioms of the system and is inherited according to the primitive rules of the system, and is not an attribute of the axiom in question. For then that property is an attribute of all the consequences of the remaining axioms, and from the fact that the axiom in question does not have that property it follows that it is not a consequence of any of the remaining axioms, and hence that it is independent of them.

The independence of an axiomatic system is, obviously, a desirable property, but is not so essential as its consistency. Sometimes the problem of independence is very difficult to solve, and only many long efforts result in proving the independence of certain axiomatic mathematical systems.

In addition to the independence of an axiom system we may also speak of the independence of the set of the primitive terms of a given system. The set of primitive terms of a system is independent if and only if none of these terms can be defined by means of the remaining primitive terms of that system. As has been mentioned above, the primitive terms of Łukasiewicz's system are independent, whereas the set of the primitive terms of the first of the axiomatic systems of the sentential calculus given above is not independent.

It can also be proved that both axiomatic systems of the sentential calculus given above are complete (in both senses of the term).[1] The conditions of completeness are satisfied only by a few, and rather

---

[1] The proof of the completeness (in both senses of the term) of Łukasiewicz's system is given in his book quoted above (pp. 81–91).

poor, deductive systems. As a rule, the proofs of completeness are fairly difficult and complicated methodological proofs concerned with deductive systems.

The equivalence of two deductive systems is usually demonstrated by proving that every axiom of one system is an axiom or thesis of the other, and that every primitive rule of one system is a primitive or derived rule of the other.

In the same way it is possible to demonstrate the equivalence of the assumptional system of the sentential calculus with each of the axiomatic systems of that calculus as given in the present book. It can easily be verified that the axioms of those systems are theses of the assumptional system, and that their primitive rules are valid in the assumptional system. For instance, Łukasiewicz's axioms occur in the assumptional system as theses T1a, T15a, T9. The rule of detachment is one of the primitive rules of the assumptional system. The rule of substitution is in that system a derived rule, which has been demonstrated in Sec. 3. Theses T29a, T34, and T14 are counterparts of definitions D1, D2, D3 in Łukasiewicz's system. From these theses, or their substitutions, and by the rule of extensionality (given *in fine* of Sec. 3) we can obtain in the assumptional system every thesis which in Łukasiewicz's system can be obtained on the strength of D1–D3 and the rule of definitional replacement. And the rule of ordinary direct proof is a derived rule of the assumptional system (cf. Sec. 2, p. 14).

In the proof of the converse theorem, and in particular in the proof showing that the rule of constructing the assumptional indirect proof is derivable in the axiomatic system, an essential role is played by what is called the deduction theorem for the sentential calculus.

Before that theorem is formulated and proved, note that each of the axiomatic systems given here is equivalent to the system in which the rule of detachment is the only primitive rule, and in which all the substitutions of axioms of a given system are axioms. For instance, the system based on the rule of detachment as the only primitive rule of inference and such that all its formulae that have one of the following three forms:

$$(\phi \to \psi) \to [(\psi \to \chi) \to (\phi \to \chi)],$$

$$(\neg\phi \rightarrow \phi) \rightarrow \phi,$$
$$\phi \rightarrow (\neg\phi \rightarrow \psi),$$

are axioms, is equivalent to Łukasiewicz's system.

For the sentential calculus in which the rule of detachment is the only primitive rule of inference, the concept of a derivation of a given formula from definite premisses and the concept of consequence are defined.

Def. I. The sequence of the formulae $\psi_1, ..., \psi_l$ is a *derivation* of the formula ϕ_n from the formulae (premisses) $\phi_1, ..., \phi_{n-1}$ if and only if $\psi_l = \phi_n$ and for every $1 \leqslant i \leqslant l$ either ψ_i is an axiom, or ψ_i is one of the formulae $\phi_1, ..., \phi_{n-1}$, or there are $k, j < i$ such that $\psi_k = (\psi_j \rightarrow \psi_i)$.

Hence the derivation of a given formula from a definite set of premisses is a finite sequence of formulae the last term of which is the formula being derived, and in which every term either is an axiom, or is one of the premisses, or is obtained from preceding terms of the sequence by the rule of detachment.

Def. II. $\phi_1, ..., \phi_{n-1} \vdash \phi_n$ if and only if there is a derivation of the formula ϕ_n from the formulae $\phi_1, ..., \phi_{n-1}$.

The formula "$\phi_1, ..., \phi_{n-1} \vdash \phi_n$" may be read as follows:

The formula ϕ_n is a consequence of the formulae $\phi_1, ..., \phi_{n-1}$;

or

The formula ϕ_n is derivable from the formulae $\phi_1, ..., \phi_{n-1}$.

A formula which is a consequence of axioms, that is, a formula for which a proof (i.e. derivation from axioms) exists, is a thesis.

These definitions lead to the following conclusions.

Cncl. 1. *If* $\vdash \phi \rightarrow \psi$, *then* $\phi \vdash \psi$.

This is so because if the sequence $\psi_1, ..., \psi_l$ is a proof of the thesis $\phi \rightarrow \psi$, then the sequence $\psi_1, ..., \psi_l, \phi, \psi$ is a derivation of the formula ψ from the formula ϕ.

Cncl. 2. *If* $\phi_1, ..., \phi_n \vdash \psi$ *and* $\psi \vdash \chi$, *then* $\phi_1, ..., \phi_n \vdash \chi$.

Proof. Assume that $\phi_1, ..., \phi_n \vdash \psi$ and $\psi \vdash \chi$. Let the sequence

$\psi_1, \ldots, \psi_l$ be a derivation of the formula ψ from the formulae $\phi_1, \ldots,$ ϕ_n, and let the sequence $\psi, \chi_1, \ldots, \chi_m$ be a derivation of the formula χ from the formula ψ. Then the sequence $\psi_1, \ldots, \psi_l, \chi_1, \ldots, \chi_m$ is the derivation of the formula χ from the formulae $\phi_1, \ldots, \phi_n$.

Cncl. 3. *If* $\phi_1, \ldots, \phi_n \vdash (\psi \to \chi)$ *and* $\phi_1, \ldots, \phi_n \vdash \psi$, *then* $\phi_1, \ldots,$ $\phi_n \vdash \chi$.

Proof. Let the sequence $\phi_1, \ldots, \phi_n, \psi_1, \ldots, \psi_l$ be a derivation of the implication $\psi \to \chi$ from the formulae $\phi_1, \ldots, \phi_n$, and let the sequence $\phi_1, \ldots, \phi_n, \chi_1, \ldots, \chi_m$ be a derivation of the formula χ from the formulae $\phi_1, \ldots, \phi_n$. Then the sequence $\phi_1, \ldots, \phi_n, \psi_1, \ldots,$ $\psi_l, \chi_1, \ldots, \chi_m, \chi$ is a derivation of the formula χ from the formulae $\phi_1, \ldots, \phi_n$.

We shall now consider systems of the sentential calculus in which the rule of detachment is the only primitive rule of inference, and in which the formulae:

(a) $\phi \to (\psi \to \phi)$,
(b) $[\phi \to (\psi \to \chi)] \to [(\phi \to \psi) \to (\phi \to \chi)]$,
(c) $\phi \to \phi$.

are theses. (Thesis (c) is deducible from theses (a) and (b) by the rule of detachment.)

For systems which satisfy these conditions the following deduction theorem is valid:

Th.1. *If* $\phi_1, \ldots, \phi_{n-1} \vdash \phi_n$, *then* $\phi_1, \ldots, \phi_{n-2} \vdash (\phi_{n-1} \to \phi_n)$.

This theorem states that if ϕ_n is a consequence of the formulae $\phi_1, \ldots, \phi_{n-1}$, then the implication $\phi_{n-1} \to \phi_n$ is a consequence of the formulae $\phi_1, \ldots, \phi_{n-2}$.

For $n = 2$, Th.1 takes on the form:

$$\textit{If } \phi_1 \vdash \phi_2, \textit{ then } \vdash (\phi_1 \to \phi_2).$$

This states that if ϕ_2 is a consequence of the formula ϕ_1, then the implication $\phi_1 \to \phi_2$ is a thesis.

Proof of Th.1. Assume that $\phi_1, \ldots, \phi_{n-1} \vdash \phi_n$. Hence there exists a sequence of formulae

(I) $$\psi_1, \ldots, \psi_t$$

which satisfy the conditions laid down in Def. I; the said sequence is a derivation of the formula ϕ_n from the formulae $\phi_1, \ldots, \phi_{n-1}$.

We shall demonstrate that the sequence of formulae

(II) $$\phi_{n-1} \to \psi_1, \ldots, \phi_{n-1} \to \psi_l,$$

after a suitable completion of its terms, is a derivation of the implication $\phi_{n-1} \to \phi_n$ from the formulae $\phi_1, \ldots, \phi_{n-2}$.

By assumption, $\psi_l = \phi_n$, so that the implication $\phi_{n-1} \to \phi_n$ is the last term of sequence (II). Let the natural number i satisfy the condition: $1 \leqslant i \leqslant l$. By assumption and by Def. I one of the following cases holds:

(1) ψ_i is an axiom,

(2) ψ_i is one of the formulae $\phi_1, \ldots, \phi_{n-1}$,

(3)   there exist $k, j < i$ such that $\psi_k = (\psi_j \to \psi_i)$.

In case (1), the formula $\phi_{n-1} \to \psi_i$ is a thesis by thesis (a) and by the rule of detachment.

In case (2), either (α) $\psi_i = \phi_{n-1}$, or (β) ψ_i is one of the formulae $\phi_1, \ldots, \phi_{n-2}$. If ($\alpha$), then the formula $\phi_{n-1} \to \psi_i$ is a thesis by thesis (c). If (β), then by thesis (a) the formula $\psi_i \to (\phi_{n-1} \to \psi_i)$ is a thesis, and hence, by Cncl. 1, the formula $\phi_{n-1} \to \psi_i$ is a consequence of the formulae $\phi_1, \ldots, \phi_{n-2}$.

In case (3), the formulae $\phi_{n-1} \to (\psi_j \to \psi_i)$, $\phi_{n-1} \to \psi_j$ precede the formula $\phi_{n-1} \to \psi_i$ in the sequence (II), and it is from them that we obtain the formula $\phi_{n-1} \to \psi_i$ on the strength of thesis (b) and the rule of detachment.

Hence, in the sequence (II) each term either is a thesis, or is a consequence of the formulae $\phi_1, \ldots, \phi_{n-2}$, or is obtained from earlier terms of the sequence by means of the rule of detachment.

The sequence (II) is completed so that every term which is a thesis is preceded by a proof of that thesis (on the strength of axioms), and every formula which is a consequence of the formulae $\phi_1, \ldots, \phi_{n-2}$ is preceded by its derivation from the formulae $\phi_1, \ldots, \phi_{n-2}$.

The sequence (II') obtained in this way is a derivation of the implication $\phi_{n-1} \to \phi_n$ from the formulae $\phi_1, \ldots, \phi_{n-2}$, since each term either is an axiom, or is one of the formulae $\phi_1, \ldots, \phi_{n-2}$, or is obtained from earlier terms of the sequence by means of the rule of detachment, and the last term of the sequence is the formula $\phi_{n-1} \to \phi_n$.

Note that if $n = 2$, then in case (2) we have only the case (α), and then the sequence (II') is a proof of the thesis $\phi_1 \to \phi_2$.

From Th.1, by its iterated application, we obtain:

Th.2. *If* $\phi_1, \phi_2, ..., \phi_{n-1} \vdash \phi_n$, *then* $\vdash \phi_1 \rightarrow \{\phi_2 \rightarrow [... (\phi_{n-1} \rightarrow \phi_n) ...]\}$.

In both axiomatic systems of the sentential calculus referred to in this Section the formula:

(d) $$(\neg\phi \rightarrow \psi) \rightarrow [(\neg\phi \rightarrow \neg\psi) \rightarrow \phi]$$

is a thesis.

If Th.1 is valid for a given system of the sentential calculus, and if that system includes thesis (d), then the following theorem is valid for that system:

Th.3. *If* $\phi_1, \phi_2, ..., \phi_{n-1}, \neg\phi_n \vdash \psi$ *and* $\phi_1, \phi_2, ..., \phi_{n-1}, \neg\phi_n \vdash \neg\psi$, *then* $\vdash \phi_1 \rightarrow \{\phi_2 \rightarrow [... \rightarrow (\phi_{n-1} \rightarrow \phi_n) ...]\}$.

Proof.
(1) $\phi_1, ..., \phi_{n-1}, \neg\phi_n \vdash \psi$ $\}$ {a.}
(2) $\phi_1, ..., \phi_{n-1}, \neg\phi_n \vdash \neg\psi$
(3) $\phi_1, ..., \phi_{n-1} \vdash (\neg\phi_n \rightarrow \psi)$ {Th.1; 1}
(4) $\phi_1, ..., \phi_{n-1} \vdash (\neg\phi_n \rightarrow \neg\psi)$ {Th.1; 2}
(5) $\phi_1, ..., \phi_{n-1} \vdash [(\neg\phi_n \rightarrow \neg\psi) \rightarrow \phi_n]$ {Cncl.1; Cncl.2; (d); 3}
(6) $\phi_1, ..., \phi_{n-1} \vdash \phi_n$ {Cncl.3; 5; 4}
$\vdash \phi_1 \rightarrow \{\phi_2 \rightarrow [... \rightarrow (\phi_{n-1} \rightarrow \phi_n) ...]\}$ {Th.2; 6}

The concept of consequence, as defined in Def. I, may be expanded by the introduction, in addition to the rule RD, also of the rules JC, OC, JD, OD, JE, OE, as the rules which permit the derivation of further proof lines according to preceding proof lines. For the concept of consequence defined in this way, the proof of Th.1 does not differ essentially from the proof given above; it merely requires the analysis of a greater number of cases and reference to the appropriate theses of the axiomatic system, namely those in which the symbols of conjunction, disjunction, and equivalence occur. Th.1, formulated for such an expanded concept of consequence, is an analogue of the rule of joining implication in the assumptional system. Th.2 is then an analogue of the rule of constructing direct proofs from assumptions, and Th.3, an analogue of the rule of constructing indirect proofs from assumptions. The rules JC, OC, JD, OD, JE,

OE, then prove to be—as the rules of joining new proof lines in proofs from assumptions—derived rules in the axiomatic system. In this way we conclude that the primitive rules of the assumptional system are (derived) rules in the axiomatic system. Since, on the other hand, the axioms of the axiomatic system are theses of the assumptional system, and the primitive rules of the axiomatic system are rules holding in the assumptional system, these two systems are equivalent.

The first axiomatic system of the sentential calculus was constructed by Frege. In his paper *Some problems in the History of Sentential Logic* Łukasiewicz describes this fact as follows: "And now we suddenly confront a phenomenon which is unprecedented in the history of logic. Without any intermediary, so that the fact cannot be explained on historical grounds, contemporary sentential logic emerges, in an almost perfect form, from the inspired brain of Gottlob Frege, the greatest logician of our times. In 1879 Frege published his small but conceptually important paper, *Begriffsschrift, eine der arithmetischen nachgebildete Formelsprache des reinen Denkens*. In this paper all sentential logic is presented for the first time in a strictly axiomatic form as a deductive system."

Frege's system, in which the symbols of implication and negation are primitive terms, is based on six axioms and the rules of substitution and detachment. The axioms are the theses given in Section 3: T10, T6 in the form of an implication, T5, T5a, the thesis $[p \rightarrow (q \rightarrow r)] \rightarrow [(p \rightarrow q) \rightarrow (p \rightarrow r)]$, called *Frege's syllogism*, and the law of commutativity (T3) in its implicational form. The last-named thesis, as was demonstrated by Łukasiewicz, is provable by means of the remaining axioms which form an independent system of axioms for the sentential calculus.

In *Principia Mathematica* by Whitehead and Russell the sentential calculus is based on the symbols of disjunction and negation as primitive terms. Referring to the result obtained by Sheffer who defined all the remaining truth functors by means of alternative disjunction (*Sheffer's stroke*), Nicod constructed a system of sentential calculus based on that truth functor and on a single axiom. Łukasiewicz's system, described in this section, dates from 1924 and is one of the many axiomatic systems of the sentential calculus constructed by him. They include systems based on implication and negation as the primitive terms and on a single axiom.

As can be seen, axiomatic systems of the sentential calculus emerged earlier than assumptional systems which, as mentioned early in Section 2, emerged only in the 1930's. This can be explained by the fact that the axiomatic method of constructing systems had long been known and used in logic and mathematics. As shown by Łukasiewicz, Aristotle's syllogistic was the first axiomatic system known in the history of human thought, to be followed by Euclid's geometry. In our times all mathematical disciplines which have developed sufficiently are constructed as axiomatic systems (although in so doing mathematicians intuitively resort to proofs from assumptions). Moreover, the methodological structure of an axiomatic system is essentially simpler than that of the assumptional systems, although, as mentioned above, the methods of proof from assumptions are much simpler than the methods of proof used in the systems built exclusively by the axiomatic method.

The sentential calculus, presented in Sections 3–5 by the assumptional, truth table, and axiomatic methods, respectively, is called the classical (or two-valued) sentential calculus. Stanisław Leśniewski (died 1939), the most eminent Polish logician next to Łukasiewicz, introduced into the sentential calculus variables that stand for truth functors and thus built a more general system, which he called protothetics. His system can be based on a single axiom which corresponds to the principle of bivalence (cf. p. 47 above).

Besides the classical sentential calculus other non-classical systems of the sentential calculus been have constructed in recent times. The best known among them are:

(1) *Modal systems*, which include modal truth functors of the sentential calculus, such as "it is necessary that", "it is possible that", etc. Towards the end of Section 4, one of the best known systems of this kind was mentioned: Lewis's system of strict implication. This system was first constructed by the axiomatic method, next truth tables were found for it (which proved to be tables with infinitely many values), and in recent years the system has also been constructed by the assumptional method.

(2) Łukasiewicz's *many-valued systems*. Beginning with his research

in 1918, Łukasiewicz constructed by the truth table method, a three-valued logic, and later many-valued logics. The building of many-valued logics was one of the greatest logical discoveries of the 20th century. Łukasiewicz constructed his three-valued logic as a modal system, and endeavours have been made to establish connections between the logic of infinitely many values and the calculus of probability. The American logician Post also constructed many-valued logics independently of Łukasiewicz, but he gave no interpretations for them. Many-valued systems of the sentential calculus were first constructed by the truth table method, and next by the axiomatic method.

(3) The *system of the intuitionist sentential calculus* constructed by Heyting by the axiomatic method, following the ideas of the Dutch mathematician Brouwer. The counterparts of some laws of the classical calculus, such as the law of the excluded middle, the theses $\neg\neg p \rightarrow p$, $(\neg p \rightarrow p) \rightarrow p$, etc., are not valid for the functors of the intuitionist sentential calculus. The intuitionists modify the sentential calculus in order to prove by the logic based on that calculus the existence of only such objects which can be "constructed". Intuitionism is one of the trends in the foundations of mathematics. The intuitionist sentential calculus was next constructed by the assumptional method, and the truth tables found for that system have proved to be of infinitely many values.

It must be emphasized that when constructing non-classical sentential calculi we need not reject the classical system. In each of the non-classical systems described above it is possible, following the introduction of appropriate definitions, to obtain the classical sentential calculus as part of such a non-classical system. Thus the construction of such systems may be interpreted as a complement of the classical system, as the construction of a more comprehensive system. In particular, as shown by Łukasiewicz, if we want to introduce modal functors into the sentential calculus we must go beyond the limits of the classical sentential calculus, in which these functors cannot be formulated.

CHAPTER 2

THE FUNCTIONAL CALCULUS

1. The Functional Calculus of the First Order. Symbols and Formulae

The *functional calculus*, also called the *predicate calculus* and *quantification theory* is based on the sentential calculus in the sense that all the symbols of the latter are included in the functional calculus, which also has new symbols added. Likewise it is assumed that the rules, both primitive and derived, of the sentential calculus are valid in the functional calculus, although they do not exhaust all the rules of the latter.

Apart from the symbols taken over from the sentential calculus, the functional calculus includes the following symbols:

1. (*Individual*) *term variables*:

$$x, y, z, x_1, y_1, z_1, \ldots$$

When the formulae of the functional calculus are used in definite reasoning, singular terms, that is names denoting one object only, are substituted for those variables. This is also quite often formulated so that the term variables represent, or stand for, singular terms. In general, we say that a variable represents, or stands for, certain constants when these constants may be substituted for the variable. In this sense, sentential variables stand for sentences.

2. *Variables standing for sentence-forming functors of term arguments*, that is, expressions which together with a certain number of terms yield sentential expressions. These functors are called *predicates*. Predicates are classified as predicates of one argument, (called the *monadic predicates*), of two arguments, etc. As examples of predicates of one argument we may quote the word "shines" in the sentence "the sun shines", the expression "is an even number" in the sentence

(1)   "4 is an even number".

Examples of predicates of two arguments are: the symbol ">" in the sentence

(2)   "$5 > 3$",

the expression "is a multiple of" in the sentence

"15 is a multiple of 5",

the expression "is the father of" in the sentence

"John is the father of Peter".

Examples of predicates of three arguments are offered by the expression "lies between ... and ..." in the sentence

"Warsaw lies between Moscow and Berlin",

the expression "is a common divisor of ... and ... " in the sentence

(3)   "4 is a common divisor of 12 and 20".

As the variables representing predicates of one argument we shall use the letters:

$$A, B, C, A_1, B_1, C_1, \ldots$$

As the variables representing predicates of two or more arguments we shall use the letters:

$$P, Q, R, S, P_1, Q_1, R_1, S_1, \ldots$$

while the context indicates how many arguments the predicate represented by a given variable has.

3. *Constants*: $\prod, \sum$. These constants are called *quantifiers*. The symbol $\prod$ is called the *universal quantifier*, and the symbol $\sum$ the *existential quantifier*. The former is read: "for every ...", the latter, "for some ..." or "there is a ... such that".

Together with the symbols taken over from the sentential calculus, these symbols exhaust the set of symbols used in that part of the functional calculus which is called the *functional calculus of the first order* or *restricted functional calculus*. In Sections 1–3 of this Chapter we shall be concerned with the functional calculus of the first order.

The following is a definition by induction of a *sentential formula* in the functional calculus of the first order:

1a. Sentential variables are sentential formulae in the functional calculus of the first order.

1b. Formulae constructed from a variable representing predicates of n arguments, followed by a bracketed sequence of n term variables $(n \geqslant 1)$, are sentential formulae of the functional calculus of the first order. Such formulae are called *atomic* or *molecular*. Thus, for instance, $A(x)$, $A(y)$, $B(x)$, $R(x_1, ..., x_n)$ are atomic. In the case of predicates of two arguments we write "xRy" instead of "$R(x, y)$", "xSy" instead of "$S(x, y)$", etc.

2a. If ϕ and ψ are formulae of the functional calculus of the first order then $\neg(\phi)$, $(\phi) \wedge (\psi)$, $(\phi) \vee (\psi)$, $(\phi) \rightarrow (\psi)$, $(\phi) \equiv (\psi)$ are formulae of that calculus.

2b. If ϕ is a sentential formula of the functional calculus of the first order and α is a term variable, then $\prod_{\alpha}(\phi)$ and $\sum_{\alpha}(\phi)$ are sentential formulae of that calculus.

3. The set of sentential formulae of the functional calculus of the first order includes only those which are sentential variables or molecular formulae or are formulae formed from these formulae by single or iterated application of rules 2a and 2b.

By 1a and 2a every formula of the sentential calculus is a sentential formula of the functional calculus.

Formulae constructed in accordance with 2b consist of three parts:

1. the quantifier "$\prod$" or "$\sum$",
2. the term variable under the quantifier,
3. the sentential formula ϕ in brackets, called the *scope* of the quantifier which precedes it.

The brackets are omitted if ϕ is a molecular formula, so that instead of "$\prod_x(A(x))$" we write "$\prod_x A(x)$".[1] The brackets are also omitted if ϕ begins again with a quantifier, so that instead of "$\prod_x(\sum_y xRy)$" we write "$\prod_x \sum_y xRy$".

---

[1] This formula is read: for every x, $A(x)$; the formula "$\sum_x A(x)$" is read: for some x, $A(x)$.

Here are some examples of formulae of the functional calculus of the first order:

(4) $$\prod_x xRx,$$

(5) $$\prod_x \sum_y xRy,$$

(6) $$\prod_x \prod_y [xRy \rightarrow \sum_z (xRz \wedge zRy)],$$

(7) $$\prod_x \prod_y \sum_z R(x, y, z).$$

These examples will be used to illustrate the concept of the scope of the quantifier, as introduced above. Thus, in formula (4) the scope of the quantifier "$\prod$" is the formula "xRx"; in formula (5), the scope of the quantifier "$\sum$" is the formula "xRy", and the scope of the quantifier "$\prod$" is the formula "$\sum_y xRy$"; in formula (6), the scope of the quantifier "$\sum$" is the formula "$xRz \wedge zRy$", the scope of the quantifier "$\prod$" with the variable "y" below is the formula in square brackets, and the scope of the quantifier "$\prod$" with the variable "x" below is the formula "$\prod_y [xRy \rightarrow \sum_z (xRz \wedge zRy)]$".

We say that a variable in a formula ϕ that is not written under a quantifier is *bound* in the formula ϕ if and only if it occurs in ϕ within the scope of a quantifier and is identical with the variable written under that quantifier. We say that a quantifier *binds* the variable which is written under it. For instance, in formula (4) the variable "x" which directly precedes the variable "R" is bound.

We say that a variable which occurs in ϕ and which does not occur under a quantifier is *free* in the formula ϕ if and only if it is not bound in ϕ. For instance, the variable "x" is free in the formula "$\sum_y xRy$"; in the formula "$\prod_x A(x) \rightarrow A(x)$" the variable "$x$" at the end of the formula is free, while the variable "x" which precedes the implication symbol in the formula is bound.

In view of conditions 2b and 1b of the definition of a sentential formula of the functional calculus of the first order, we may describe such a system as that part of the entire functional calculus in which quantifiers bind only term variables and in which molecular formulae include only sentence-forming functors of term arguments (and do not include sentence-forming functors whose arguments are functors).

The important concept of a sentential function can now be introduced: A *sentential function* is a formula which includes free variables and which yields sentences on substitution of constants for all these variables.

Among sentential functions we distinguish those of one free variable (e.g., "x is an even number"), those of two free variables (e.g., "$x > y$"), those of three free variables (e.g., "x is a divisor of y and z"), etc. By substituting in these sentential functions the names of appropriate numbers for the variables we obtain sentences (1)—(3) given earlier in this section. Equations are familiar examples of sentential functions. The term "condition" is often used to denote sentential functions.

The concept of a sentential function must, of course, be distinguished from that of a function in the mathematical sense of the term. Sentential functions are formulae of a certain kind, whereas mathematical functions are not formulae but certain relations (cf. Part II, Chap. I, Sec. 6).

Formulae of the functional calculus are interpreted so that the formula "$A(x)$" represents sentential functions containing the free variable "x", and the formula "$R(x_1, \ldots, x_n)$", sentential functions containing the free variables $x_1, \ldots, x_n$. In other words: sentential functions containing the free variable "x" may be substituted for "$A(x)$", and sentential functions containing the free variables $x_1, \ldots, x_n$, for "$R(x_1, \ldots, x_n)$".[1]

For instance, if the sentential function "$x = y+1$" is substituted for the formulae "xRy", then formula (5) yields the sentence

$$\prod_x \sum_y x = y+1,$$

and if the sentential function "$z = x+y$" is substituted for the formula "$R(x, y, z)$", then formula (7) yields the sentence

$$\prod_x \prod_y \sum_z (z = x+y).$$

---

[1] A precise formulation of this rule of substitution, called the rule of substitution for functional formulae, is given in Sec. 5.

The following interpretation of the formulae of the functional calculus is often adopted:

Predicates of one argument denote *properties,* and predicates of two or more arguments denote *relations* (between two or more elements). Molecular formulae of the form "$A(x)$" are correspondingly read: "the property A is an attribute of the object x", or "the object x has the property A"; formulae of the form "$R(x_1, ..., x_n)$" are read: "the relation R holds between the objects $x_1, ..., x_n$", formulae of the form "$x R y$" are also read: "the object x is in the relation R to the object y". The terms "property" and "relation" may, in this interpretation, be understood so broadly that every sentential function with one free term variable refers to a certain property, and every sentential function with n ($n > 1$) free term variables refers to a certain relation between n elements.

Beside ordinary quantifiers discussed above the functional calculus also includes what are called *quantifiers with a limited range.*

Here are examples of formulae containing such quantifiers

$$(8) \qquad \qquad \prod_{x>0} |x| = x,$$

$$(9) \qquad \qquad \sum_{x \neq 0} a . x = a.$$

$\displaystyle\prod_{x>0}$ is read:

$$\text{``for every } x \text{ greater than 0'',}$$

$\displaystyle\sum_{x \neq 0}$ is read:

$$\text{``for some } x \text{ other than 0''.}$$

Formula (8) is an abbreviation of

$$\prod_{x} (x > 0 \rightarrow |x| = x);$$

formula (9) is an abbreviation of

$$\sum_{x} (x \neq 0 \wedge a . x = a).$$

In general, the formula $\displaystyle\prod_{\phi(\alpha)} \psi(\alpha)$ is an abbreviation of

$$\prod_{\alpha} [\phi(\alpha) \rightarrow \psi(\alpha)]$$

and the formula $\sum\limits_{\phi(\alpha)} \psi(\alpha)$ is an abbreviation of

$$\sum_{\alpha}[\phi(\alpha) \wedge \psi(a)],$$

where $\phi(\alpha)$ and $\psi(\alpha)$ are sentential functions in which α is a free variable.

If the formula $\phi(\alpha)$ is in the scope of a quantifier with a limited range and includes more than one free variable, doubt may arise as to which variable is bound by the quantifier. For instance, is the formula

(10)
$$\prod_{|x|>|y|} x^2 > y^2$$

an abbreviation of $\prod\limits_{x}(|x| > |y| \to x^2 > y^2)$, or an abbreviation of $\prod\limits_{y}(|x| > |y| \to x^2 > y^2)$?

To avoid this doubt we assume that a quantifier with a limited range binds the variable which is written directly under it. Hence in formula (10) the variable x is bound, and in $\prod\limits_{|y|<|x|} x^2 > y^2$ the variable y is bound.

Quantifiers, especially those with a limited range, are often used in writing down mathematical definitions and theorems. We shall now give several examples of this kind (in which we assume that the variables x, y, z range over the set of real numbers, and the variables k, m, n, over the set of natural numbers).

$$x > y \equiv \sum_{z>0} x = y + z,$$

$$x < y \equiv \sum_{z} x < z < y,$$

$$n/m \equiv \sum_{k} m = n.k.$$

Definition of the limit of a sequence:

$$\lim_{n\to\infty} a_n = l \equiv \prod_{\varepsilon>0} \sum_{m} \prod_{n>m} |a_n - l| < \varepsilon.$$

Definition of a continuous function:

f is a continuous function

$$\equiv \prod_{\varepsilon>0} \prod_{x} \prod_{y} \sum_{\delta>0} (|x-y| < \delta \to |f(x) - f(y)| < \varepsilon).$$

Definition of a uniformly continuous function:

f is a uniformly continuous function

$$\equiv \prod_{\varepsilon>0} \sum_{\delta>0} \prod_x \prod_y (|x-y| < \delta \rightarrow |f(x)-f(y)| < \varepsilon).$$

Frege was the first logician to introduce quantifiers, but his work, quoted in the preceding chapter and published in 1879, remained unnoticed at that time, perhaps because of the very difficult symbolism he used. *Circa* 1885, Peirce introduced the symbols $\prod$ and $\sum$ to represent quantifiers. He also noticed that the sentence $\prod_x A(x)$ may be treated as the conjunction:

$$A(x_1) \wedge A(x_2) \wedge A(x_3) \wedge \ ...,$$

and that the sentence $\sum_x A(x)$ may be treated as the disjunction:

$$A(x_1) \vee A(x_2) \vee A(x_3) \vee \ ...$$

But in this connection he emphasizes that these are only analogues, since the number of objects may be infinite, and, as is known, conjunctions and disjunction may be built of a finite number of elements only.

The term "propositional function" (in this book replaced by "sentential function") was introduced by Russell in 1903, while sentential functions of many variables were studied as early as 1892–3 by Peirce and Frege.

A comprehensive and systematic treatment of the functional calculus was given in Vol. I of *Principia Mathematica* by Whitehead and Russell (1910). In their work this section is entitled "Theory of Apparent Variables", because bound variables were then called *apparent variables* (while free variables were called *real variables*).

Exercises

1. With the help of quantifiers write down the sentences:
    (a) The numbers 3 and 10 have no common divisor other than 1.
    (b) No natural number is less than 0.
    (c) For every natural number there exists a natural number greater than it.
    (d) There exists a natural number such that no other natural number is less than it.

(e) The system of equations:
$$x+2 = 5,$$
$$2x+4 = 6$$
is contradictory.

(f) Every pupil of the second form has read a book.

(g) There is a book which has been read by every second-form pupil.

2. Write down, with the help of quantifiers, the definitions of:

    (a) continuity of a function on the set S;

    (b) uniform continuity of a function on the set S;

    (c) convergence of the sequence of functions $\{f_n\}$ at every point of the set S;

    (d) uniform convergence of the sequence of functions $\{f_n\}$ on the set S.

3. Indicate which variables in the following formulae are free, and which are bound:

$$\prod_x [A(x) \to B(x)], \quad \prod_x A(x) \to B(x),$$
$$\prod_x [\sum_y (xRy) \to (yRx)], \quad \prod_{xRy} [A(x) \lor A(y)],$$
$$\prod_{xRy} [A(x) \to A(y)].$$

4. Rewrite the following formulae without resorting to quantifiers with a limited range:

$$\prod_{A(x)} \sum_{B(y)} [A(x) \to B(y)], \quad \prod_{xPy} \sum_{yQx} xRy.$$

2. Primitive Rules of the Functional Calculus of the First Order

The primitive rules of the functional calculus of the first order are:

1. *All the primitive rules of the sentential calculus*, generalized so as to cover not only formulae of the sentential calculus, but also the sentential formulae in the functional calculus.

We therefore assume that in the schemata of the rules RD, JC, OC, JD, OD, JE, OE the symbols ϕ and ψ stand for any sentential formulae of the functional calculus. The formulae $\phi_1, \ldots, \phi_n$, which are component formulae of formula (I), referred to in connection with defining the rule for constructing an indirect proof from assumptions (cf. Chap. I, Sec. 2, pp. 13–15), are interpreted in an analogous way. The rule of joining to the proof theses which have previously been proved is extended to cover not only theses of the sentential calculus, but also those of the functional calculus.

2. *The rules for joining and omitting the universal and the extential quantifier.*

In writing down the schemata of the rules for joining and omitting quantifiers we shall use the symbol $\phi(\alpha/\xi)$ which stands for the formula obtained from ϕ by the substitution of the expression ξ for the term variable α. When such a substitution is performed, the following conditions must be satisfied:

(1) The variable α in the formula ϕ is replaced only where it is free. If α occurs more than once as a free variable it is replaced by the expression ξ as often as it occurs in free form.

(2) If the variable α occurs within the scope of the quantifier which binds the variable β, no expression containing the free variable β may be substituted for the variable α. In other words: no substitution may be performed which has the effect of binding variables which were free.

The second condition is necessary, since otherwise in some cases substitution would lead from true to false sentences. For instance, the formula "$\sum_m m > n$", in which the variables m and n range over the set of natural numbers, is true since it states that for any natural number n there exists a number greater than it. But it is not permitted to substitute in that formula the expression "$m+1$" for the free variable "n", since the variable "m", free in the expression "$m+1$", would become bound in the result of the substitution, i.e. in the sentence "$\sum_m m > m+1$". This sentence is also obviously false.

It is assumed that if α is a free variable in ϕ, then the symbol $\phi(\alpha/\xi)$ is meaningful only if ξ may be substituted for α in the formula ϕ in a way which satisfies both conditions. Hence, if, for instance, ϕ is the formula "$\sum_m m > n$", then the symbol $\phi(n/m)$ is meaningless and does not denote any formula.

It is also assumed that if α is not a free variable in ϕ, then $\phi(\alpha/\xi)$ is identical with the formula ϕ.

The rule for omitting the universal quantifier may be written in the form of the schema:

$$\text{O}\Pi \qquad\qquad \frac{\prod_\alpha \phi}{\phi(\alpha/\xi).}$$

An example of inference by the rule OΠ is

$$\frac{\prod_x x = x}{2 = 2.}$$

The rule of joining the universal quantifier:

JΠ
$$\frac{\phi}{\prod_\alpha \phi}$$

may be used in proofs from assumptions only with the proviso that the variable α is not a free variable in the assumptions of the proof.

This limitation will be explained by means of the example of the following theorem from the arithmetic of real numbers:

$$x > 0 \rightarrow x + y > y.$$

A proof from assumptions of this theorem would begin with writing down the assumption

$$x > 0.$$

In the conclusions drawn from this assumption, for instance in the conclusion obtained on the strength of the rule JD:

$$x > 0 \ \lor \ x = 0,$$

the variable x does not represent an arbitrary real number, but stands only for positive numbers. Hence the conclusion that

$$\prod_x (x > 0 \ \lor \ x = 0)$$

would not be correct.

An example of inference by the rule JΠ is

$$\frac{x.1 = x}{\prod_x (x.1 = x).}$$

The rule for joining the existential quantifier is defined by the schema

JΣ
$$\frac{\phi\,(\alpha/\xi)}{\sum_\alpha \phi.}$$

Like the rule OΠ, this rule may be applied without restriction.

Note that the rule JΣ permits the formula $\sum_\alpha \phi$ to be obtained from the formula $\phi\,(\alpha/\xi)$, which must satisfy both conditions specified above in the definition of that symbol. Hence, for instance, from the formula "$\prod_n n+1 > n$", which is true in the domain of natural numbers, we may not, by that rule, obtain the formula "$\sum_k \prod_n k > n$", since the formula "$\prod_n n+1 > n$" is not a substitution for the formula "$\prod_n k > n$", since in the latter formula the expression "$n+1$" may not be substituted for the variable k, in accordance with the second condition formulated in the definition of the symbol $\phi(\alpha/\xi)$.

An example of inference by the rule $J\Sigma$ is

$$\frac{2 \text{ is an even and a prime number}}{\sum_x (x \text{ is an even and a prime number}).}$$

The rule $J\Sigma$ serves to prove the existence of objects of a certain kind by giving an example of such an object.

The formulation of the rule for omitting the existential quantifier as a schema will be preceded by an explanation of its intuitive meaning. This rule formalizes a certain step in reasoning often used in mathematical proofs. The following inference is often made in such proofs: It has been established that at least one object satisfies the condition ϕ. Let a stand for one (arbitrary) such object. Consequently, a satisfies the condition ϕ. The symbol "a" is here not a variable, but a constant denoting a definite object which satisfies the condition ϕ. In formalizing the inference we first introduce not only the term variables $x, y, z, \ldots$, but also the term constants $a, b, c, \ldots$, which will occur only in lines of proof and not in theses of the functional calculus.

The rule for omitting the existential quantifier ($O\Sigma$) permits us to join to the proof, for instance, the formula "$A(a)$", if "$\sum_x (Ax)$" belongs to the proof. However this rule must be applied with certain restrictions, the intuitive sense of which will now be explained. If the following lines: (1) $\sum_x \phi(x)$, and (2) $\sum_x \psi(x)$ (or $\sum_y \psi(y)$) are included in the proof and if by applying the rule $O\Sigma$ to (1) we have obtained $\phi(x/a)$, we may not, when applying the rule $O\Sigma$ to (2), join $\psi(x/a)$ to the proof. The proof lines (1) and (2) state only that there exists an object satisfying the condition ϕ and that there exists an object satisfying the condition ψ. These objects may, of course, be different, and it can happen that they must be different if the conditions ϕ and ψ are mutually exclusive. Hence if we apply the rule OΣ more than once in the same proof, we must each time introduce a new constant which has so far not occurred in the proof.

To understand the intuitive meaning of the second restriction on the application of the rule OΣ let us consider the following example.

Formula (3), $\sum_{y} y > x$, the variables of which range over the set of natural numbers, states that for any natural number x there is a natural number y which is greater than x. This, of course, is true. And yet we cannot deduce from (3) the formula "$a > x$", since the latter would mean that a certain natural number a is greater than any natural number x, which is obviously false. In order to be able to apply the rule OΣ in this case we provide "a" with a subscript "x" and deduce from (3) the formula "$a_x > x$", which states that for any natural number x there exists a natural number a_x greater than x and in some way determined by it (it may, for instance, be the number $x+1$).

Likewise, from formula (4), $\sum_{y} L(x, y, z)$, stating that for any two different points x and z there is a point y lying between x and z, by the rule OΣ we may deduce the formula $L(x, a_{x,z}, z)$ stating that between any two different points x and z there lies a definite point $a_{x,z}$ determined by the points x and z.

Taking all this into account we formulate the rule for omitting the existential quantifier in the form of the following schema:

$$\text{O}\Sigma \qquad \frac{\sum_{\alpha} \phi}{\phi(\alpha/\sigma_{\beta_1,\dots,\beta_n})} \ ,$$

where $\beta_1, \dots, \beta_n$ are all the free term variables other than α, in formula ϕ, and the formula $\phi(\alpha/\sigma_{\beta_1,\dots,\beta_n})$ is the result of a substitution for the variable α in the ϕ of the constant σ with the subscripts $\beta_1, \dots, \beta_n$. Note that the variables $\beta_1, \dots, \beta_n$ in the formula $\sigma_{\beta_1,\dots,\beta_n}$ are treated as free, so that the expression $\sigma_{\beta_1,\dots,\beta_n}$ may be substituted in the formula ϕ for the variable α if and only if that variable does not occur in the scope of any quantifier binding the variables $\beta_1, \dots, \beta_n$.

It is to be emphasized that by the rule $O\Sigma$ we may from the formula $\sum_{\alpha} \phi$ obtain only such a formula $\phi(\alpha/\sigma_{\beta_1,\ldots,\beta_n})$ which satisfies the conditions specified in the definition of the symbol $\phi(\alpha/\xi)$. For instance, if ϕ is the formula "$xRy \wedge \sum_{y} xSy$", then from the formula $\sum_{x} \phi$ we may not, by the rule $O\Sigma$, obtain the formula "$a_y Ry \wedge \sum_{y} a_y Sy$", since the latter is not obtained by the substitution of a_y for x in ϕ, because such a substitution would infringe condition (2) in the definition of the symbol $\phi(\alpha/\xi)$. If the variables x and y range over a set of at least two elements, then the formula "$\sum_{x} (x \neq y \wedge \sum_{y} x = y)$" is true. In conformity with the explanations given above, we may not apply the rule $O\Sigma$ to obtain from that formula the formula "$a_y \neq y \wedge \sum_{y} a_y = y$", which leads to a contradiction and hence is undoubtedly false.

Quantifiers with a limited range may be introduced either by definition or by the primitive rules of joining and omitting, with the schemata

$$O\Pi^* \qquad \frac{\prod_{\phi(\alpha)} \psi(\alpha)}{\phi(\alpha/\xi) \to \psi(\alpha/\xi),}$$

$$J\Pi^* \qquad \frac{\phi(\alpha) \to \psi(\alpha)}{\prod_{\phi(\alpha)} \psi(\alpha)}$$

(if α does not occur in the assumptions of the proof as a free variable).

$$J\Sigma^* \qquad \frac{\phi(\alpha/\xi)}{\psi(\alpha/\xi)}{\sum_{\phi(\alpha)} \psi(\alpha),}$$

$$O\Sigma^* \qquad \frac{\sum_{\phi(\alpha)} \psi(\alpha)}{\phi(\alpha/\sigma_{\beta_1,\ldots,\beta_n})}$$
$$\psi(\alpha(\sigma_{\beta_1,\ldots,\beta_n})$$

(with restrictions analogous to the case of the rule $O\Sigma$).

Exercises

1. Assume that the term constants of the functional calculus stand only for the three constant terms a, b, c. In this case the formula $\prod_x \phi(x)$ may be interpreted as the conjunction of the formulae

$$\phi(x/a), \quad \phi(x/b) \quad \text{and} \quad \phi(x/c),$$

and the formula $\sum_x \phi(x)$ as a disjunction of these formulae. Find the meaning, in this interpretation, of the rules $O\Pi$ and $O\Sigma$.

2. Use the rules $O\Pi^*$, $J\Pi^*$, $O\Sigma^*$ and $J\Sigma^*$ to prove the following equivalences

(I) $$\prod_{\phi(\alpha)} \psi(\alpha) \equiv \prod_{\alpha} [\phi(\alpha) \rightarrow \psi(\alpha)],$$

(II) $$\sum_{\phi(\alpha)} \psi(\alpha) \equiv \sum_{\alpha} [\phi(\alpha) \wedge \psi(\alpha)].$$

3. Taking the equivalences (I) and (II) above as definitions, prove the rules $O\Pi^*$, $J\Pi^*$, $O\Sigma^*$ and $J\Sigma^*$.

3. Theses and Derived Rules
of the Functional Calculus of the First Order

From the fact, stated at the beginning of the preceding Section, that all the primitive rules of the sentential calculus are primitive rules of the functional calculus, it follows that every thesis of the sentential calculus is a thesis of the functional calculus, and that every derived rule of the sentential calculus is a derived rule of the functional calculus. The derived rules of the sentential calculus are generalized so as to cover the sentential formulae in the functional calculus, in a way analogous to that in which the primitive rules were generalized in the preceding Section. It is to be emphasized that when applying the rule $J\Pi$ to the additional assumptions of a proof or to the formulae obtained in the proof from these additional assumptions, we may not bind with a universal quantifier any variable that occurs free in an additional assumption of the proof.

T1. $$\prod_x A(x) \rightarrow A(y).$$

Proof. (1) $\prod_x A(x)$ {a.}

         $A(y)$ {$O\Pi$: 1}

T2. $$A(y) \rightarrow \sum_x A(x).$$

Proof. (1) $A(y)$ $\{a.\}$

$\qquad\qquad \sum_{x} A(x)$ $\{J\Sigma: 1\}$

As in the sentential calculus, the theses of the functional calculus given in this Section have as their analogues certain metatheses the proofs of which are analogous to the proofs of the theses. The following metatheses correspond to T1 and T2:

M1. $\vdash \prod_{\alpha} \phi \rightarrow \phi(\alpha/\xi),$

M2. $\vdash \phi(\alpha/\xi) \rightarrow \sum_{\alpha} \phi.$

Their proofs are entirely analogous to the proofs of T1 and T2.
By M2, the following formulae are theses:

$$A(x) \rightarrow \sum_{x} A(x),$$

$$\neg A(x) \rightarrow \sum_{x} \neg A(x), \text{ etc.}$$

When referring below to T1 and T2 we shall also mean those formulae which are theses on the strength of Metatheses M1 and M2.

T3. $\neg \prod_{x} A(x) \equiv \sum_{x} \neg A(x).$

Proof. (a) (1) $\neg \prod_{x} A(x)$ $\{a.\}$

(2) $\neg \sum_{x} \neg A(x)$ $\{a.i.p.\}$

(3) $\neg A(x) \rightarrow \sum_{x} \neg A(x)$ $\{T2\}$

(4) $A(x)$ $\{\text{Toll.}: 3; 2\}$

(5) $\prod_{x} A(x)$ $\{J\Pi: 4\}$

contr. $\{1; 5\}$

(b) (1) $\sum_{x} \neg A(x)$ $\{a.\}$

(2) $\prod_{x} A(x)$ $\{a.i.p.\}$

(3) $\neg A(a)$ $\{O\Sigma: 1\}$

(4) $A(a)$ $\{O\Pi: 2\}$

contr. $\{3; 4\}$

In line (5) of part (a) of the proof the rule JΠ is used correctly, since the variable x does not occur in the assumptions of the proof as a free variable.

Thesis T3 may be read as follows:

It is not true that every object has a given property if and only if there exist objects which do not have that property.

The theses given in this Section, beginning with T3, have the property that none of them contains molecular formulae with the same predicate and different term variables. (Note that T1 and T2 do not share that property, since each of them contains two molecular formulae $A(x)$ and $A(y)$, which have the same predicate A but differ by the name variables which are the arguments of that predicate.) These theses can be divided into two kinds. In the first, which includes T3–T10 and T19–T25, sentential variables do not occur; in the second, which includes T11–T18, sentential variables do occur.

For the theses of the first kind, the schema of the formula which occurs in the corresponding metathesis, is constructed as follows:

The term variables occurring in a given thesis are replaced by the variables $\alpha, \beta, \gamma, \ldots$, and the molecular formulae of the functional calculus by the variables $\phi, \psi, \chi, \ldots$, with the provision that the same variables (the same molecular formulae) are replaced by the same variables, and different variables (different molecular formulae) by different variables.

In view of the above, Thesis T3 has as its analogue the following metathesis

M3. $$\vdash \neg \prod_{\alpha} \phi \equiv \sum_{\alpha} \neg \phi,$$

and the thesis $A(x) \to \sum_{x} A(x)$ has as its analogue the metathesis

$$\vdash \phi \to \sum_{\alpha} \phi.$$

By T3 (M3) the following conditions are equivalent:

$$\neg \prod_{x} x+3 = 5,$$
$$\sum_{x} \neg (x+3 = 5).$$

The proof of the theorem that for every thesis of the kind now under discussion its corresponding metathesis is also provable, is by induction and is analogous to the proof of the corresponding theorem of the sentential calculus.

As in the case of derived rules of the sentential calculus we shall speak, in the functional calculus, of rules which are derived from certain theses, instead of speaking of rules derived from the meta-theses corresponding to these theses. For instance, we shall say that the following rule of negating the universal quantifier, derived from M3, is derived from T3:

$$N\Pi \qquad \frac{\neg \prod_\alpha \phi}{\sum_\alpha \neg \phi.}$$

Thesis T3 will be called the *law of negating the universal quantifier*. The next thesis is the *law of negating the existential quantifier*. The laws of negating the quantifiers are also called *De Morgan's laws in the functional calculus*.

T4. $$\neg \sum_x A(x) \equiv \prod_x \neg A(x).$$

Proof. (a) (1) $\neg \sum_x A(x)$ {a.}

(2) $A(x) \to \sum_x A(x)$ {T2}

(3) $\neg A(x)$ {Toll. 2; 1}

$\prod_x \neg A(x)$ {JΠ: 3}

(b) (1) $\prod_x \neg A(x)$ {a.}

(2) $\sum_x A(x)$ {a.i.p.}

(3) $A(a)$ {OΣ: 2}

(4) $\neg A(a)$ {OΠ: 1}

     contr.   {3; 4}

By T4, the following conditions are equivalent:

$$\neg \sum_x x+2 = x,$$

$$\prod_x \neg(x+2 = x).$$

T4 leads to the rule of negating the existential quantifier:

NΣ
$$\frac{\neg \sum_{\alpha} \phi}{\prod_{\alpha} \neg \phi.}$$

On the strength of T3 and T4 we state that the negation of a formula that begins with a string of universal and existential quantifiers is equivalent to the formula which is obtained from the former by replacing each universal quantifier by an existential quantifier, and vice versa, and by negating the formula that follows the quantifiers. For instance,

$$\neg \prod_x \sum_y \prod_z R(x, y, z) \equiv \sum_x \prod_y \sum_z \neg R(x, y, z).$$

We shall now consider the distributive laws for the quantifiers. They permit us to distribute in a definite way the quantifiers which precede compound formulae, over the elements of those formulae, or to shift quantifiers which precede elements of a compound formula before the entire (compound) formula.

T5.
$$\prod_x (A(x) \to B(x)) \to (\prod_x A(x) \to \prod_x B(x)).$$

Proof. (1) $\prod_x (A(x) \to B(x))$ ⎫
     (2) $\prod_x A(x)$ ⎬   {a.}
     (3) $A(x) \to B(x)$ {OΠ: 1}
     (4) $A(x)$ {OΠ: 2}
     (5) $B(x)$ {RD: 3; 4}
        $\prod_x B(x)$ {JΠ: 5}

An example of reasoning by the derived rule based on T5 is

$$\frac{\prod_x (x = x \to x \leqslant x)}{\prod_x x \leqslant x.}$$

T6.
$$\prod_x (A(x) \to B(x)) \to (\sum_x A(x) \to \sum_x B(x)).$$

Proof. (1) $\prod\limits_{x}(A(x) \to B(x))$ ⎤

    (2) $\sum\limits_{x} A(x)$ ⎦   {a.}

    (3) $A(a)$ {OΣ: 2}

    (4) $A(a) \to B(a)$ {OΠ: 1}

    (5) $B(a)$ {RD: 4; 5}

        $\sum\limits_{x} B(x)$ {JΣ: 5}

An example of inference by the derived rule based on T6 is

$$\prod_{x}(x+2 = 5 \to x+2 \leqslant 5)$$
$$\sum_{x} x+2 = 5$$
$$\overline{\sum_{x} x+2 \leqslant 5.}$$

T7. $\prod\limits_{x}(A(x) \wedge B(x)) \equiv \prod\limits_{x} A(x) \wedge \prod\limits_{x} B(x).$

The proof of this thesis is left to the reader.

By T7 the following conditions are equivalent:

$$\prod_{x}(f(x) = 0 \wedge g(x) = 0),$$
$$\prod_{x} f(x) = 0 \wedge \prod_{x} g(x) = 0.$$

T8. $\prod\limits_{x} A(x) \vee \prod\limits_{x} B(x) \to \prod\limits_{x}(A(x) \vee B(x)).$

Proof. (1) $\prod\limits_{x} A(x) \vee \prod\limits_{x} B(x)$ {a.}

    (1.1) $\prod\limits_{x} A(x)$ {ad. a.}

    (1.2) $A(x)$ {OΠ: 1.1}

    (1.3) $A(x) \vee B(x)$ {JD: 1.2}

    (1.4) $\prod\limits_{x}(A(x) \vee B(x))^{(1)}$ {JΠ: 1.3}

    (2.1) $\prod\limits_{x} B(x)$ {ad. a.}

    (2.2) $B(x)$ {OΠ: 2.1}

---

    [1] Rule JΠ may be applied here because the free variable x (which is not free in the assumption of the proof) does not occur in the additional assumption f the proof (1.1), from which we have obtained (1.3).

(2.3) $A(x) \lor B(x)$ {JD: 2.2}

(2.4) $\prod_x (A(x) \lor B(x))$ {JΠ: 2.3}

$\prod_x (A(x) \lor B(x))$ {1.1 → 1.4; 2.1 → 2.4; 1}

An example of inference by the derived rule based on T8 is

$$\frac{\prod_x f(x) > 0 \lor \prod_x f(x) < 0}{\prod_x (f(x) > 0 \lor f(x) < 0).}$$

The implication converse to T8 is false, as is shown by the formula

$$\prod_x (x = 0 \lor x \neq 0) \to \prod_x x = 0 \lor \prod_x x \neq 0$$

in which the antecedent is true, and the consequent false.

T9. $\sum_x (A(x) \land B(x)) \to \sum_x A(x) \land \sum_x B(x).$

Proof. (1) $\sum_x (A(x) \land B(x))$ {a.}

(2) $A(a) \land B(a)$ {OΣ: 1}

(3) $A(a)$ ⎫
(4) $B(a)$ ⎭ {OC: 2}

(5) $\sum_x A(x)$ {JΣ: 3}

(6) $\sum_x B(x)$ {JΣ: 4}

$\sum_x A(x) \land \sum_x B(x)$ {JC: 5; 6}

An example of inference by the derived rule based on T9:

$$\frac{\sum_x (x+2 = 5 \land 0 < x < 4)}{\sum_x x+2 = 5 \land \sum_x 0 < x < 4.}$$

The implication converse to T9 is false, as can be seen from the following example:

$$\sum_x (x+1 = 10) \land \sum_x (0 < x < 4) \to \sum_x (x+1 = 10 \land 0 < x < 4),$$

where the antecedent is true and the consequent false.

T10. $\sum\limits_{x} \left(A(x) \lor B(x) \right) \equiv \sum\limits_{x} A(x) \lor \sum\limits_{x} B(x).$

Proof. (a) (1) $\sum\limits_{x} \left(A(x) \lor B(x) \right)$ {a.}

       (2) $A(a) \lor B(a)$ {OΣ: 1}

       (1.1) $A(a)$ {ad. a.}

       (1.2) $\sum\limits_{x} A(x)$ (JΣ: 1.1}

       (1.3) $\sum\limits_{x} A(x) \lor \sum\limits_{x} B(x)$ {JD: 1.2}

       (2.1) $B(a)$ {ad. a.}

       (2.2) $\sum\limits_{x} B(x)$ {JΣ: 2.1}

       (2.3) $\sum\limits_{x} A(x) \lor \sum\limits_{x} B(x)$ {JD: 2.2}

             $\sum\limits_{x} A(x) \lor \sum\limits_{x} B(x)$ {1.1 → 1.3; 2.1 → 2.3; 2}

   (b) (1) $\sum\limits_{x} A(x) \lor \sum\limits_{x} B(x)$ {a.}

       (1.1) $\sum\limits_{x} A(x)$ {ad. a.}

       (1.2) $A(a)$ {OΣ: 1.1}

       (1.3) $A(a) \lor B(a)$ {JD: 1.2}

       (1.4) $\sum\limits_{x} \left(A(x) \lor B(x) \right)$ {JΣ: 1.3}

       (2.1) $\sum\limits_{x} B(x)$ {ad. a.}

       (2.2) $B(a)$ {OΣ: 2.1}

       (2.3) $A(a) \lor B(a)$ {JD: 2.2}

       (2.4) $\sum\limits_{x} \left(A(x) \lor B(x) \right)$ {JΣ: 2.3}

             $\sum\limits_{x} \left(A(x) \lor B(x) \right)$ {1.1 → 1.4; 2.1 → 2.4; 1}

By T10, the following conditions are equivalent:

$$\sum_{x} 2x^2 \leqslant 0, \qquad \sum_{x} 2x^2 < 0 \lor \sum_{x} 2x^2 = 0.$$

We shall now study a number of laws of the functional calculus in which sentential variables occur. Any sentential formulae may be substituted for these variables, but with the restriction analogous to condition 2 concerning substitution for term variables: If a given sentential variable occurs within the scope of the quantifier which binds the term variable α, then it is not permitted to substitute for that sentential variable any sentential formulae containing the free variable α.

The theses to be discussed now are called the *laws of shifting quantifiers*.

T11. $$\prod_x (p \to A(x)) \equiv p \to \prod_x A(x).$$

Only part (b) of the proof will be given.

Proof. (b) (1) $p \to \prod_x A(x)$ {a.}

     (1.1) p {ad. a.}

     (1.2) $\prod_x A(x)$ {RD: 1; 1.1}

     (1.3) $A(x)$ {OΠ: 1.2}

     (2) $p \to A(x)$ {1.1 → 1.3}

        $\prod_x (p \to A(x))$ {JΠ: 2}

For the theses of the second kind, singled out on p. 103, which include T11, we construct schemata of formulae occurring in the corresponding metatheses; we do so in a way analogous to the case of the theses of the first kind discussed there, with the provision that sentential variables are replaced by the variables $\phi, \psi, \chi, \ldots$, distinct from those which are used to replace the molecular formulae of the functional calculus occurring in a given thesis (of course, different sentential variables are replaced by different variables, and the same sentential variables by the same variables).

The following restriction is also adopted.

If a sentential variable, for instance the variable p, occurs in a given thesis within the scope of the quantifier which binds the variable α, and has as its analogue in the corresponding metathesis the variable ϕ, then in that corresponding metathesis we must introduce the condition stating that α is not a free variable in formula ϕ (since on the strength of the restriction analogous to the restriction formulated on p. 96 formulae containing the free variable α may not be substituted for the sentential variable p).

In accordance with the above, Thesis T11 has as its analogue the metathesis

M11. $\quad \vdash \prod_\alpha (\phi \to \psi) \equiv \phi \to \prod_\alpha \psi \quad$ *if α is not free in ϕ.*

By T11 (M11) the following conditions are equivalent:

$$\prod_x (y > 0 \to |x|+y > 0),$$

$$y > 0 \to \prod_x |x|+y > 0.$$

T12. $\qquad\qquad \sum_x (p \to A(x)) \equiv p \to \sum_x A(x).$

Only part (b) of the proof is given.

Proof. (b) (1) $\quad p \to \sum_x A(x)$ $\qquad\qquad$ {a.}

$\qquad$ (1.1) $\ p$ $\qquad\qquad\qquad\qquad\qquad$ {ad. a.}

$\qquad$ (1.2) $\ \sum_x A(x)$ $\qquad\qquad\qquad\quad$ {RD: 1; 1.1}

$\qquad$ (1.3) $\ A(a)$ $\qquad\qquad\qquad\qquad\quad$ {OΣ: 1.2}

$\qquad$ (2) $\quad p \to A(a)$ $\qquad\qquad\qquad$ {1.1 $\to$ 1.3}

$\qquad\qquad\quad \sum_x (p \to A(x))$ $\qquad\qquad$ {JΣ: 2}

Hence the following conditions are equivalent:

$$\sum_x (b^2-4ac > 0 \to ax^2+bx+c = 0),$$

$$b^2-4ac > 0 \to \sum_x (ax^2+bx+c = 0).$$

The proofs of the following two theses are left to the reader.

T13. $\qquad\qquad \prod_x (A(x) \to p) \equiv \sum_x A(x) \to p.$

Hence the following conditions are equivalent:

$$\prod_x (ax^2 > 0 \to a > 0),$$

$$\sum_x ax^2 > 0 \to a > 0.$$

T14. $\qquad\qquad \sum_x (A(x) \to p) \equiv \prod_x A(x) \to p.$

By T14, the following conditions are equivalent:

$$\sum_{x} (ax^2+bx+c \neq 0 \rightarrow b^2-4ac < 0),$$

$$\prod_{x} (ax^2+bx+c \neq 0) \rightarrow b^2-4ac < 0.$$

T15. $\prod_{x}(p \vee A(x)) \equiv p \vee \prod_{x} A(x).$

Proof. (a) (1) $\prod_{x}(p \vee A(x))$ {a.}

    (2) $p \vee A(x)$ {OΠ: 1}

    (1.1) p {ad. a}

    (1.2) $p \vee \prod_{x} A(x)$ {JD: 1.1}

    (2.1) $\neg p$ {ad. a}

    (2.2) $A(x)$ {OD: 2; 2.1}

    (2.3) $\prod_{x} A(x)$ {JΠ: 2.2}

    (2.4) $p \vee \prod_{x} A(x)$ {JD: 2.3)

        $p \vee \prod_{x} A(x)$ {1.1 → 1.2; 2.1 → 2.4}

Part (b) of the proof is analogous to the proof of T8.
The proofs of the following three theses are left to the reader.

T16. $\sum_{x}(p \wedge A(x)) \equiv p \wedge \sum_{x} A(x).$

T17. $\prod_{x}(p \wedge A(x)) \equiv p \wedge \prod_{x} A(x).$

T18. $\sum_{x}(p \vee A(x)) \equiv p \vee \sum_{x} A(x).$

We shall now discuss the laws for changing the order of quantifiers.

T19. $\prod_{x}\prod_{y} xRy \equiv \prod_{y}\prod_{x} xRy.$

Proof. (a) (1) $\prod_{x}\prod_{y} xRy$ {a.}

    (2) $\prod_{y} xRy$ {OΠ: 1}

    (3) xRy {OΠ: 2}

(4) $\displaystyle\prod_x xRy$ {JΠ: 3}

$\displaystyle\prod_y\prod_x xRy$ {JΠ: 4}

Part (b) of the proof is analogous to part (a).

T20. $\displaystyle\sum_x\sum_y xRy \equiv \sum_y\sum_x xRy.$

Proof. (a) (1) $\displaystyle\sum_x\sum_y xRy$ {a.}

(2) $\displaystyle\sum_y aRy$ {OΣ: 1}

(3) aRb {OΣ: 2}

(4) $\displaystyle\sum_x xRb$ {JΣ: 3}

$\displaystyle\sum_y\sum_x xRy$ {JΣ: 4}

Part (b) of the proof is analogous to part (a).

Thesis 20 has its analogue in the metathesis

M20. $\displaystyle\vdash\sum_\alpha\sum_\beta \phi \equiv \sum_\beta\sum_\alpha \phi.$

By M20, the following formula is a thesis:

$$\sum_x\sum_y R(x, y, z) \equiv \sum_y\sum_x R(x, y, z).$$

Part (a) of the proof of this formula is as follows:

Proof. (a) (1) $\displaystyle\sum_x\sum_y R(x, y, z)$ {a.}

(2) $\displaystyle\sum_y R(a_z, y, z)$ {OΣ: 1}

(3) $R(a_z, b_z, z)$ {OΣ: 2}

(4) $\displaystyle\sum_x R(x, b_z, z)$ {JΣ: 3}

$\displaystyle\sum_y\sum_x R(x, y, z)$ {JΣ: 4}

In this proof we have for the first time used terms dependent on variables (constants with indices), namely the expressions "a_z" and "b_z". The last two steps of the proof are correct since the formula

"$R(a_z, b_z, z)$" is a substitution of the formula "$R(x, b_z, z)$", and the formula "$\sum\limits_{x} R(x, b_z, z)$" is a substitution of the formula "$\sum\limits_{x} R(x, y, z)$".

T21. $$\sum\limits_{x}\prod\limits_{y} xRy \rightarrow \prod\limits_{y}\sum\limits_{x} xRy.$$

Proof. (1) $\sum\limits_{x}\prod\limits_{y} xRy$ {a.}

(2) $\prod\limits_{y} aRy$ {OΣ: 1}

(3) aRy {OΠ: 2}

(4) $\sum\limits_{x} xRy$ {JΣ: 3}

    $\prod\limits_{y}\sum\limits_{x} xRy$ {JΠ: 4}

Here is an example of inference by a derived rule based on T21:

$$\frac{\sum\limits_{x}\prod\limits_{y} x.y = y}{\prod\limits_{y}\sum\limits_{x} x.y = y.}$$

The implication which is converse to T21 is false, as is shown by the following formula from the arithmetic of natural numbers:

$$\prod\limits_{y}\sum\limits_{x} x > y \rightarrow \sum\limits_{x}\prod\limits_{y} x > y.$$

The formula is false since its antecedent is true (as it states that for every natural number there is a number that is greater than it), and the consequent is false (as it states that there exists a natural number greater than any natural number).

The implication converse to T21 cannot be proved in a way analogous to the proof of T21, in view of the limitations imposed on the rules for handling the quantifiers. By the rule OΠ, the assumption "$\prod\limits_{y}\sum\limits_{x} xRy$" yields the formula "$\sum\limits_{x} xRy$", and the latter, by the rule OΣ, yields "$a_y Ry$". If the rule JΠ is applied to the latter formula, we obtain "$\prod\limits_{y} a_y Ry$". But from the latter we cannot obtain the formula "$\sum\limits_{x}\prod\limits_{y} xRy$" by the rule J$\Sigma$, since in the formula "$\prod\limits_{y} xRy$" we may not substitute the term "a_y" for the variable "x", since such a substitution would bind the free index variable "y" occurring in "a_y".

T22. $$\prod_x (A(x) \equiv B(x)) \rightarrow (\prod_x A(x) \equiv \prod_x B(x)).$$

Proof. (1) $\prod_x (A(x) \equiv B(x))$ {a.}

       (2) $A(x) \equiv B(x)$ {OΠ: 1}

       (3) $A(x) \rightarrow B(x)$ ⎫

                                  {OE: 2}

       (4) $B(x) \rightarrow A(x)$ ⎭

       (5) $\prod_x (A(x) \rightarrow B(x))$ {JΠ: 3}

       (6) $\prod_x (B(x) \rightarrow A(x))$ {JΠ: 4}

       (7) $\prod_x A(x) \rightarrow \prod_x B(x)$ {RD: T5; 5}[1]

       (8) $\prod_x B(x) \rightarrow \prod_x A(x)$ {RD: T5; 6}

            $\prod_x A(x) \equiv \prod_x B(x)$ {JE: 7; 8}

An example of inference by a rule derived from T22 is

$$\frac{\prod_x (2(x+3)-5 > 2x \equiv 2x+1 > 2x)}{\prod_x (2(x+3)-5 > 2x) \equiv \prod_x (2x+1 > 2x).}$$

T23. $$\prod_x (A(x) \equiv B(x)) \rightarrow (\sum_x A(x) \equiv \sum_x B(x)).$$

Proof. (1) $\prod_x (A(x) \equiv B(x))$ {a.}

       (2) $A(x) \equiv B(x)$ {OΠ: 1}

       (3) $A(x) \rightarrow B(x)$ ⎫

                                  {OE: 2}

       (4) $B(x) \rightarrow A(x)$ ⎭

       (5) $\prod_x (A(x) \rightarrow B(x))$ {JΠ: 3}

       (6) $\prod_x (B(x) \rightarrow A(x))$ {JΠ: 4}

       (7) $\sum_x A(x) \rightarrow \sum_x B(x)$ {RD: T6; 5}

---

[1] We make use here of the rule of joining to the proof those theses which have been proved previously. For brevity's sake the thesis referred to is not written in full in the proof, but only indicated by its number.

$$(8) \quad \sum_x B(x) \rightarrow \sum_x A(x) \qquad\qquad \{\text{RD: T6; } 6\}$$

$$\sum_x A(x) \equiv \sum_x B(x) \qquad\qquad \{\text{JE: 7; } 8\}$$

An example of inference by the rule derived from T23 is

$$\prod_x (3(x-1)+1 = 2(x-3) \equiv x+4 = 0)$$
$$\overline{\sum_x (3(x-1)+1 = 2(x-3)) \equiv \sum_x (x+4 = 0)}.$$

On the strength of the rule of extensionality from the sentential calculus[1] and Theses T22 and T23 it can be proved analogously that the rule of extensionality is also valid in the functional calculus where it has the schemata:

$$\prod_\alpha (\phi \equiv \psi)$$
$$\overline{\chi \equiv \chi(\phi//\psi)},$$

$$\prod_\alpha (\phi \equiv \psi)$$
$$\overline{\dfrac{\chi}{\chi(\phi//\psi)}}.$$

From Theses T3 and T4, by the law of double negation (T5b, Sec. 5, Chap. I) and the law permitting us to join negation to both sides of an equivalence (T33a, Sec. 3, Chap. I), we obtain the theses:

T24. $\qquad\qquad \prod_x A(x) \equiv \neg \sum_x \neg A(x).$

T25. $\qquad\qquad \sum_x A(x) \equiv \neg \prod_x \neg A(x).$

This shows that the universal quantifier can be defined by means of negation and the existential quantifier. Likewise, the existential quantifier can be defined by means of negation and the universal quantifier.

Many of the theses given in this Section are valid in an analogous form for quantifiers with a limited range. Such theses for quantifiers with a limited range will be symbolized by the same numbers which are attached to the corresponding theses for ordinary quantifiers, but the numbers will have an asterisk appended to them. Proofs of these theses, carried out by means of the rules $O\Pi^*$, $J\Pi^*$, $O\Sigma^*$,

---

[1] Cf. pp. 43–44.

$J\Sigma^*$, are partly analogous to the proofs of the corresponding theses formulated with ordinary quantifiers. As an example we give the proof of the thesis corresponding to T21.

T21*. $$\sum_{A(x)} \prod_{B(y)} xRy \rightarrow \prod_{B(y)} \sum_{A(x)} xRy$$

Proof. (1) $\displaystyle\sum_{A(x)} \prod_{B(y)} xRy$ {a.}

(2) $A(a)$ $\left.\begin{array}{l} \\ \\ \end{array}\right\}$ {$O\Sigma^*$: 1}

(3) $\displaystyle\prod_{B(y)} aRy$

(4) $B(y) \rightarrow aRy$ {$O\Pi^*$: 3}

(1.1) $B(y)$ {ad. a.}

(1.2) aRy {RD: 4; 1.1}

(1.3) $\displaystyle\sum_{A(x)} xRy$ {$J\Sigma^*$: 2; 1.2}

(5) $B(y) \rightarrow \displaystyle\sum_{A(x)} xRy$ {1.1 → 1.3}

$\displaystyle\prod_{B(y)} \sum_{A(x)} xRy$ {$J\Pi^*$: 5}

Here is an example of inference by the rule derived from T21:

$$\frac{\displaystyle\sum_{x \text{ is a book}} \prod_{y \text{ is a pupil in the second form}} y \text{ has read } x}{\displaystyle\prod_{y \text{ is a pupil in the second form}} \sum_{x \text{ is a book}} y \text{ has read } x.}$$

In other words, from the premiss: "there is a book which has been read by every second-form pupil" we may draw the conclusion: "every second-form pupil has read a certain book".

The implication converse to T21 is false, which is shown by the following example:

$$\prod_{x \text{ is a state}} \sum_{y \text{ is a city}} y \text{ is the capital of } x \rightarrow$$

$$\sum_{y \text{ is a city}} \prod_{x \text{ is a state}} y \text{ is the capital of } x.$$

The antecedent of this implication is true (as every state has its capital), but the consequent is false (as it states that a city exists which is the capital of every state, which is evidently false).

Apart from T21*, the most frequently applied theses for quantifiers with a limited range are theses T3*, T4*, T19*, T20*, T22*, T23*, which are given in the list of theses at the end of the book.

In the system presented above the following rules can be derived.

(1) The rule for joining the universal quantifier to the antecedent of an implication:

R1.
$$\frac{\phi \to \psi}{\prod_{\alpha} \phi \to \psi.}$$

(2) The rule for joining the universal quantifier to the consequent of an implication:

R2.
$$\frac{\phi \to \psi}{\phi \to \prod_{\alpha} \psi}$$

(provided that the variable α is free neither in ϕ nor in the assumptions of the proof).

(3) The rule for joining the existential quantifier to the consequent of an implication:

R3.
$$\frac{\phi \to \psi}{\phi \to \sum_{\alpha} \psi.}$$

(4) The rule for joining the existential quantifier to the antecedent of an implication:

R4.
$$\frac{\phi \to \psi}{\sum_{\alpha} \phi \to \psi}$$

(provided that the variable α is free neither in ϕ nor in the assumptions of the proof).

(5) The rule for omitting the universal quantifier in the consequent of an implication:

R1'.
$$\frac{\phi \to \prod_{\alpha} \psi}{\phi \to \psi.}$$

(6) The rule for omitting the existential quantifier in the antecedent of an implication:

R3'.
$$\frac{\sum_{\alpha} \phi \to \psi}{\phi \to \psi.}$$

(The proofs of these rules are left to the reader.)

(7) The rule of substitution for free term variables:

$$\text{If } \vdash \phi, \text{ then } \vdash \phi(\alpha/\xi).$$

Proof. (1) $\vdash \phi$

(2) $\vdash \prod_{\alpha} \phi$ {JΠ: 1}

$\vdash \phi(\alpha/\xi)$ {OΠ: 2}

(8) The rule of substitution for variables representing predicates:

$$\text{If } \vdash \phi, \text{ then } \vdash \phi(\varphi_1/\varphi_2),$$

where φ_1 and φ_2 are variables representing predicates.

The proof of this rule is by induction on the order of the thesis. For the theses of the first order we construct the proof of the thesis $\phi(\varphi_1/\varphi_2)$ by replacing the variable φ_1 by the variable φ_2 everywhere in the proof of the thesis ϕ. We next assume the validity of the rule for theses of orders lower than n and prove it for theses of order n, in a way analogous to the first step of the induction.

(9) The rule for relettering bound variables:

With the provision that the variable β is not free in ψ this rule permits us, in every formula ϕ

(a) to replace an occurrence of the formula $\prod_{\alpha} \psi$ by the formula

$$\prod_{\beta} \psi(\alpha/\beta),$$

(b) to replace an occurrence of the formula $\sum_{\alpha} \psi$ by the formula

$$\sum_{\beta} \psi(\alpha/\beta).$$

Thus, for instance, the formula "$\prod_{x} A(x)$" may be replaced by the formula "$\prod_{y} A(y)$", and "$\prod_{x} xRy$" may be replaced by "$\prod_{z} zRy$". Likewise, the formula "$\sum_{x} A(x)$" may, for instance, be replaced by the formula "$\sum_{y} A(y)$", or "$\sum_{x} xRy$" by "$\sum_{z} zRy$".

The assumption that the variable β is not free in ψ is essential here, since otherwise the replaced formula might not be equivalent, as in the case of the formulae "$\prod_{x} xRy$" and "$\prod_{y} yRy$", or the formulae "$\sum_{x} xRy$" and "$\sum_{y} yRy$".

The rule now under discussion is proved by means of the rule of extensionality and the metatheses:

(a) $\vdash \prod_\alpha \psi \equiv \prod_\beta \psi(\alpha/\beta)$, if the variable β is not free in ψ,

(b) $\vdash \sum_\alpha \psi \equiv \sum_\beta \psi(\alpha/\beta)$, if the variable β is not free in ψ.

We shall now give the proof of (a). By OΠ the formula $\prod_\alpha \psi$ yields the formula $\psi(\alpha/\beta)$, whence we obtain the formula $\prod_\beta \psi(\alpha/\beta)$ by the rule JΠ which can be applied here since, by assumption, the variable β is not free in the formula $\prod_\alpha \psi$.

On the other hand, the formula $\prod_\beta \psi(\alpha/\beta)$ yields, by OΠ, the formula $(\psi(\alpha/\beta))(\beta/\alpha)$, which, if it is assumed that β is not free in ψ, is identical with the formula ψ. From this we obtain, by JΠ, the formula $\prod_\alpha \psi$. (This rule may be applied here, since the variable α is not free in the formula $\prod_\beta \psi(\alpha/\beta)$.)

The proof of (b) is analogous.

The rule of substitution for formulae of the functional calculus is also a derived rule, but its proof by induction is fairly complicated. It is easier to prove it by the rule of joining definitions to a proof, to be discussed in Sec. 5, where we shall also formulate and analyse this rule of substitution.

The rules specified above are primitive rules in certain formulations of the first order functional calculus, which we shall now mention briefly.

The first order functional calculus can be constructed by joining to the axioms and primitive rules of the sentential calculus (Chap. I, Sec. 6) the rule of substitution for free term variables and the rules R1 to R4. Rules R1′ and R3′ may be adopted in place of R1 and R3.

The first order functional calculus can also be constructed by the axiomatic method, if to the axioms and primitive rules of the sentential calculus (Chap. I, Sec. 6) we join as axioms of the functional calculus Theses T1 and T11, and as primitive rules, the rule for joining the universal quantifier to theses, the rule of substitution for free term variables and for formulae of the functional calculus and the rule for relettering variables bound by the universal quantifier. The existential quantifier is introduced by the definition:

$$\sum_\alpha \phi \overline{\underset{\text{df}}{=}} \neg \prod_\alpha \neg \phi,$$

(cf. T25).

Systems built in this way are equivalent to the system of the first order functional calculus as presented in this textbook.

In the proof of the equivalence of these systems with the assumptional system an essential role is played, as in the case of the sentential calculus, by the deduction theorem for the first order functional calculus.

Consider, for instance, the system of the first order functional calculus based on the sentential calculus which uses implication and negation as its primitive terms; the system can be based on the rules R1–R4 and the deduction theorem (cf. Chap. I, Sec. 6) is valid for it. The definition Def. I of Chap. I, Sec. 6, is expanded by defining, for the system in question, a derivation of the formula ϕ_n from the formulae $\phi_1, \ldots, \phi_{n-1}$ as a finite sequence of formulae, such that the last term of the sequence is the formula ϕ_n and each term of the sequence either is an axiom, or is one of the formulae $\phi_1, \ldots, \phi_{n-1}$, or is obtained from preceding terms of the sequence by the rule RD or one of the rules R1–R4, with the observance of the reservations given in the formulation of the rules R2 and R4 (the variable α, introduced by R2 or R4, may not be free in any of the formulae $\phi_1, \ldots, \phi_{n-1}$, etc.). By means of the concept of derivation, defined in this way, we define the concept of consequence analogously to Df. II in Chap. I, Sec. 6. In the system in question, Th. 1, Th. 2, and Th. 3, Chap. I, Sec. 6, can be proved if the concept of consequence is so defined. The deduction theorem is proved analogously to Th.1 in Chap. I, Sec. 6, with the proviso that—in conformity with the definition of the concept of derivation—we consider four other possible cases (in which further proof lines are obtained from preceding proof lines by means of one of the rules R1–R4). In so doing we refer to the appropriate theses of the first order functional calculus.

Having proved Th.1, Th.2, and Th. 3 for the system under consideration we next prove that the rule OΣ is derivable in that system

in the sense that if a formula can be proved by means of the rule
OΣ, it can also be proved by means of the primitive rules (RD, R1–
R4) and theses of the system.

Like the sentential calculus, the system of the first order functional
calculus gives rise to issues of consistency, independence, syn-
tactical and semantical completeness.

It can be demonstrated that this system is consistent, and that the
primitive rules on which it is based are independent. As opposed
to the sentential calculus the first order functional calculus is not
syntactically complete: it can be demonstrated that, for instance,
the formula "$\sum_x A(x) \to \prod_x A(x)$" is neither a thesis nor, if joined
to the theses, does it lead to a contradiction. On the other hand, the
first order functional calculus is semantically complete, i.e. every
true formula of the calculus is a thesis. This was proved by K. Gödel
in 1930.

The concept of truth, as used here, is defined by the concept of
satisfaction. The intuitive sense of the latter concept will be explained
by the following examples.

We say that that the number 2 satisfies the equation $x + 3 = 5$,
that the numbers 1 and 3 satisfy the condition $x + y = 4$. On the
other hand, the number 1 does not satisfy the equation $x + 3 = 5$,
nor do the numbers 1 and 2 satisfy the condition $x + y = 4$. If
every sequence of objects belonging to a non-empty set, or, as is
usually said, to a non-empty domain, satisfies a given formula,
then we say that that formula is *true* in that domain. For instance,
the formula "$x < y \to \sum_z x < z < y$" is true in the domain of all
rational numbers (or all real numbers), but is not true in the domain
of all natural numbers, since, e.g., the natural numbers 2 and 3 do
not satisfy it. A formula which is true in every non-empty domain
(i.e., a domain to which at least one object belongs) is called a *true
formula*.[1] In this sense, every thesis of the first order functional
calculus is true and every true formula of that calculus is a thesis.

[1] Precise definitions of the concepts of satisfaction and truth were given
by A. Tarski in *The Concept of Truth in Formalized Languages*, 1956 (first published
in Polish in 1933).

The condition that the domain must be non-empty is essential, for some theses of the first order functional calculus are not true in the empty domain. For instance, the formula "$\neg \sum_x \neg A(x) \rightarrow \sum_x A(x)$" is a thesis, but in the empty domain it is not true. Its antecedent is then true (for since there exist no objects in this domain, there exist no objects which do not have the property A), but its consequent is false, since it states the existence in a given domain of objects having the property A.

Although every true formula of the first order functional calculus can be proved, there is no general method which enables us to decide in a finite number of definite steps whether a given formula is true. Thus, as opposed to the sentential calculus, the first order functional calculus is an undecidable system.[1] That is why we could not give for it a verification method like the zero-one method for the sentential calculus. Yet certain parts of the system now under discussion, for instance the first order monadic functional calculus (which includes all the theses given in this Section except for the laws of exchanging the order of quantifiers), are decidable, and more comprehensive textbooks give the various verification methods for them.

As we have seen in Chap. I, Sec. 3, and in the present Section, theorems of the sentential calculus and of the functional calculus may be formulated in two ways: 1) as theses of the system, 2) as theorems of a metasystem, stating that formulae of a definite structure are theses. The latter approach will be called metalogical. In this connection, when we apply the sentential calculus or the first order functional calculus to a field of science, we may proceed in two ways.

In the first, the system of theorems of a given field of science is based on the sentential calculus or on the first order functional calculus in the same way as the first order functional calculus has been based on the sentential calculus. In this case, all the symbols of the sentential calculus or the first order functional calculus, in particular the sentential variables or the variables standing for predic-

---

[1] This theorem was proved by A. Church in 1936. The concept of decidability is made precise by means of the concept of recursive functions.

ates, belong to the given system, and all the theses of the sentential calculus or the first order functional calculus are theses of the system.

In the second approach, we resort to a metalogical formulation of the theorems of the sentential calculus or the first order functional calculus. In this case, neither sentential variables nor variables standing for predicates are introduced into the system. The Greek letters ϕ, ψ, χ, ..., which occur in the metalogical formulations of theorems of the sentential calculus or the first order functional calculus, then stand for any formulae of the given system. The sentential calculus (the first order functional calculus) treated in this way is sometimes called an *applied sentential calculus* (*applied first order functional calculus*).

The primitive rules of the sentential calculus and of the first order functional calculus, given in the respective Chapters, permit us to construct these calculi both in the first and in the second way.

Exercises

1. Prove the theses:

(a) $\displaystyle\sum_x (A(x) \to B(x)) \equiv \prod_x A(x) \to \sum_x B(x),$

(b) $\displaystyle\sum_x \prod_y (A(x) \to B(y)) \equiv \prod_y \sum_x (A(x) \to B(y)),$

(c) $\displaystyle\prod_x (A(x) \to p) \to \left(\prod_x A(x) \to p\right),$

(d) $\displaystyle\left(\sum_x A(x) \to p\right) \to \sum_x (A(x) \to p),$

(e) $\displaystyle\sum_x \sum_{A(y)} xRy \equiv \sum_{A(y)} \sum_x xRy,$

(f) $\displaystyle\prod_x \prod_{A(y)} xRy \equiv \prod_{A(y)} \prod_x xRy.$

Demonstrate by means of examples that the implications converse to (c) and (d) are false.

2. Prove the following laws of negating, distributing, transferring and exchanging the order of quantifiers with a limited range:

(a) $\displaystyle\neg \prod_{A(x)} B(x) \equiv \sum_{A(x)} \neg B(x),$

(b) $\displaystyle\neg \sum_{A(x)} B(x) \equiv \prod_{A(x)} \neg B(x),$

(c) $\displaystyle\prod_{A(x)} (B(x) \to C(x)) \to \left(\prod_{A(x)} B(x) \to \prod_{A(x)} C(x)\right),$

(d) $\prod\limits_{A(x)} (B(x) \to C(x)) \to \left(\sum\limits_{A(x)} B(x) \to \sum\limits_{A(x)} C(x) \right),$

(e) $\prod\limits_{A(x)} B(x) \vee \prod\limits_{A(x)} C(x) \to \prod\limits_{A(x)} (B(x) \vee C(x)),$

(f) $\sum\limits_{A(x)} (B(x) \vee C(x)) \equiv \sum\limits_{A(x)} B(x) \vee \sum\limits_{A(x)} C(x),$

(g) $\sum\limits_{A(x)} (B(x) \wedge C(x)) \to \sum\limits_{A(x)} B(x) \wedge \sum\limits_{A(x)} C(x),$

(h) $\prod\limits_{A(x)} (p \to B(x)) \equiv p \to \prod\limits_{A(x)} B(x),$

(i) $\sum\limits_{A(x)} (p \to B(x)) \to \left(p \to \sum\limits_{A(x)} B(x) \right),$

(j) $\prod\limits_{A(x)} (B(x) \to p) \equiv \sum\limits_{A(x)} B(x) \to p,$

(k) $\sum\limits_{A(x)} (B(x) \to p) \to \left(\prod\limits_{A(x)} B(x) \to p \right),$

(l) $\prod\limits_{A(x)} \prod\limits_{B(y)} xRy \equiv \prod\limits_{B(y)} \prod\limits_{A(x)} xRy,$

(m) $\sum\limits_{A(x)} \sum\limits_{B(y)} xRy \equiv \sum\limits_{B(y)} \sum\limits_{A(x)} xRy,$

(n) $\sum\limits_{A(x)} (B(x) \wedge p) \equiv \sum\limits_{A(x)} B(x) \wedge p.$

Demonstrate by means of examples that the implications converse to (i) and (k) are not theses. Prove these converse implications preceded by the condition $\sum\limits_{x} A(x).$

4. Identity

In addition to the constants of the sentential calculus and the functional calculus, the logical terms include the symbol "$=$", which is called the *symbol of identity* or *equality*. This symbol is used in most mathematical formulae. The formula "$x = y$" is read: "x is identical with y" or "x equals y".

When introducing the symbol of identity in logic we adopt the axiom

A1. $\qquad\qquad\qquad\qquad x = x.$

This is the only axiom which we introduce into the logical systems analysed in the present book. The axiom is quite intuitive.

Equally intuitive is the rule for joining new lines of proof to a proof, which has the schema

EI $\qquad\qquad \dfrac{\begin{array}{c} \alpha = \beta \\ \phi \end{array}}{\phi(\beta//\alpha).}$

The letters α and β here are term variables, ϕ is any sentential formula, and $\phi(\beta//\alpha)$ is the formula obtained from ϕ by the replacement of the free variable α, which is not within the scope of a quantifier that binds the variable β, by the variable β. If the variable α occurs more than once in ϕ, it need not be replaced everywhere by the variable β: if ϕ stands for the formula "$x = y \rightarrow y = x$" then $\phi(t//x)$ stands both for the formula "$t = y \rightarrow y = t$" and for the formulae "$t = y \rightarrow y = x$" and "$x = y \rightarrow y = t$".

The rule introduced above is called the *rule of extensionality for identity*. It is analogous to the *rule of extensionality for equivalence*. But while the latter is a derived rule of the sentential calculus, the rule of extensionality for identity is a primitive rule.

By joining to the first order functional calculus the axiom A1 and the rule EI, and by generalizing the rules of that calculus so that they cover sentential formulae constructed by means of the symbol of identity, we obtain what is called *first order functional calculus with identity*. Some of the simplest theses of this calculus are given below.

T1. $$x = y \rightarrow y = x.$$

Proof. (1) $x = y$ {a.}
      (2) $x = x$ {A1}
          $y = x$ {EI: 1; 2}

The following rule, which is also symbolized EI, is derived from EI and T1:

$$\alpha = \beta$$
EI $$\frac{\phi}{\phi(\alpha//\beta).}$$

T2. $$x = y \wedge y = z \rightarrow x = z,$$

Proof. (1) $x = y$ $\left.\vphantom{\begin{matrix}x=y\\y=z\end{matrix}}\right\}$
      (2) $y = z$ {a.}
          $x = z$ {EI: 1; 2}

Axiom A1 and Theses T1 and T2 state that identity is reflexive, symmetric and transitive[1].

---

[1] A relation R is reflexive if $\prod\limits_{x} xRx$; it is symmetric if $\prod\limits_{x}\prod\limits_{y} (xRy \rightarrow yRx)$; it is transitive if $\prod\limits_{x}\prod\limits_{y}\prod\limits_{z} (xRy \wedge yRz \rightarrow xRz)$.

T1 and T2 yield

T2a. $$x = z \wedge y = z \to x = y.$$

The rule of extensionality EI is restricted by two conditions:
(1) only free variables can be replaced,
(2) the variable α, which we replace, does not occur in ϕ within the scope of a quantifier that binds the variable by which we replace α.

We shall now demonstrate that both these restrictions are essential. For this purpose let us note that the formulae

(α) $$\prod_x x = x,$$

(β) $$\prod_x x = y \to z = y$$

are theses of the first order functional calculus with identity. Formula (α) is obtained from A1 by applying the rule of joining the universal quantifier to theses; formula (β) is obtained by the rule OΠ.

The formula

$$x = y \to z = y,$$

in which the variables stand for names of natural numbers, is, of course, false. If we did not observe restrictions (1) and (2), we could prove it as follows:

Proof. (1) $x = y$ {a.}

(2) $\prod_x x = x$ {α}

(3) $\prod_x x = y \to z = y$ {β}

(4) $\prod_x x = x \to z = y$ {EI: 1; 3}

$z = y.$ {RD: 4; 2}

To obtain formula (4) we have, contrary to restriction (2), replaced the variable "y", which occurs within the scope of the quantifier binding the variable x, by "x".

Here is another proof—also incorrect—of the formula above:

Proof. (1′) $x = y$ {a.}

(2′) $\prod_x x = x$ {α}

(3′) $\prod_x x = y \to z = y$ {β}

$$(4') \prod_x x = y \qquad \qquad \{\text{EI: } 1'; 2'\}$$

$$z = y. \qquad \qquad \{\text{RD: } 3'; 4'\}$$

To obtain formula (4') we have, contrary to restriction (1), replaced the bound variable "x" by the variable "y".

T3. $\qquad \qquad x = y \land A(x) \to A(y).$

The easy proof of this thesis is left to the reader.

By one of the laws of compound transposition T3 yields

T3a. $\qquad \qquad A(x) \land \neg A(y) \to x \neq y.$

The formula "$x \neq y$" is used here as an abbreviation of the formula "$\neg(x = y)$". Thesis T3 may be read as follows: *If a certain property is an attribute of a certain object, then it is also an attribute of any object identical with the former.*

By T3a, *if a certain property is an attribute of the object x, and is not an attribute of the object y, then x is different from y.*

A formulation of Theses T3 and T3a can be found in Aristotle's works, which also include some other theorems concerned with identity (e.g., T2a).

T4. $\qquad \qquad A(x) \equiv \sum_{y=x} A(y).$

Proof. (a) (1) $A(x)$ $\qquad \qquad \qquad \{\text{a.}\}$

(2) $x = x$ $\qquad \qquad \qquad \{\text{A1}\}$

$\displaystyle \sum_{y=x} A(y)$ $\qquad \qquad \{\text{J}\Sigma^*\text{: } 2; 1\}$

(b) (1) $\displaystyle \sum_{y=x} A(y)$ $\qquad \qquad \{\text{a.}\}$

(2) $a_x = x$ $\left. \vphantom{\begin{matrix}a\\a\end{matrix}} \right\}$

(3) $A(a_x)$ $\qquad \qquad \{\text{O}\Sigma^*\text{: } 1\}$

$A(x)$ $\qquad \qquad \qquad \{\text{EI: } 2; 3\}$

Thesis T4 has been called by H. Scholz the *generalized fundamental theorem of Descartes.* In his well-known reasoning, intended to obtain a theorem that would be beyond all doubt, Descartes concludes that such a theorem is "I think". And he continues: I think, hence I exist. By T4, the sentence "I think" is equivalent to the sentence "there exists an x such that x is identical with me and x thinks" (in other words: "there exists a thinking being which is identical with me"). From the latter follows the sentence "there exists an x such that x is identical with me" (or: "I exist"). The name of the proper fundamental theorem of Descartes was given by Scholz to the thesis: $A(x) \to \sum_{y} y = x$, on which the reasoning described above can be based directly.

Sentences of the form "$A(x)$", in which a certain singular term occurs in place of the variable x, are called *singular sentences*. Thesis T4 permits us to replace a singular sentence by a logically equivalent sentence which begins with the existential quantifier.

The next thesis, the proof of which is left to the reader, permits us *to replace a singular sentence by a logically equivalent sentence that begins with the universal quantifier.*

T5. $$A(x) = \prod_{y=x} A(y).$$

When we apply the first order functional calculus with identity to systems in which compound terms (such as $a+b, a.b$, etc.) may be substituted for term variables, we use the derived rule of extensionality for identity, which permits us to replace compound terms. One of the formulations of this derived rule is given below.

Let ξ and η stand for terms. The rule with the schema

$$\frac{\xi = \eta}{\phi} \frac{}{\phi(\eta//\xi)}$$

permits us to replace ξ by η, provided that the following two conditions (which are generalized forms of conditions (1) and (2) given in the formulation of Rule EI) are satisfied:

(1) no free variable of the term ξ is bound in ϕ in the place where ξ is replaced;

(2) no free variable of the term η may become bound in $\phi(\eta//\xi)$ in the place where ξ has been replaced by η (i.e., the term ξ does not occur in ϕ, in the place where it has been replaced, within the scope of any quantifier binding any variable which is free in the term η).

The proof of this rule is left to the reader.

By this rule, for instance, on the strength of the thesis $(a+b)^2$ $= a^2+2ab+b^2$, we obtain from the formula "$(a+b)^2 = x$" the formula "$a^2+2ab+b^2 = x$".

In the first order functional calculus with identity we may define the unit quantifier:

D1. $$\vdash \sum_{\alpha} {}_1 \phi(\alpha) \equiv \sum_{\alpha} \phi(\alpha) \wedge \prod_{\phi(\alpha)} \prod_{\phi(\beta)} \alpha = \beta;$$

$\phi(\alpha)$ is here a sentential function in which α occurs as a free variable, β is not free in $\phi(\alpha)$, $\phi(\beta)$ is obtained from $\phi(\alpha)$ by the substitution of the variable β for the variable α.

The formula "$\sum_{x} {}_1 A(x)$" is read: there is one and only one x such that $A(x)$.

We may use the unit quantifier to give the following theorem from the arithmetic of real numbers:

$$\sum_{x} x^3 = 2 \wedge \prod_{x^3=2} \prod_{y^3=2} x = y$$

the much simpler form of

$$\sum_{x} {}_1 x^3 = 2.$$

On the strength of the law of negating a conjunction and the law of negating a universal quantifier with a limited range we obtain from D1 the thesis

T6. $$\daleth \sum_{x} {}_1 A(x) \equiv \daleth \sum_{x} A(x) \vee \sum_{A(x)} \sum_{A(y)} x \neq y.$$

By T6, *the sentence "it is not true that there exists one and only one object which has the property A" is equivalent to the sentence "there are no objects which have the property A or there are two different objects which have the property A".*

The rule of extensionality of the functional calculus and the definition D1 can be used to prove the following thesis:

T7. $$\prod_{x} [A(x) \equiv B(x)] \rightarrow [\sum_{x} {}_1 A(x) \equiv \sum_{x} {}_1 B(x)].$$

This thesis permits us to apply the rule of extensionality in the functional calculus also to those formulae in which the singular quantifier occurs.

In the first order functional calculus with identity we may introduce descriptions. A description is a meaningful term $\iota\,\phi(\alpha)$, consisting of the description operator ι, the term variable α, and the sentential function $\phi(\alpha)$, in which α is a free variable. The sentential function is called the *scope* of the description operator in a given expression. As in the case of formulae with quantifiers we make a distinction between the free and bound variables within the scope of the description operator. The description "$\iota\,A(x)$" is read: the only x such that $A(x)$. If there is exactly one object which satisfies the sentential function $\phi(\alpha)$, then the expression $\iota\,\phi(\alpha)$ is meaningful, otherwise the expression is meaningless. Suppose, for instance, that the variable x ranges over the set of the natural numbers. In such a case the expression "$\iota(2 < x < 4)$" is meaningful, for there is exactly one natural number between 2 and 4. On the other hand, the expressions "$\iota(2 < x < 3)$" and "$\iota(2 < x < 5)$" are meaningless, since neither of the sentential functions "$2 < x < 3$" and "$2 < x < 5$" is satisfied by exactly one natural number: in the former case, such a natural number does not exist at all, and in the latter, there are two different natural numbers which satisfy the sentential function in question. If the sentential function $\phi(\alpha)$, which forms the scope of a description operator, includes free variables other than α, which is bound by that operator, then certain substitutions of the expression $\iota\,\phi(\alpha)$ may be meaningful, while others may be meaningless. For instance, from the expression "$\iota(y < x < z)$", in which variables range over the set of the natural numbers, we may by substitution obtain both the meaningful term "$\iota(2 < x < 4)$" and the meaningless expression "$\iota(2 < x < 5)$". The expression "$\iota(y < x < z)$" is considered to be meaningful with the assumption $\sum_1(y < x < z)$, that is, when formulating theorems including the expression "$\iota(y < x < z)$" we shall precede it by the condition $\sum_1(y < x < z)$.

In conformity with the above, in the first order functional calculus with identity the descriptions, interpreted as meaningful terms of the form $\iota \underset{\alpha}{\phi}(\alpha)$, will not include the expression "$\iota \underset{x}{(x \neq x)}$", because $\neg \sum_{x} x \neq x$, but will include the term "$\iota \underset{x}{(x = y)}$", because $\sum_{1} x = y$; they will also include the expression "$\iota \underset{x}{A(x)}$", with the assumption that $\underset{x}{\sum_{1}} A(x)$. Descriptions, being terms, may be substituted for term variables by applying either the rule of substitution in theses, or the rule of omitting the universal quantifier. But, in conformity with what has been said above concerning the meaningfulness of the expression $\iota \underset{\alpha}{\phi}(\alpha)$, we apply the following general restriction concerning the theses that include descriptions:

Every thesis obtained from the theses obtained so far and containing the description $\iota \underset{\alpha}{\phi}(\alpha)$ must be preceded by the condition $\underset{\alpha}{\sum_{1}} \phi(\alpha)$. This condition may be omitted if it is a thesis or a consequence of other conditions that form the antecedent of the formula in question.

By this restriction, the axiom A1 yields by substitution the thesis:

(a) $\underset{x}{\sum_{1}} A(x) \to \iota \underset{x}{A(x)} = \iota \underset{x}{A(x)}$.

Likewise, the thesis $\underset{y}{\sum} y = x$ yields by substitution the thesis:

(b) $\underset{x}{\sum_{1}} A(x) \to \underset{y}{\sum} y = \iota \underset{x}{A(x)}$.

In conformity with the above restriction, from thesis (b) we may not obtain by transposition the formula: $\neg \underset{y}{\sum} y = \iota \underset{x}{A(x)} \to \neg \underset{x}{\sum_{1}} A(x)$ as a thesis. But by applying transposition to thesis (b) and taking the above restriction into consideration, we obtain the thesis:

(c) $\underset{x}{\sum_{1}} A(x) \to [\neg \underset{y}{\sum} y = \iota \underset{x}{A(x)} \to \neg \underset{x}{\sum_{1}} A(x)]$.

As a special case of thesis (a) we obtain the thesis

(a') $\underset{x}{\sum_{1}} x = y \to [\iota \underset{x}{(x = y)} = \iota \underset{x}{(x = y)}]$.

Since, however, the formula $\underset{x}{\sum_{1}} x = y$ is a thesis (cf. Exercise 2c), we can apply the rule of detachment to that thesis and to (a′) to obtain the thesis

(d) $\iota_x(x = y) = \iota_x(x = y).$

We also adopt the following primitive rule of joining the description operator:

$$J\iota \qquad \frac{\sum_{\alpha}{}_1 \phi(\alpha)}{\phi(\alpha / \iota_\alpha \phi(\alpha)).}$$

We shall now give, by way of example, the proofs for several theses, in which use shall be made of the rule $J\iota$ as a rule for joining new proof lines to a proof.

T8. $\qquad \sum_x {}_1 A(x) \rightarrow [A(y) \equiv y = \iota_x A(x)].$

Proof. (1) $\quad \sum_x {}_1 A(x)$ $\qquad\qquad$ {a.}

(2) $\quad A(\iota_x A(x))$ $\qquad\qquad$ {Jι: 1}

(1.1) $\quad A(y)$ $\qquad\qquad$ {ad. a.}

(1.2) $\quad A(y) \wedge A(\iota_x A(x)) \rightarrow y = \iota_x A(x)$

$\qquad\qquad\qquad\qquad\qquad$ {J1; 1}

(1.3) $\quad y = \iota_x A(x)$ $\qquad\qquad$ {1.2; 1.1; 2}

(2.1) $\quad y = \iota_x A(x)$ $\qquad\qquad$ {ad. a.}

(2.2) $\quad A(y)$ $\qquad\qquad$ {EI: 2.1; 2}

$\qquad A(y) \equiv y = \iota_x A(x)$ $\qquad$ {1.1 → 1.3; 2.1 → 2.2}

T9. $\qquad \sum_x {}_1 A(x) \rightarrow [B(\iota_x A(x)) \equiv \sum_{B(x) \wedge A(x)} \prod_{A(y)} y = x].$

Proof. (1) $\quad \sum_x {}_1 A(x)$ $\qquad\qquad$ {a.}

(2) $\quad A(\iota_x A(x))$ $\qquad\qquad$ {Jι: 1}

(1.1) $\quad B(\iota_x A(x))$ $\qquad\qquad$ {ad. a.}

(1.2) $\quad A(y) \rightarrow y = \iota_x A(x)$ $\qquad$ {RD, OE: T8; 1}

(1.3) $\quad \prod_{A(y)} y = \iota_x A(x)$ $\qquad$ {JΠ*: 1.2}

$$(1.4) \quad \sum_{B(x) \wedge A(x)} \prod_{A(y)} y = x \qquad \{JC, \ J\Sigma^*: \ 1.1; \ 2; \ 1.3\}$$

$$(2.1) \quad \sum_{B(x) \wedge A(x)} \prod_{A(y)} y = x \qquad \{ad. \ a.\}$$

$$(2.2) \quad \left. \begin{array}{l} B(a) \\ (2.3) \quad A(a) \end{array} \right\} \qquad \{O\Sigma^*, \ OC: \ 2.1\}$$

$$(2.4) \quad A(a) \rightarrow a = \underset{x}{\iota A(x)} \qquad \{RD, \ OE: \ T8; \ 1\}$$

$$(2.5) \quad a = \underset{x}{\iota A(x)} \qquad \{RD: \ 2.4; \ 2.3\}$$

$$(2.6) \quad \underset{x}{B(\iota A(x))} \qquad \{EI: \ 2.5; \ 2.2\}$$

$$\underset{x}{B(\iota A(x))} \equiv \sum_{B(x) \wedge A(x)} \prod_{A(y)} y = x$$

$$\{1.1 \rightarrow 1.4; \ 2.1 \rightarrow 2.6\}$$

T10. $\qquad \underset{x}{\sum_1 A(x)} \wedge \underset{x}{\prod [A(x) \equiv B(x)]} \rightarrow \underset{x}{\iota A(x)} = \underset{x}{\iota B(x)}.$

Proof.

$$(1) \quad \left. \begin{array}{l} \underset{x}{\sum_1 A(x)} \\[2mm] (2) \quad \underset{x}{\prod [A(x) \equiv B(x)]} \end{array} \right\} \qquad \{a.\}$$

$$(3) \quad \underset{x}{\sum_1 B(x)} \qquad \{RD, \ RD_E: \ T7; \ 2; \ 1\}$$

$$(4) \quad \underset{x}{A(\iota A(x))} \qquad \{J\iota: \ 1\}$$

$$(5) \quad \underset{x}{A(\iota A(x))} \equiv \underset{x}{B(\iota A(x))} \qquad \{O\Pi: \ 2\}$$

$$(6) \quad \underset{x}{B(\iota A(x))} \qquad \{RD_E: \ 5; \ 4\}$$

$$(7) \quad \underset{x}{B(\iota A(x))} \equiv \underset{x}{\iota A(x)} = \underset{x}{\iota B(x)} \qquad \{RD: \ T8; \ 3\}$$

$$\underset{x}{\iota A(x)} = \underset{x}{\iota B(x)} \qquad \{RD_E: \ 7; \ 6\}$$

In the antecedent of T10 we have omitted the condition $\underset{x}{\sum_1 B(x)}$ for, as can be seen from the proof, it is a consequence of the remaining conditions that form the antecedent of T10. Thesis T10 permits us to expand the rule of extensionality of the functional calculus so as to cover formulae containing descriptions.

Exercises

1. Prove the theses:

$$(a) \quad \underset{x}{\prod} \underset{y}{\sum} x = y,$$

(b) $x = y \equiv \prod_{z} (x = z \equiv y = z),$

(c) $u = z \wedge v = t \equiv \prod_{x} \prod_{y} (x = u \wedge y = v \equiv x = z \wedge y = t),$

(d) $u = z \wedge v = t \rightarrow \prod_{x} \prod_{y} (x = u \vee y = v \equiv x = z \vee y = t),$

(e) $xRy \equiv \sum_{z=x}^{\prime} \sum_{u=y} zRu,$

(f) $xRy \equiv \prod_{z=x} \prod_{u=y} zRu.$

2. Prove the theses:

 (a) $\vdash \sum_{\alpha} {}_{1} \phi(\alpha) \equiv \sum_{\phi(\alpha)} \prod_{\phi(\beta)} \beta = \alpha,$

 (b) $\vdash \sum_{\alpha} {}_{1} \phi(\alpha) \equiv \sum_{\alpha} \prod_{\beta} (\beta = \alpha \equiv \phi(\beta)),$

 (c) $\sum_{x} {}_{1} x = y,$

 (d) $\iota_{x} (x = y) = y.$

Demonstrate that by adopting each of equivalences (a), (b) as a definition of the unit quantifier we may derive from it the equivalence given in Dl.

3. Define the numerical quantifiers occurring in the following formulations:
 (a) there are two and only two x such that $A(x)$,
 (b) there are at least two x such that $A(x)$,
 (c) there are at most two x such that $A(x)$,
 (d) there are three and only three x such that $A(x)$.

5. Definitions

 In systems based on the first order functional calculus (with identity) we usually resort to one more rule, not discussed so far, namely the rule for joining definitions to a proof. The formulation of this rule will be preceded by a few general remarks about definitions.

 In nearly every discipline definitions of new terms are introduced by means of terms previously defined. In constructing an axiomatic system we make a point of reducing the number of primitive terms to a minimum. Other terms of such a system are defined by means of primitive terms.

 By means of the definitions we can replace every formula containing a defined term by an equivalent formula not containing that term, or, as it is often put, we can eliminate a defined term from any context.

 Definitions are formulated either as rules (belonging to a meta-system) which permit us to replace definite formulae of the system by other formulae, or as formulae of a special form, included among the theses of the system. The first method was discussed above in connection with definitions in the axiomatic system of the sentential calculus (Chap. I, Sec. 6). Of the definitions of the second kind we shall be concerned here with normal, conditional and inductive definitions.

Normal definitions have the form of equivalences or equalities. The term being defined occurs on the left-hand side of such a definition. That part of a normal definition which includes the term being defined is called the *definiendum*, and the other part is called the *definiens*.

As examples we may quote:

(1) $$\text{Prim}(x) \equiv x > 1 \wedge \prod_{y \neq 1} (y/x \rightarrow y = x),$$

(2) $$x \geqslant y \equiv x > y \vee x = y,$$

(3) $$1 = \text{seq}(0),$$

(4) $$\text{seq}(n) = n + 1,$$

(5) $$|x| = y \equiv (x \geqslant 0 \wedge y = x \vee x < 0 \wedge y = -x),$$

(6) $$x - y = z \equiv z + y = x.$$

In Definitions (1) and (4) the variables range over the set of all natural numbers, and in the remaining definitions, over the set of all real numbers.

In Definition (1) the term defined is a predicate of one argument, in Definition (2), a predicate of two arguments; in Definition (3) the term defined is a constant, in Definitions (4) and (5), a term-forming functor of one term argument, in Definition (6), a term-forming functor of two term arguments.

To be correct, normal definitions must satisfy the following conditions:

1. The definiens may include only primitive terms or terms which have previously been correctly defined. The term being defined must be different from all the terms already occurring in the system. Hence the definiens may not include either the term being defined or any term which has been defined by means of the term being defined.

If the last condition is not satisfied, we speak of a vicious circle in the definition.

2. In the definiendum, every variable may occur only once.

If this condition is not satisfied we are not in a position to eliminate the term being defined from certain contexts by means of the given definition.

For instance, by means of the equivalence

$$\log_a a = b \equiv a^b = a,$$

which does not satisfy condition 2, we cannot eliminate the symbol "log", e.g. from the context "$\log_{10} 100 = 2$".

3. The condition of homogeneity: every variable which occurs free on one side of a definition should also occur free on its other side.

A definition in which this condition is not satisfied may lead to a contradiction. For instance, from the definition

(7) $a \parallel b \equiv a$ and b lie in the plane α and have no point in common,

we may obtain a contradiction in the following way:

Let a_1 and a_2 have no point in common, lie on the plane α_1 but not in the plane α_2. By substituting in (7): a/a_1, b/b_1, α/α_1, and detaching the right-hand side of this substitution we obtain $a \parallel b$. But by substituting a/a_1, b/b_1, α/α_2 in (7), negating both sides of this substitution and detaching the right-hand side of the equivalence thus obtained, we obtain $\neg(a \parallel b)$.

Conditions 1, 2, 3 are sufficient conditions for the correctness of normal definitions in which the definiendum apart from the term being defined includes variables only (as in Definitions (1)–(4)).

In Definitions (5) and (6) the definiendum apart from the term being defined includes the symbol of equality. The general form of such definitions is as follows:

(a) $$f(x_1, \ldots, x_n) = y \equiv \phi(x_1, \ldots, x_n, y)$$

where "f" is the term-forming functor of n terms or functor arguments which is being defined, and the definiens is a sentential function in which the variables $x_1, \ldots, x_n, y$, and only these variables, are free.

A normal definition of the form (a), in addition to conditions 1, 2, 3, should also satisfy what is called the *condition of existence and uniqueness*. We say that a normal definition satisfies this condition if the formula

(8) $$\prod_{x_1} \cdots \prod_{x_n} \sum_y {}_1 \phi(x_1, \ldots, x_n, y)$$

is a thesis.

If, on the contrary, formula (8) is false, a definition of form (a) may lead to a contradiction. For instance, the definition

(9) $$\sqrt{x} = y \equiv y^2 = x$$

does not satisfy the condition of existence and uniqueness, as it is not true that $\prod_x \sum_y {}_1 y^2 = x$. If in (9) we first substitute $x/4$, $y/2$, and then substitute $x/4$, $y/-2$, we have to conclude that $2 = -2$, which contradicts the thesis $2 \neq -2$.

Conditions 1, 2, 3, and the condition of existence and uniqueness are sufficient conditions for the correctness of normal definitions of form (a).

As an example of a conditional definition we may give:

(10) $$y \neq 0 \rightarrow \left(\frac{x}{y} = z \equiv y \cdot z = x \right).$$

Should we omit the antecedent from Definition (10), the definition would lead to a contradiction, since then the definiens would not satisfy the condition of existence and uniqueness, as $0.2 = 0$ and $0.3 = 0$, so that it is not true that $\prod_x \prod_y \sum_z {}_1 y \cdot z = x$.

The general form of conditional definitions is the following:

(b) $$\psi(x_{i_1}, \ldots, x_{i_k}) \rightarrow [f(x_1, \ldots, x_n) = y \equiv \phi(x_1, \ldots, x_n, y)],$$

where the variables $x_{i_1}, \ldots, x_{i_k}$ are some (sometimes all) of the variables $x_1, \ldots x_n$, $\psi (x_{i_1}, \ldots, x_{i_k})$ is a sentential formula in which the variables $x_{i_1}, \ldots, x_{i_k}$, and only these variables, are free, and the consequent of formula (b) is a formula of form (a).

A definition of form (b) is correct if its consequent satisfies conditions 1, 2, 3, required of normal definitions, and the condition of existence and uniqueness for objects which satisfy the antecedent of the definition in question. Thus, the condition of existence and uniqueness here has the form

(11) $$\prod_{x_1} \cdots \prod_{x_n} [\psi(x_{i_1}, \ldots, x_{i_k}) \rightarrow \sum_y {}_1 \phi(x_1, \ldots, x_n, y)].$$

For instance, for Definition (10) this condition has the form

$$\prod_{x} \prod_{y \neq 0} \sum_{z}{}_1 y . z = x.$$

When we define terms by means of conditional definitions we also adopt certain restrictions as to the meaningfulness of formulae containing those terms. We consider all and only those formulae to be meaningful from which the term being defined can be eliminated by means of its conditional definition.

In particular, those formulae are meaningful which are obtained from the expression "$f(x_1, \ldots, x_n)$" by the substitution for the variables $x_1, \ldots, x_n$ of values for which the condition of the given definition is true.

For instance, in the case of the term defined by means of Definition (10) the expressions: $\dfrac{6}{3}$, $\dfrac{x}{3}$, etc., are meaningful, since the condition $y \neq 0$ is true for 3.

Meaningful, too, are formulae of the form of an implication in which the consequent contains only substitutions of the expression "$f(x_1, \ldots, x_n)$" in which some of the variables $x_{i_1}, \ldots, x_{i_k}$ are replaced by expressions which include variables, and in which the antecedent is formed by the appropriate substitution from the formula $\psi(x_{i_1}, \ldots, x_{i_k})$.

For instance, in the example under consideration, the formula

$$a+b \neq 0 \to \frac{x}{a+b} + 1 = z$$

is meaningful.

If in the consequent the term being defined occurs more than once in different contexts, then the antecedent must be a conjunction of the appropriate conditions or a formula from which such a conjunction follows.

For instance, in the example under consideration the formulae

$$y \neq 0 \wedge t \neq 0 \to \left(\frac{x}{y} = \frac{z}{t} \equiv x . t = y . z \right),$$

$$a . b \neq 0 \to \frac{x}{a \cdot b} = \frac{x}{a} \cdot \frac{1}{b}$$

are meaningful.

On the other hand, we do not consider meaningful formulae which have been obtained from the expression "$f(x_1, ..., x_n)$" by substituting for the variables $x_{i_1}, ..., x_{i_k}$ constants for which the condition of a given definition does not hold.

For instance, in the example under consideration the expression "$\frac{x}{0} = z$" is not meaningful, since the condition $y \neq 0$ does not hold for 0.

The definition:

$$(12) \qquad \begin{cases} a.1 = a \\ a.(n+1) = a.n + a \end{cases}$$

which belongs to the arithmetic of natural numbers, is an example of an inductive definition.

This definition consists of two equations. In the former, the term being defined occurs only on the left-hand side of the definition, and one of its arguments is the constant "1". In the latter, the term being defined occurs on both sides, but on the right-hand side one of its arguments is "n", and on the left-hand side the corresponding argument is "$n + 1$".

Inductive definitions also take on much more complicated forms.

Inductive definitions are reducible to the normal form. For instance, for definition (12) each of the following formulae is such a form

$$(13) \quad a.b = c \equiv \sum_{f} \{\prod_{a} f(a,1) = a \wedge \prod_{a}\prod_{n} [f(a,n+1) = f(a,n)+a]$$
$$\wedge\, f(a,b) = c\},$$

$$(13') \quad a \cdot b = c \equiv \prod_{f} \{\prod_{a} f(a,1) = a \wedge \prod_{a}\prod_{n} [f(a,n+1) = f(a,n)+a]$$
$$\rightarrow f(a,b) = c\}.^{[1]}$$

The formula "$\prod_{a} f(a,1) = a \wedge \prod_{a}\prod_{n}[f(a,n+1) = f(a,n)+a]$", occurring in (13) and (13'), is obtained by forming a conjunction of the two equations that form definition (12), preceding these equations by universal quantifiers, and replacing the functor ".", which is being defined, by a variable.

---

[1] We disregard here the proof demonstrating that each of the formulae (13) and (13') is a consequence of the formulae (12), and vice versa, that formulae (12) are consequences of each of the formulae (13) and (13').

The procedure is analogous in other cases.

Formulae (13) and (13′) include quantifiers binding a variable which stands for functors. Thus here we go beyond the limits of the first order functional calculus.

The definition of a thesis of the nth order, given on p. 18, is an inductive definition. By analogy, we can formulate a definition of a sentential formula of the nth order. These definitions, too, can be reduced to a normal form. We shall use the definition of a sentential formula of the sentential calculus (cf. pp. 2–3 above) to make the reader familiar with a certain formulation of the normal form for such definitions, a formulation which is often used in analogous cases. In this formulation we use the concept of a set closed under certain operations, and the concept of the smallest set having a certain property. The definitions of these concepts are as follows:

The set S is *closed under the operation O* if and only if the application of the operation O to any element (s) of the set S yields as a result an element of the set S.

The set S is the smallest set having the property P if and only if the set S has the property P and the set S is included in every set that has the property P.

With these concepts we can reformulate as follows the definition given on pp. 2–3:

The set of sentential formulae of the sentential calculus is the smallest set which includes sentential variables, and which is closed under the operations of forming the negation, conjunction, disjunction, implication, and equivalence.

By analogy, we define the set of theses of a given axiomatic system as the smallest set which includes the axioms of that system and is closed under the rules of inference of that system.

If we wrote down such definitions in symbols, they would include quantifiers binding not only variables which stand for individuals, which in this case are formulae, but also for sets of such individuals.

The rule of joining definitions to a system permits us

(1) to introduce, as theses of a system, correct definitions of new terms, constructed by means of primitive terms or terms previously defined in a correct manner;

(2) to join to an assumptional proof, as new proof lines, correct definitions of new terms, constructed by means of terms of the system in question (primitive terms or terms previously correctly defined) and by means of constants introduced to the proof on the strength of the rule $O\Sigma$.

In this formulation we make no restrictions whatever as to the forms of definitions, so that we may introduce, in assumptional proofs, not only normal definitions, but also conditional and inductive definitions. For instance, having introduced the constants "a" and "b", standing for certain numbers, into a proof by means of the rule $O\Sigma$, we may define by induction the sequence $\{c_n\}$:

$$c_1 = a,$$
$$c_{n+1} = c_n \cdot b.$$

We have been concerned here with definitions and the rule of joining definitions principally because in Part II we shall avail ourselves of definitions of the types discussed above. Moreover, even in Part I we have introduced definitions of various kinds, and hence we have to understand the conditions which they must satisfy in order to be correct.

The rule of joining definitions is generally used in systems based on the functional calculus of the first order. Within that calculus itself there is no need to introduce that rule as a new primitive rule since the rules previously introduced suffice to prove all the true formulae of the system.

However, we shall demonstrate here how we can prove the rule of substitution for functional formulae, which is one of the most complicated logical rules and whose proof by induction is also complicated, by means of the rule of joining definitions. The way we do this makes it possible to understand the necessity of introducing the restrictions which occur in the formulation of the rule now to be considered.

The basic idea of the rule of substitution for functional formulae is based on the interpretation of the molecular formulae of the functional calculus of the first order, by which, e.g., the formula "$A(x)$" stands for any sentential function with the free variable "x" and these sentential functions may be substituted for the formula "$A(x)$" (cf. p. 91 above). When formulating the rule in a precise manner we must, however, first take into account the fact that in a given formula, for instance in Thesis T1 (Sec. 3, Chap. II), a variable standing for predicates may occur in various places and with various arguments, and secondly, we must introduce certain restrictions concerning free and bound variables, without which the rule would lead from true formulae to false formulae. Since, in view of all this, the formulation of the rule is fairly complicated, we shall first give an example of its application.

EXAMPLE 1. By this rule, from the thesis

(1) $$\prod_x A(x) \to A(y)$$

we obtain the thesis

(2) $$\prod_x R(z, x, u) \to R(z, y, u),$$

by making the following substitutions in (1):

$$A(x)/R(z, x, u),$$
$$A(y)/R(z, y, u).$$

The substituted formulae differ from one another only in the fact that the variable "x" appears in the formula substituted for "$A(x)$" in the same place where the variable "y" occurs in the formula substituted for "$A(y)$". To symbolize this substitution in a general way we introduce the variable "t", assuming that it occurs neither in the thesis in which the substitution is made nor in the formula which is being substituted. Hence we have

$$A(t)/R(z, t, u).$$

We now formulate the rule of substitution for functional formulae in its general form:

Let ϕ be any thesis of the functional calculus of the first order, in which the variable φ stands for predicates. We shall now define how to obtain from ϕ a new thesis by replacing in ϕ all the formulae constructed from the predicate φ and some of its arguments by formulae which differ from a certain formula ψ only by variables. In the example given above the role of the formula ψ was played by the formula "$R(z, t, u)$".

Let $\varphi(\beta_1, ..., \beta_k)$ be part of the formula ϕ and let $\psi(\alpha_1, ..., \alpha_k)$ be any sentential formula containing the free variables $\alpha_1, ..., \alpha_k$, which we assume to occur in ϕ neither as free nor as bound variables. The variables $\alpha_1, \ldots, \alpha_k$ need not be all the variables occurring in the formula ψ. We substitute for the formula $\varphi(\beta_1, \ldots, \beta_k)$ the formula which is obtained from $\psi(\alpha_1, \ldots, \alpha_k)$ by the substitution of β_1 for α_1, β_2 for α_2, ..., β_k for α_k. We proceed in a similar way with every part of the formula ϕ which has the form $\varphi(\gamma_1, \ldots, \gamma_k)$.

We assume that:

(i) None of the variables $\alpha_1, \ldots, \alpha_k$ occurs within the scope of a quantifier binding the variables which we have substituted in formula ψ for the variables $\alpha_1, \ldots, \alpha_k$, for instance the variables $\beta_1, \ldots, \beta_k$ or $\gamma_1, \ldots, \gamma_k$.

(ii) If one of the formulae $\varphi(\beta_1, \ldots, \beta_k), \varphi(\gamma_1, \ldots, \gamma_k)$, etc., occurs in ϕ within the scope of the quantifier binding any variable γ, then in the formula $\psi(\alpha_1, \ldots, \alpha_k)$ the variable γ may not occur as a free variable.

We shall now give examples of incorrect substitutions, infringing one of the restrictions (i), (ii).

EXAMPLE 2.

(1)
$$\prod_x A(x) \to A(y),$$

$$A(t) / \sum_y t R y,$$

(2)
$$\prod_x \sum_y x R y \to \sum_y y R y.$$

This substitution violates restriction (i), since the variable t occurs in the formula "$\sum_y t R y$" within the scope of the quantifier binding the variable y, which, contrary to the general restrictions on the rule of substitution, we have substituted for the variable t.

Formula (2) is false, as can be seen from the following substitution:

$$\prod_x \sum_y x < y \to \sum_y y < y$$

in which the antecedent is true and the consequent false.

EXAMPLE 3.

(1)
$$\prod_x A(x) \to A(y),$$

$$A(t) / x R t,$$

(2) $\prod_x xRx \to xRy.$

This substitution violates restriction (ii), since the formula "$A(x)$" occurs in formula (1), in which we substitute, within the scope of a quantifier binding the variable x, which at the same time is a free variable in the formula "xRt." The falsehood of formula (2) is demonstrated by the falsehood of its substitution:

$$\prod_x x = x \to x = y,$$

in which the antecedent is true and the consequent false.

We shall now show how in Example 1 formula (2) can be derived from formula (1) by means of the rule of substitution for variables (both term variables and variables standing for predicates) and the rule of joining definitions.

In this proof we introduce the definition

$$f_1\langle R, z, u \rangle (t) \equiv R(z, t, u).$$

In this definition the expression "$f_1\langle R, z, u \rangle$" is a predicate of one argument. Expressions of this type are called *predicates dependent on parameters*. In the example under consideration the variables "R", "z", "u" are parameters.

EXAMPLE 1.

(1) $\prod_x A(x) \to A(y)$

(2) $\prod_t [f_1\langle R, z, u \rangle(t) \equiv R(z, t, u)]$ {Df.}

(3) $\prod_x f_1\langle R, z, u \rangle(x) \to f_1\langle R, z, u \rangle(y)$ {(1): $A/f_1\langle R, z, u \rangle$}

(4) $f_1\langle R, z, u \rangle(x) \equiv R(z, x, u)$ ⎱
(5) $f_1\langle R, z, u \rangle(y) \equiv R(z, y, u)$ ⎰   {OΠ: 2}

      $\prod_x R(z, x, u) \to R(z, y, u)$ {3; 4; 5}[1]

The procedure is analogous in other examples of this type: we introduce a definition of the appropriate predicate dependent on parameters, we substitute it for the variable standing for predicates,

---

[1] We have made use here of the rule of extensionality of the functional calculus (cf. p. 115).

and then we replace the substitutions of the left-hand side of the definition by the corresponding substitutions of the right-hand side.

To explain the role of restrictions (i), (ii), given in the formulation of the rule of substitution for functional formulae, we shall examine why in Examples 2 and 3 an analogous proof cannot be given.

EXAMPLE 2.

(1) $\prod_x A(x) \to A(y)$

(2) $\prod_t [f_2\langle R\rangle(t) \equiv \sum_y tRy]$ {Df.}

(3) $\prod_x f_2\langle R\rangle(x) \to f_2\langle R\rangle(y)$ {(1): $A/f_2\langle R\rangle$}

(4) $f_2\langle R\rangle(x) \equiv \sum_y xRy$ {OΠ: 2}

Here the proof breaks down because the substitution in (2) of "y" for "t" would violate the restriction on the rule of substitution for variables, since the free variable y would become bound in the expression resulting by substitution.

EXAMPLE 3.

(1) $\prod_x A(x) \to A(y)$

(2) $\prod_t [f_3\langle R, x\rangle(t) \equiv xRt]$ {Df.}

Here the proof breaks down because in (1) we may not substitute "$f_3\langle R, x\rangle$" for "A", since the free variable x would then become bound in the expression resulting by the substitution.

To state this in a general form:

If restriction (i) is not satisfied, that is, for instance, if the variable α_i occurs in ψ within the scope of the quantifier which binds the variable β_j, then in the definition

$$f\langle \delta_1, \ldots, \delta_m\rangle(\alpha_1, \ldots, \alpha_k) \equiv \psi(\delta_1, \ldots, \delta_m, \alpha_1, \ldots, \alpha_k)$$

the variable β_j may not be substituted for the variable α_i, so that, as in Example 2, we may not replace the formula $\varphi(\beta_1, \ldots, \beta_j, \ldots, \beta_k)$ by the formula $\psi(\delta_1, \ldots, \delta_m, \beta_1, \ldots, \beta_j, \ldots, \beta_k)$.

And if restriction (ii) is not satisfied, that is, for instance, if $\varphi(\beta_1, \ldots, \beta_k)$ occurs in ϕ within the scope of the quantifier binding the variable γ, and the variable γ is free in the formula $\varphi(\alpha_1, \ldots, \alpha_k)$, then on introducing the definition

$$f\langle \gamma, \delta_1, ..., \delta_m \rangle (\alpha_1, ..., \alpha_k) \equiv \psi(\alpha_1, ..., \alpha_k, \gamma, \delta_1, ..., \delta_m),$$

as in Example 3 we may not substitute in the formula ϕ the expression "$f\langle \gamma, \delta_1, ..., \delta_m \rangle$", for the variable γ would then become bound.

It is now clear that if restrictions (i), (ii) are satisfied, then the thesis which is obtained by substitution for functional formulae can also be obtained by an appropriate definition and substitution for variables. For if condition (ii) is satisfied, then on introducing an appropriate definition we may substitute for the variable standing for predicates, and if condition (i) is also satisfied, then the left-hand side of an appropriate substitution of the definition may be replaced everywhere by its right-hand side.

We have discussed the rule of substitution for functional formulae in great detail chiefly because we perform such substitutions in applying formulae of the functional calculus to any field of science.

That is why we must clearly realize the restrictions to be observed in this connection.

Note also that formula (2), which we obtain (correctly) in Example 1, is a thesis in accordance with Metathesis M1 (cf. p. 102 above) corresponding to Thesis (1), since the formula "$R(z, y, u)$" can be obtained from the formula "$R(z, x, u)$" through the substitution of "y" for "x".

On the other hand, formulae (2), which we obtain (incorrectly) in Examples 2 and 3, do not come under the schema of Metathesis M1, since the formula "$\sum\limits_{y} yRy$" cannot be obtained from the formula "$\sum\limits_{y} xRy$" through the substitution of "y" for "x", nor is the formula "xRy" a substitution of the formula "xRx".

In this case, as in many others, it is easier to apply metatheses of the functional calculus than to apply the corresponding theses which require substitution for functional formulae.

Exercises

1. Formulate inductive definitions of addition and raising to a power of natural numbers, and then transform them into normal definitions.

2. In the thesis: $\sum\limits_{x} \prod\limits_{y} R(x, y) \to \sum\limits_{x} R(x, x)$ substitute $R(t_1, t_2)/\sum\limits_{u} R(t_1, z, t_2, u)$.

Deduce the thesis obtained through this substitution by means of an appropriate definition.

3. In the thesis given in Ex. 2 substitute $R(t, t_1)/t = t.t_1$.

6. Functional Calculi of Higher Orders

In the preceding Sections of the present Chapter we have discussed the functional calculus of the first order. The essential properties of that calculus will now be repeated.

In the functional calculus of the first order, the only variables, apart from sentential variables, are term variables and variables standing for predicates. In the formulae of that calculus, quantifiers bind term variables only. Molecular formulae of the functional calculus of the first order are sentential variables and formulae consisting of a predicate[1] and its arguments, that is such formulae as

$$A(x), B(y), xRy, R(x, y, z).$$

We shall introduce a more general concept of a *molecular formula*. This concept is fairly difficult, and therefore its definition will be preceded by a number of examples. The equivalence

DI. $$\alpha(A) \equiv \sum_{x} {}_1 A(x)$$

defines the symbol α. If we interpret predicates of one argument as formulae indicating properties of individual objects, we may read formulae of the type $\alpha(A)$: A is a property which is an attribute of exactly one object.

DII. $$A\rho B \equiv \neg \sum_{x} (A(x) \equiv B(x)).$$

The formula $A\rho B$ may be read: the properties A and B are not simultaneously attributes of any object.

DIII. $$x\sigma A \equiv A(x) \wedge \alpha(A).$$

The formula thus defined is read: x is the only object that has the property A. The first argument of the functor σ is a term, and the second a predicate.

---

[1] Strictly speaking: of a predicate or a variable standing for a predicate. This and analogous abbreviations will often be used in this Section.

Predicates are otherwise called *functors of the first order*, and molecular formulae of the functional calculus of the first order are called *molecular formulae of the first order*. *Sentence-forming* functors of arguments which are terms or functors of the first order, if at least one such argument is a functor of the first order, are functors of the second order.

A formula constructed of a functor of the second order and its arguments is called a *molecular formula of the second order*. Thus, the symbols α, ρ, σ, defined above, are functors of the second order, and the formulae $\alpha(A)$, $A\rho B$, $x\sigma A$ are molecular formulae of the second order.

We now assume that the bold-faced letters $\mathbf{A}, \mathbf{B}, \mathbf{C}, \ldots$ stand for second order functors of one argument, and the bold-faced letters $\mathbf{R}, \mathbf{S}, \mathbf{T}$, for second-order functors of several arguments.

The following formulae are molecular formulae of the second order, symbolized exclusively by means of the variables

$$\mathbf{A}(A), \; \mathbf{B}(A), \; A\mathbf{R}B, \; x\mathbf{R}A, \; \mathbf{S}(A, x, y).$$

In the same way as the functional calculus of the first order has been constructed as a superstructure on the sentential calculus, we may construct over the former a richer system having as molecular formulae sentential variables and molecular formulae of the first and second orders. This system is called the *functional calculus of the second order*. In this calculus quantifiers bind sentential variables and variables standing for predicates, but do not bind variables standing for functors of the second order.

We give below, by way of example, a number of theses belonging to the functional calculus of the second order. The first three of them are analogues of certain theses belonging to the functional calculus of the first order; in particular, TIII corresponds to T21 (Sec. 3, Chap. II), and TII corresponds to the simple implication contained in T7 (Sec. 3, Chap. II).

TI. $$\prod_A \mathbf{A}(A) \to \sum_A \mathbf{A}(A).$$

TII. $$\prod_A [\mathbf{A}(A) \wedge \mathbf{B}(A)] \to \prod_A \mathbf{A}(A) \wedge \prod_A \mathbf{B}(A).$$

TIII. $$\sum_A \prod_B (A\mathbf{R}B) \to \prod_B \sum_A (A\mathbf{R}B).$$

TIV. $$\prod_x \sum_A A(x).$$

Thesis TIV may be read: *for every object there is a property which is its attribute.* This thesis does not contain molecular formulae of the second order, but it does not belong to the functional calculus of the first order because the existential quantifier binds in it a variable which stands for predicates.

TV. $$\sum_R (xRy \equiv yRx).$$

By this thesis there exists a relation which holds between the objects x and y if and only if it holds between y and x.

The rules for handling quantifiers which bind variables that stand for predicates are fully analogous to the rules of the functional calculus of the first order. We shall illustrate this by the example of TII.

Proof. TII. (1) $\prod_A [A(A) \wedge B(A)]$ {a.}

(2) $A(A) \wedge B(A)$ {OΠ: 1}

(3) $A(A)$ ⎫

(4) $B(A)$ ⎭   {OC: 2}

(5) $\prod_A A(A)$ {JΠ: 3}

(6) $\prod_A B(A)$ {JΠ: 4}

$\prod_A A(A) \wedge \prod_A B(A)$ {JC: 5; 6}

In proving theses of the functional calculus of the second order we often make use of definitions. For instance, the proof of TIV is based on the following definition:

(a) $*\langle x \rangle(y) \equiv x = y.$

The formula "$*\langle x \rangle (y)$", introduced in this definition, is a predicate of one argument dependent on the parameter "x". This formula might be interpreted as a formula denoting the property which is an attribute of an object if and only if that object is identical with the object x, hence, denoting a property characteristic of the object x.

Definition (a) and Axiom A1 (Sec. 4, Chap. II) easily yield

(b) $*\langle x \rangle(x).$

This formula will be employed in later proofs.

Proof. TIV. (1) $*\langle x\rangle(x)$ {b}

(2) $\sum_A A(x)$ {JΣ: 1}

$\prod_x \sum_A A(x)$ {JΠ: 2}

TV is proved on the strength of the thesis: $x = y \equiv y = x$. Another important thesis will now be proved.

TVI. $x = y \equiv \prod_A (A(x) \equiv A(y))$.

Proof. (a) (1) $x = y$ {a.}

(2) $A(x) \equiv A(x)$ {T11a. Sec. 3, Chap. I}

(3) $A(x) \equiv A(y)$ {EI: 1; 2}

$\prod_A (A(x) \equiv A(y))$ {JΠ: 3}

(b) (1) $\prod_A (A(x) \equiv A(y))$ {a.}

(2) $*\langle x\rangle(x) \equiv *\langle x\rangle(y)$ {OΠ: 1}

(3) $*\langle x\rangle(y)$ {RD$_E$: 2; b}

$x = y$ {RD$_E$: a; 3}

By TVI, *the objects x and y are identical if and only if every property which is an attribute of one of them is an attribute of the other.* Hence no property can be an attribute of the object x without simultaneously being an attribute of the object y identical with x.

Thesis TVI may be considered the *definition of identity*. Its basic idea can be found even in the works of Aristotle who stated that only identical objects have all the same properties as attributes. As St. Thomas Aquinas (13th century) put it, objects are identical if whatever is an attribute of one of them is an attribute of any other. Best known was the formulation of Leibniz (17th century), stating that objects are identical which do not differ by any property (*principium identitatis indiscernibilium*). A symbolic formulation of Thesis TVI as a definition of identity was first given by Peirce in 1885. This definition became known after it was included in Whitehead and Russell's *Principia Mathematica*. It is sometimes called *Russell's* or *Leibniz's and Russell's definition of identity.*

Sentence-forming functors which have terms and functors of the first and/or second order as arguments are called *functors of the third order* provided at least one argument is a functor of the second order. Molecular formulae of the third order are defined in a way analogous to the definition of molecular formulae of the second order.

We call a system which is constructed as a superstructure on the functional calculus of the second order in the same way as the latter is constructed as a superstructure on the functional calculus of the first order *a functional calculus of the third order*; in this system, molecular formulae are sentential variables and molecular formulae of the first, second, and third order and quantifiers bind term variables and variables standing for functors of the first and second orders, but do not bind variables standing for functors of the third order. In a similar way we construct functional calculi of the fourth, fifth, etc., orders. It is obvious that the higher the order of the functional calculus the more restricted is the application, in the various disciplines, of the theses of this calculus which are not theses of the calculi of lower orders. Yet one aspect of progress in contemporary mathematics consists in the fact that the various branches of mathematics are based on functional calculi of increasingly high orders.

We infer from TVI that the identity of individuals can be defined in the functional calculus of the second order. Likewise the relation

$$A = B,$$

which we may read: the property A is identical with the property B, is definable in the functional calculus of the third order. But we may also introduce the relation of identity of properties into the functional calculus of the second order by adopting the axiom

A1'. $A = A,$

analogous to Axiom A1, and by adopting a rule analogous to Rule EI.

The issues raised in this Section are also commented upon in Sec. 3 of the *Supplement*.

Exercises

1. Prove the theses:

(a) $\sum_A \prod_B [\prod_x (A(x) \to B(x)) \land \prod_x (A(x) \to \neg B(x))],$

(b) $\sum_A \prod_x A(x),$

(c) $\sum_A \prod_x \neg A(x)$,

(d) $A(x) \equiv \sum_B [\prod_x (A(x) \equiv B(x)) \wedge B(x)]$.

2. Prove AI and Rule EI by adopting Thesis TVI as the definition of identity.

3. Show that the symbol of identity can also be introduced by means of the following primitive rules of omitting and joining:

OI
$$\frac{x = y}{A(x) \equiv A(y);}$$

JI
$$\frac{A(x) \equiv A(y)}{x = y.}$$

It is assumed in JI that "A" is not a free variable in the assumptions of the proof.

Prove Thesis TVI by means of Rules OI and JI, and prove Rules OI and JI by means of Thesis TVI.

4. Prove the equivalence of Thesis TVI and the thesis:

$$x = y \equiv \prod_A (A(x) \rightarrow A(y)).$$

5. Prove the theses:

(a) $x \neq y \equiv \sum_A (A(x) \wedge \neg A(y) \vee A(y) \wedge \neg A(x))$,

(b) $x \neq y \equiv \sum_A (A(x) \wedge \neg A(y))$.

7. Examples of Formalized Mathematical Proofs

EXAMPLE 1. Proof of the existence of a left-sided identity for multiplication, based on the following axioms of group theory:

AI. $x \cdot (y \cdot z) = (x \cdot y) \cdot z$.

AII. $\sum_z (y \cdot z = x)$.

AIII. $\sum_z (z \cdot y = x)$.

Thesis: $\sum_y \prod_x (y \cdot x = x)$.

Proof. (1) $\sum_z (z \cdot y = y)$ {AIII}

(2) $a_y \cdot y = y$ {OΣ: 1}

(3) $y \cdot b_{x,y} = x$ {OΣ: AII}

(4) $a_y \cdot (y \cdot b_{x,y}) = (a_y \cdot y) \cdot b_{x,y}$ {AI}

(5) $a_y \cdot x = (a_y \cdot y) \cdot b_{x,y}$ {EI: 3; 4}

(6) $a_y \cdot x = y \cdot b_{x,y}$ {EI: 2; 5}

(7) $a_y \cdot x = x$ {EI: 3; 6}

(8) $\prod\limits_{x} (a_y \cdot x = x)$ {JΠ: 7}

$\sum\limits_{y} \prod\limits_{x} (y \cdot x = x)$ {JΣ: 8}

EXAMPLE 2. Proof of the theorem on the existence of the bisector.[1] We introduce the following abbreviations to be used in the notation of the proof:

The letter X stands for the set of variables O, A, A', B, B', which represent points shown in Fig. 1.

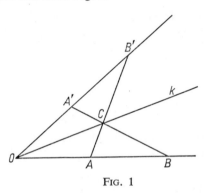

FIG. 1

The formula "$\Phi(X)$" stands for the sentence "the points A, B lie on one side of the angle, the points A', B', on the other side of the angle, and the segments $OA, AB, OA', A'B'$ are equal".

The formula "$\Psi(X, C)$" stands for the sentence "the point C lies on the straight lines AB' and $A'B$".

The formula "$\Theta(C, k)$" stands for the sentence "k is the straight line OC".

The formula "$\varDelta(k)$" stands for the sentence "the straight line k is the bisector of the given angle".

When applying the rule OΣ in the proof we shall use as constants the letters X, C, k with subscripts: X_1, C_1, k_1.

---

[1] Cf. A. Mostowski, *Logika matematyczna* (Mathematical Logic), 1948, pp. 68–9.

The following theorems are accepted without proof:

I. $$\prod_X [\Phi(X) \to \sum_C \Psi(X, C)].$$

In the verbal formulation: If it is assumed that $\Phi(X)$, *the straight lines AB' and A'B intersect.*

II. $$\prod_{X,C,k} [\Phi(X) \land \Psi(X, C) \land \Theta(C, k) \to \Delta(k)].$$

(If it is assumed that $\Phi(X), \Psi(X, C)$ and $\Theta(C, k)$, then *the angles COA and COA' are equal.*)

III. $$\sum_X \Phi(X).$$

(*On each side of an angle we can mark segment OA, AB, OA', A'B' which are equal.*)

IV. $$\prod_C \sum_K \Theta(C, k).$$

(*A straight line always passes through the points O and C.*)

Thesis:

$$\sum_k \Delta(k) \quad \text{(there exists a bisector of a given angle).}$$

Proof.
(1)	$\Phi(X_1)$		$\{O\Sigma: \text{III}\}$
(2)	$\Phi(X_1) \to \sum_C \Psi(X_1, C)$		$\{O\Pi: \text{I}\}$
(3)	$\sum_C \Psi(X_1, C)$		$\{\text{RD}: 2; 1\}$
(4)	$\Psi(X_1, C_1)$		$\{O\Sigma: 3\}$
(5)	$\sum_k \Theta(C_1, k)$		$\{O\Pi: \text{IV}\}$
(6)	$\Theta(C_1, k_1)$		$\{O\Sigma: 5\}$
(7)	$\Phi(X_1) \land \Psi(X_1, C_1) \land \Theta(C_1, k_1) \to \Delta(k_1)$		$\{O\Pi: \text{II}\}$
(8)	$\Phi(X_1) \land \Psi(X_1, C_1) \land \Theta(C_1, k_1)$		$\{\text{JC}: 1; 4; 6\}$
(9)	$\Delta(k_1)$		$\{\text{RD}: 7; 8\}$
	$\sum_k \Delta(k)$		$\{\text{J}\Sigma: 9\}$

EXAMPLE 3. Proof of the ordinary principle of induction:

I. $$A(0) \land \prod_a [A(a) \to A(a+1)] \to A(a).$$

It is assumed here that the variables a, b, c range over the set of natural numbers, which is taken to include the number 0. In the proof we make use of the following theorems:

II. $$\prod_a \{[\prod_{b<a} A(b)] \to A(a)\} \to A(a).^{(1)}$$

T1. $$a+b = b+a.$$
T2. $$a+0 = a.$$
T3. $$a \neq 0 \wedge a \neq 1 \to 1 < a.$$
T4. $$a < b \equiv \sum_{c \neq 0} a+c = b.$$
T5. $$1 \neq 0.$$

Thesis: $$A(0) \wedge \prod_a [A(a) \to A(a+1)] \to A(a).$$

Proof.
(1) $A(0)$ $\quad\quad\quad\quad\quad\quad\quad$ } {a.}
(2) $\prod_a [A(a) \to A(a+1)]$ }

(3) $\neg A(a)$ $\quad\quad\quad\quad\quad$ {a.i.p.}
(4) $A(0) \to A(0+1)$ $\quad\quad$ {O$\prod$: 2}
(5) $0+1 = 1$ $\quad\quad\quad\quad\quad$ {T1; T2}
(6) $A(1)$ $\quad\quad\quad\quad\quad\quad\quad$ {RD, EI: 4; 1; 5}

(7) $\neg\prod_a \{[\prod_{b<a} A(b)] \to A(a)\}$ {Toll: II; 3}

(8) $\sum_a \{[\prod_{b<a} A(b)] \wedge \neg A(a)\}$ {N$\prod$, T21. Sec. 3, Chap. I: 7}

(9) $\prod_{b<a_1} A(b)$ }
$\quad\quad\quad\quad\quad\quad\quad\quad\quad$ {O$\sum$, OC: 8}
(10) $\neg A(a_1)$ }
(11) $a_1 \neq 0$ $\quad\quad\quad\quad\quad\quad$ {10; 1; T3a. Sec. 4. Chap. II}
(12) $a_1 \neq 1$ $\quad\quad\quad\quad\quad\quad$ {10; 6; T3a. Sec. 4. Chap. II}
(13) $1 < a_1$ $\quad\quad\quad\quad\quad\quad$ {RD: T3; 11; 12}
(14) $\sum_{c \neq 0} (1+c = a_1)$ $\quad\quad$ {RD$_{\text{E}}$: T4; 13}

(15) $1+c_1 = a_1$ }
$\quad\quad\quad\quad\quad\quad\quad\quad$ {O$\sum$*: 14}
(16) $c_1 \neq 0$ }
(17) $c_1+1 = a_1$ $\quad\quad\quad\quad$ {EI: T1; 15}
(18) $c_1 < a_1$ $\quad\quad\quad\quad\quad\quad$ {RD$_{\text{E}}$, J$\sum$*: 17; T5; T4}

---

(1) This theorem is also often called the *principle of induction*.

(19) $c_1 < a_1 \to A(c_1)$ $\{O\Pi^*\colon 9\}$

(20) $A(c_1)$ $\{RD\colon 19;\ 18\}$

(21) $A(c_1) \to A(c_1+1)$ $\{O\Pi\colon 2\}$

(22) $A(a_1)$ $\{RD,\ EI\colon 21;\ 20;\ 17\}$

     contr. $\{10;\ 22\}$

PART TWO

Elements of Set Theory

CHAPTER 1

GENERAL SET THEORY

1. The Algebra of Sets

The concept of a *set* is one which is used in various disciplines as commonly as the constants of the sentential calculus and the quantifiers. It has also been widely used in logic, although for some time it was not subjected to close analysis. What are called *extensions of terms* (concepts) are sets of objects denoted by given terms (classified under given concepts), and relations between extensions are certain relations between sets. The concept of a set used is particularly often in the various branches of mathematics. For instance, in the calculus we examine sets of numbers and functions; in algebra, sets of polynomials and equations; in geometry, sets of points, straight lines, and planes. That branch of mathematics which is concerned with the properties of sets, regardless of what are the elements of those sets, is called *set theory*.

The fundamental concept of set theory is that of belonging to a set (also called the *membership* relation). We denote that the object *a belongs* to (is an element of) the set A by writing

$$a \in A.$$

This formula is also often read: the set A *contains* the element a, or briefly: a is in A.

If by $\mathbf{N}$, $\mathbf{M}$, and $\mathbf{R}$ we denote respectively the set of all natural numbers (i.e., positive integers), the set of all rational numbers, and the set of all real numbers, then the formulae

$$2 \in \mathbf{N}, \quad 10 \in \mathbf{N}, \quad 2/3 \in \mathbf{M}, \quad \sqrt{2} \in \mathbf{R}, \quad \pi \in \mathbf{R},$$

are true, while the formulae

$$-2 \in \mathbf{N}, \quad 0 \in \mathbf{N}, \quad 2\tfrac{1}{2} \in \mathbf{N}, \quad \sqrt{2} \in \mathbf{M}, \quad \sqrt{-5} \in \mathbf{R},$$

are false.

The set of objects investigated by a given discipline—deductive or natural—is usually defined. We shall call it the *universal set* and represent it by the symbol 1. In the arithmetic of real numbers the set of all real numbers is the universal set; in the arithmetic of natural numbers it is the set of all natural numbers; in anthropology, the set of all human beings.

In set theory we can define many important mathematical concepts, such as the concepts of a function, sequence, etc. The concept of a number has also come to be better analysed in terms of set theory. Thus set theory is the fundamental mathematical theory.

In the various branches of mathematics we often deal with sets containing an infinite number of elements. Set theory is concerned with the characteristic properties of such sets and their various kinds. The origin and development of set theory is directly connected with the study of infinite sets. This branch of mathematics was founded by Georg Cantor (1845–1918). His first works met with incomprehension and opposition on the part of many contemporary mathematicians, since even eminent mathematicians believed that the notion of infinity would never become a mathematical concept. This view was shared for instance by Gauss, who was active even before Cantor's first works appeared. However, these preconceptions have proved unfounded, and set theory has developed rapidly, in a short time becoming the fundamental mathematical discipline applicable to the various branches of mathematics.

The creation and development of Cantor's set theory was preceded by the working out of certain ideas concerning sets in what is known as the algebra of sets.[1] In this Section we adopt as the primitive terms for constructing the algebra of sets: the symbol ∈ for the membership relation, and the symbol 1 for the universal set.

The following conventions are adopted concerning symbolism:
Lower-case letters:

$$x, y, z, t, u, \ldots$$

will stand for any objects which are not sets. Such objects will be called *individuals*. Any sets of individuals will be denoted by capital

---

[1] The algebra of sets was given firm foundations by the English mathematician George Boole. His first work in the field was published in 1854. Many theorems of the algebra of sets were known to Leibniz (fl. 17th cent.).

letters:
$$X, Y, Z, T, U, \dots$$

Bold-faced capitals:
$$\boldsymbol{X, Y, Z, T, U}, \dots$$

will stand for any sets all of whose elements are themselves sets. Such sets will be called *families* of sets.

We adopt the axiom

A1.1. $x \in 1$.

Thus we consider only those individuals which are elements of a defined universal set. Further axioms of set theory will be introduced as required.

We now define two fundamental concepts of the algebra of sets:

D1.1. $X \doteq Y \equiv \prod_{x} (x \in X \equiv x \in Y)$.

The formula $X \doteq Y$ is read: *the set X is extensionally equal to the set Y*. Hence every individual which is an element of one of two sets which are extensionally equal is also an element of the other.

D1.2. $X \subset Y \equiv \prod_{x} (x \in X \rightarrow x \in Y)$.

The formula $X \subset Y$ is read: *the set X is included in the set Y*, or: *the set X is a part of the set Y*, or: *the set X is a subset of the set Y*.

Thus, the set X is a part of the set Y if and only if every element of the set X is an element of the set Y, which may (but need not!) also contain elements that do not belong to the set X.

We shall now prove a number of theorems of the algebra of sets. The proofs of these theorems will be set out in the same way as the proofs of logical theses, but the logical rules used will not as a rule be quoted explicitly. Proofs will sometimes be abbreviated but not in such a way as to make it difficult for the reader to complete them himself.

T1.1. $X \subset 1$.

This theorem states that *every set of individuals is a subset of the universal set*.

Proof. (1) $x \in 1 \to (x \in X \to x \in 1)$ $\{p \to (q \to p)\}^{(1)}$

     (2) $x \in X \to x \in 1$ $\{1, \text{Al.1}\}$

     (3) $\prod_x (x \in X \to x \in 1)$ $\{2\}$

        $X \subset 1$ $\{\text{D1.2}, 3\}$

T1.2a. $X \doteq X,$

  b. $X \doteq Y \to Y \doteq X,$

  c. $X \doteq Y \wedge Y \doteq Z \to X \doteq Z.$

This theorem may be formulated in words as follows: *the relation of extensional equality of sets is reflexive, symmetric, and transitive.* Only the proof of formula c. is given, the proofs of the remaining two being left to the reader.

Proof. c. (1) $X \doteq Y$⎫ $\{\text{a.}\}$
        (2) $Y \doteq Z$⎭

       (3) $\prod_x (x \in X \equiv x \in Y)$ $\{\text{D1.1}, 1\}$

       (4) $\prod_x (x \in Y \equiv x \in Z)$ $\{\text{D1.1}, 2\}$

       (5) $x \in X \equiv x \in Y$ $\{3\}$

       (6) $x \in Y \equiv x \in Z$ $\{4\}$

       (7) $x \in X \equiv x \in Z$ $\{(p \equiv q) \to [(q \equiv r) \to$
                                   $(p \equiv r)], 5, 6\}$

       (8) $\prod_x (x \in X \equiv x \in Z)$ $\{7\}$

          $X \doteq Z$ $\{\text{D1.1}, 8\}$

T1.3a. $X \doteq Y \to X \subset Y,$

  b. $X \subset Y \wedge Y \subset X \to X \doteq Y.$

We shall prove only formula a., leaving the proof of b. to the reader.

Proof. a. (1) $X \doteq Y$ $\{\text{a.}\}$

       (2) $x \in X \equiv x \in Y$ $\{\text{D1.1}, 1\}$

       (3) $x \in X \to x \in Y$ $\{2\}$

       (4) $\prod_x (x \in X \to x \in Y)$ $\{3\}$

          $X \subset Y$ $\{\text{D1.2}, 4\}$

---

[1] Instead of giving a reference we have written out the logical thesis involved. A similar procedure will be adopted throughout.

T1.4a. $$X \subset X, \qquad \{T1.3a, T1.2a\}$$
b. $$X \subset Y \wedge Y \subset Z \to X \subset Z.$$

Thus the relation $\subset$ is reflexive and transitive, but not symmetric since for instance the set of prime numbers is included in the set of natural numbers, but not vice versa. The proof of this theorem is quite straightforward.

L1.1. $$x = y \to \prod_X (x \in X \equiv y \in X).$$

Proof. (1) $x = y$ $\{a.\}$
       (2) $x \in X \equiv x \in X$ $\{p \equiv p\}$
       (3) $x \in X \equiv y \in X$ $\{EI: 1, 2\}$
           $\prod_X (x \in X \equiv y \in X)$ $\{3\}$

D1.3a. $$x \in \{y\} \equiv x = y,$$
b. $$x \in \{y_1, y_2, ..., y_n\} \equiv x = y_1 \vee x = y_2 \vee ... \vee x = y_n.$$

Thus the symbol $\{y\}$ denotes the set of which y is the only element; the symbol $\{y_1, y_2, ..., y_n\}$ stands for the set of which the individuals $y_1, y_2, ..., y_n$ and no others are the elements. The set $\{y\}$ is called a *unit set*.

L1.2. $$\prod_X (x \in X \equiv y \in X) \to x = y.$$

Proof. (1) $\prod_X (x \in X \equiv y \in X)$ $\{a.\}$
       (2) $x \in \{y\} \equiv y \in \{y\}$ $\{1\}$
       (3) $x = y \equiv y = y$ $\{D1.3a, 2\}^{(1)}$
           $x = y$ $\{y = y, 3\}$

T1.5 $$x = y \equiv \prod_X (x \in X \equiv y \in X). \qquad \{L1.1, L1.2\}$$

Thus two individuals are identical if and only if for any set they either both are, or both are not, elements of that set. T1.5 can serve as a definition of the identity of individuals. Frequently, however, the symbol of identity is treated as a primitive term in the algebra of sets. Theorem T1.5 is clearly analogous to Thesis TVI, given in

----

(1) We have made use here of the rule of extensionality of the sentential calculus (cf. Part One, Chap. 1, Sec. 3).

Sec. 6, Chap. II of Part One. They differ only in that in the former reference is made to sets, and in the latter to properties. There is also a closer relationship between them, since either of them can be proved by means of the other (cf. Sec. 5, p. 186).

T1.6. $X = Y \to X \doteq Y$.

> *Proof.* (1) $X = Y$ {a.}
> $X \doteq Y$ {EI: 1, T1.2a}

Thus identical sets are extensionally equal.

The question arises whether the converse theorem is also true, that is whether extensionally equal sets are identical. Intuitively this does not give rise to any doubts, yet this theorem cannot be proved on the strength of the assumptions adopted so far. We therefore introduce the new axiom:

A1.2. $X \doteq Y \to X = Y$.

This axiom, which is called the *axiom of extensionality for sets*, states that sets which have the same elements are identical. Hence a set is fully defined if it is stated which objects belong to it.

Consequently, the following equivalence is true:

T1.7. $X \doteq Y \equiv X = Y$.   {A1.2, T1.6}

In other words every formula of the type $X \doteq Y$ may be replaced by a formula of the type $X = Y$. Hence from now on we shall use only the symbol "$=$".

Note that T1.2 follows immediately from T1.7 and from the fact that the relation of identity is reflexive, symmetric, and transitive. However had T1.2 not been proved we could not adopt Axiom A1.2 without risking a contradiction.

We shall now introduce some further definitions from the algebra of sets. In doing so we shall make use of the symbol $\notin$. The formula $x \notin X$ is equivalent to the formula $\neg(x \in X)$.

D1.4. $x \in X \cup Y \equiv x \in X \lor x \in Y$.
D1.5. $x \in X \cap Y \equiv x \in X \land x \in Y$.
D1.6. $x \in X - Y \equiv x \in X \land x \notin Y$.
D1.7. $x \in \bar{X} \equiv x \notin X$.
D1.8. $x \in \varnothing \equiv x \in \bar{1}$.

The first of these definitions defines the *union* of sets, the second the *intersection* of sets, the third the *difference* of sets, the fourth the *complement* of the set X, and the fifth the *null set* represented by $\emptyset$. The first three definitions are related to arithmetical processes; this can most easily be seen in the case of the difference of sets: the union and the intersection of sets are sometimes called the *sum* and *product* respectively, and the symbols "$+$" and "$.$" are used by some authors instead of "$\cup$" and "$\cap$".

We see that those and only those elements belong to the union of two sets which belong to at least one of the seperate sets; those and only those elements belong to the intersection of two sets which belong to both of them; those and only those elements belong to the difference of two sets which belong to the first set but do not belong to the second; those and only those elements belong to the complement of a set which do not belong to that set; and finally, those and only those elements belong to the null set which belong to the complement of the universal set (cf. T1.9 below).

By way of example, we give some theorems employing the concepts defined above.

T1.8.　　$\{x_1, x_2, ..., x_n\} = \{x_1\} \cup \{x_2\} \cup ... \cup \{x_n\}.$ 　{D1.3a, b, D1.4,

D1.1}

T1.9.　　　　　　　　　$x \notin \emptyset.$

That is to say that *no object is an element of the null set.*

Proof. (1)　$x \in \emptyset$ 　　　　　　　　　　　{a.i.p.}

　　　(2)　$x \in \bar{1}$ 　　　　　　　　　　　{D1.8, 1}

　　　(3)　$x \notin 1$ 　　　　　　　　　　　{D1.7, 2}

　　　　　contr. 　　　　　　　　　　　{A1.1, 3}

T1.10a.　　　　　　　　$\emptyset = \bar{1},$

　　b.　　　　　　　　$1 = \bar{\emptyset}.$

Thus *each of the two sets $\emptyset$ and 1 is the complement of the other.*

Proof. a. (1)　$\prod_{x} (x \in \emptyset \equiv x \in \bar{1})$ 　　　　　{D1.8}

　　　(2)　$\emptyset \doteq \bar{1}$ 　　　　　　　　　{D1.1, 1}

　　　　　$\emptyset = \bar{1}$ 　　　　　　　　　{A1.2, 2}

In future proofs we shall omit those parts of proofs which lead from formulae of the type $X \doteq Y$ to formulae of the type $X = Y$.

Proof. b. (1) $x \in \overline{\varnothing}$ {D1.7, T1.9}

      (2) $x \in 1 \equiv x \in \overline{\varnothing}$ {$p \rightarrow [q \rightarrow (p \equiv q)]$, A1.1, 1}

          $1 = \overline{\varnothing}$ {D1.1, 2}

T1.11a. $X \cup \varnothing = X.$

    b. $X \cap 1 = X.$

    c. $X \cup 1 = 1.$

    d. $X \cap \varnothing = \varnothing.$

Formula a. states that the null set is the identity for the addition of sets, and formula b. that the universal set is the identity for multiplication of sets.

Proof. a. (1) $(x \in X \lor x \in \varnothing) \rightarrow x \in X$ {$\neg q \rightarrow ((p \lor q) \equiv p)$,

                                                         T.19}

      (2) $x \in X \cup \varnothing \equiv x \in X$ {D1.4, 1}

          $X \cup \varnothing = X$ {D1.1, 2}

The proofs of the remaining formulae are similar; in the proof of:

formula b. we refer to A1.1, D1.5 and the Thesis $q \rightarrow (p \land q \equiv p)$,

formula c. we refer to A1.1, D1.4 and the Thesis $q \rightarrow (p \lor q \equiv q)$,

formula d. we refer to T1.9, D1.5 and the Thesis $\neg q \rightarrow (p \land q \equiv q)$.

Next we define the relation $X \underset{\neq}{\subset} Y$ as follows:

D1.9. $X \underset{\neq}{\subset} Y \equiv X \subset Y \land X \neq Y.$

The relation defined above is to be interpreted: *X is a proper subset of the set Y.*

Note also that instead of writing $x \in X \land y \in X$ we usually write $x, y \in X$. Sets whose intersection is the null set are called *disjoint sets.*

We shall now describe a graphical method of verifying formulae in the algebra of sets. These pictorial representations are called Venn diagrams. Sets other than the universal set are symbolized by circles. When we are dealing with two sets, X and Y, we draw two intersecting circles, one standing for X, and the other for Y. This gives the diagram

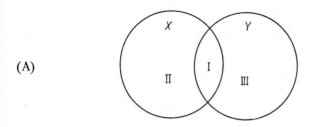

(A)

in which the product $X \cap Y$ is represented by area I, the difference $X-Y$ by area II, and the sum $X \cup Y$ by the area consisting of I, II and III.

To symbolize the complement of a set we introduce a rectangle which stands for the universal set. When the circle representing the set X is drawn inside this rectangle, we obtain the diagram

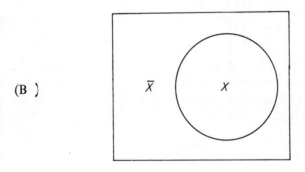

(B)

in which the part of the rectangle outside the circle represents the complement of X, i.e., $\overline{X}$.

The set X is included in the set Y if and only if the area representing the set X is part of, or coincides with, the area representing the set Y.

Bearing this in mind it can be seen from Diagram (A) that the following formulae are true: $X \cap Y \subset X$, $X \cap Y \subset Y$, $X \subset X \cup Y$, $Y \subset X \cup Y$, $X-Y \subset X$.

The set X is equal to the set Y if and only if both are represented by the same area. Thus, for instance, it can be seen from Diagram (A) that the formulae $X \cap Y = Y \cap X$, $X \cup Y = Y \cup X$ are true, while Diagram (B) shows that the formulae $X \cap 1 = X$, $X \cup \overline{X} = 1$ are also true.

We verify the formula

$$X \cup (Y \cap Z) = (X \cup Y) \cap (X \cup Z)$$

by drawing the diagram

(C)

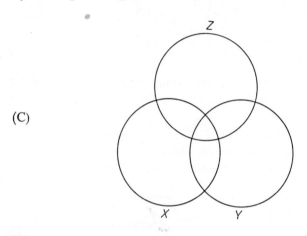

and by marking on it first the area representing the set $X \cup (Y \cap Z)$, and next the area representing the set $(X \cup Y) \cap (X \cup Z)$.

For the left-hand side of the equation we then obtain the diagram

(C₁)

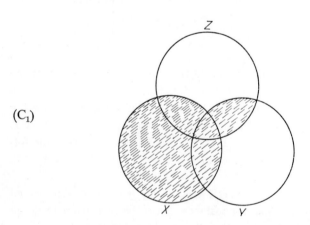

in which the shaded area stands for the set $X \cup (Y \cap Z)$. For the right-hand side we obtain the diagram

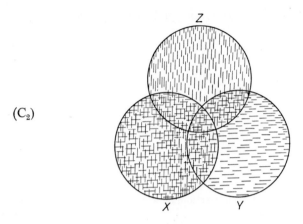

(C_2)

Here the area shaded horizontally stands for the set $X \cup Y$, the area shaded vertically for the set $X \cup Z$, and the common part of these two areas, shaded crosswise, stands for the product $(X \cup Y) \cap (X \cup Z)$. Since the areas representing the sets $X \cup (Y \cap Z)$ and $(X \cup Y) \cap (X \cup Z)$ coincide, the sets are equal, and hence the formula is true.

In the example just discussed we have made use of broken lines in shading, since continuous lines are used to indicate that a given set is empty (i.e., is the null set). Thus, for instance, the truth of the formula $X \subset Y$ is marked on Diagram (A) by shading the area II with continuous lines. In accordance with the explanations given above, the formula $X \subset Y$ is true if and only if the area standing for the set X is part of, or coincides with, with the area that stands for the set Y; however this occurs if and only if the set $X-Y$, represented by the area II, is empty. Thus the truth of the formula $X \subset Y$ is symbolized by the diagram

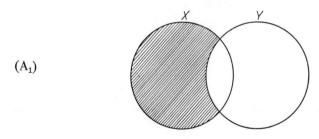

(A_1)

The falsehood of such a formula is symbolized by the diagram

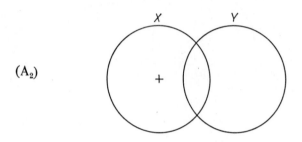

(A₂)

The symbol "+" placed in a particular area indicates that the set represented by that area is not empty.

In view of the above and of D1.9 the truth of the formula $X \underset{\neq}{\subset} Y$ can be represented by the diagram

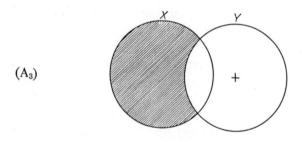

(A₃)

To verify by Venn diagrams those formulas of the algebra of sets which have the form of an implication we use the abbreviated zero-one method of verifying formulae of the sentential calculus. Thus, for instance, we assume that the antecedent of a given implication is true, we mark on an appropriate diagram the conditions for the truth of the antecedent, and then we examine whether the consequent of that implication can be false. In other cases we assume that the consequent of a given implication is false, we mark on a diagram the conditions for this to be so, and then we examine whether the antecedent of the implication can be true.

In this way we verify, for instance, Formula T1.4b by means of the following diagram

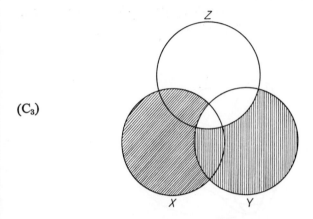

(C₃)

It can be seen from Diagram (C_3) that if the antecedent of T1.4b is true, i.e., if $X \subset Y$ and $Y \subset Z$, then it is also true that the set X is included in the set Z (since when we have marked the conditions for the truth of the antecedent the whole area of circle X outside the area of circle Z is shaded). Hence Formula T1.4b is true.

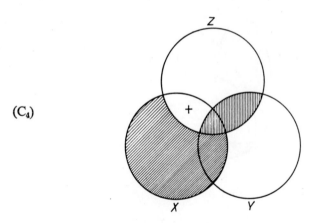

(C₄)

The falsehood of the formula

$$X \subset Z \wedge Y \cap Z = \emptyset \to X \subset Y$$

is shown by the diagram (C_4) from which it can be seen that when the antecedent of the implication is true its consequent can be false, since the set $X - Y$ can be not empty.

Formulae with four variables are verified by means of the following diagram consisting of four intersecting ellipses:[1]

(D)

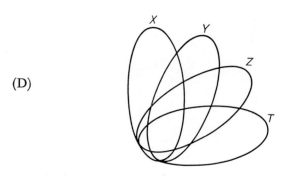

The truth of the formula

$$X \subset Z \land Y \subset T \to X \cap Y \subset Z \cap T$$

is shown by the diagram

(D₁)

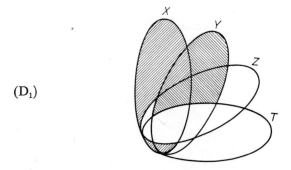

The falsehood of the formula

$$X \subset Y \land Z \subset T \to X \cap Y \subset Z \cap T$$

---

[1] A generalized graphical method of verification by means of Venn diagrams is given by S. Łuszczewska-Romahnowa, *Analiza i uogólnienie metody sprawdzania formuł logicznych przy pomocy diagramów Venna* (An Analysis and Generalization of Venn's Diagrammatic Decision Procedure), in *Studia Logica*, Vol. 1, 1953.

is shown by the diagram

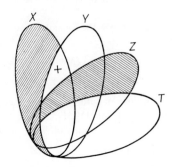

(D$_2$)

Exercises

1. Prove the following formulae:

   (a) $\bar{X} = 1 - X$,

   (b) $X - Y = X \cap \bar{Y}$,

   (c) $\emptyset \subset X$,

   (d) $X = Y \equiv \bar{X} = \bar{Y}$,

   (e) $X \subset Y \equiv \bar{Y} \subset \bar{X}$,

   (f) $X \subset Y \equiv X \cup Y = Y$,

   (g) $X \subset Y \equiv X \cap Y = X$,

   (h) $X \subset Y \equiv X - Y = \emptyset$,

   (i) $Z \cap (X - Z) = \emptyset$.

2. Find examples which prove the falsehood of the following formulae:

   (a) $(X - Y) \cup Y = X$,

   (b) $(X \cup Y) - Y = X$.

3. Show that Theorem T1.11 can be deduced from formulae (e), (f), (g) of Exercise 1, from T1.1, and from the following two formulae which will be proved in Sec. 3:
$$X \cup Y = Y \cup X, \quad X \cap Y = Y \cap X.]$$

4. The sum (union) and the product (intersection) of sets can also be defined as follows:
$$Z = X \cup Y \equiv X \subset Z \wedge Y \subset Z \wedge \prod_U (X \subset U \wedge Y \subset U \to Z \subset U),$$
$$Z = X \cap Y \equiv Z \subset X \wedge Z \subset Y \wedge \prod_U (U \subset X \wedge U \subset Y \to U \subset Z).$$

By these definitions the sum of two sets is the "smallest" set containing both component sets, and the product of two sets is the "largest" set contained in both component sets.

(a)   From these definitions prove the formulae given in Exercise 3, and also the formulae:

$$(X \cup Y) \cup Z = X \cup (Y \cup Z), \quad (X \cap Y) \cap Z = X \cap (Y \cap Z),$$

$$(X \cup Y) \cap Z = (X \cap Z) \cup (Y \cap Z), \quad (X \cap Y) \cup Z = (X \cup Z) \cap (Y \cup Z).$$

(b)   Show that the definitions of sum and of product given above are equivalent to definitions D1.4 and D1.5.

5. Verify the following formulae by means of Venn diagrams:

$$X \subset Z \wedge Y \subset Z \rightarrow X \cup Y \subset Z,$$

$$X \subset Y \wedge X \subset Z \rightarrow X \subset Y \cap Z,$$

$$Y \cap Z = \emptyset \wedge X \cap Z \neq \emptyset \rightarrow X - Y \neq \emptyset,$$

$$Y \cap Z = \emptyset \wedge X - Y = \emptyset \rightarrow X \cap Z = \emptyset.$$

6. The basic formulae of Aristotelian logic can be introduced by the following definitions:

$$SaP \equiv S \subset P,$$

$$SeP \equiv S \cap P = \emptyset,$$

$$SiP \equiv S \cap P \neq \emptyset,$$

$$SoP \equiv S - P \neq \emptyset.$$

Verify graphically the following laws of direct inference and moods of syllogisms:

$$SaP \rightarrow \neg SoP,$$

$$SiP \rightarrow \neg SeP,$$

$$\neg SiP \rightarrow SoP,$$

$$MaP \wedge SaM \rightarrow SiP,$$

$$MeP \wedge SiM \rightarrow SoP,$$

$$MaP \wedge SeM \rightarrow SeP,$$

$$PeM \wedge SaM \rightarrow SeP,$$

$$PaM \wedge SeM \rightarrow SoP.$$

Which of these laws are true only if it is assumed that the sets are not empty?

7. Verify graphically the formulae

$$X \subset Z \wedge Y \subset T \rightarrow X \cup Y \subset Z \cup T,$$

$$X \subset Y \wedge T \subset Z \wedge Y \cup Z \subset X \cup T \wedge X \cap Y = \emptyset \wedge Z \cap T \subset Y$$
$$\rightarrow X \cup Y \cup Z \cup T = Y \cap Z \cap T - X,$$

$$X \subset Y \wedge Z \subset T \wedge X \cup Z = 1 \wedge Y \cap T = \emptyset \rightarrow Y \subset X \wedge T \subset Z.$$

When verifying the last formula place Diagram (D) in a rectangle representing the universal set.

8. Prove the following theorems:
   (a) $Y \subset X \to (X-Y) \cup Y = X$,
   (b) $X \cap Y = \emptyset \to (X \cup Y) - Y = X$,
   (c) $Y \subset X \land Z = X - Y \to X = Y \cup Z$,
   (d) $X \underset{\neq}{\subset} Y \land Y \cap Z = \emptyset \to X \cup Z \underset{\neq}{\subset} Y \cup Z$,
   (e) $Z = X - Y \to Y \cap Z = \emptyset$,
   (f) $X \subset Y \land Y \cap Z = \emptyset \to X \cap Z = \emptyset$,
   (g) $X \underset{\neq}{\subset} Y \to Y - X \neq \emptyset$,
   (h) $X \underset{\neq}{\subset} Y \land Y \subset Z \to X \not\subset Z$,
   (i) $Z \cap X = \emptyset \to (Z \cup Y) \cap (X-Y) = \emptyset$,
   (j) $Z \subset X \cup Y \to X \cup Z \cup (Y-Z) = X \supset Y$.

   Verify these formulae by means of Venn diagrams.

9. The formula $X \doteq Y = (X-Y) \cup (Y-X)$ defines an operation on sets which differs from the operations introduced so far. The set $X \doteq Y$ is called the *symmetric difference* of the sets X and Y. Prove the formulae
   (a) $X \doteq X = \emptyset$,
   (b) $X \doteq Y = Y \doteq X$,
   (c) $X \doteq Y = (X \cup Y) - X \cap Y$,
   (d) $X - Y = X \doteq X \cap Y$,
   (e) $X \cup Y = (X \doteq Y) \doteq X \cap Y$.

   Draw a diagram for symmetric difference and use it to verify (b), (c) and (d)

10. Prove the following theorems:
   (a) $x \in X \doteq \overline{Y} \equiv \neg (x \in X \equiv x \in Y)$,
   (b) $X = Y \equiv X \doteq Y = \emptyset$,
   (c) $X \cap Y = \emptyset \to X \doteq Y = X \cup Y$,
   (d) $X \subset Y \to Y \doteq X = Y - X$.

2. Relationship between the Sentential Calculus and the Algebra of Sets

In the present Section it is convenient to assume that the variables of the sentential calculus are the symbols

$$p_1, p_2, p_3, \ldots$$

It is also assumed that

$$Z_1, Z_2, Z_3, \ldots$$

stand for any sets, and that the symbols $\alpha, \beta, \gamma, \ldots$ denote any formulae of the sentential calculus in which only the constants

$$\vee, \wedge, \neg$$

occur.

I. We denote by α^* the formula which is obtained from α by the substitution for the variables $p_1, p_2, p_3, \ldots$ of the formulae $x \in Z_1$, $x \in Z_2$, $x \in Z_3$, ..., respectively.

II. We denote by Z_α the formula which is obtained from the formula α by the substitution for the variables $p_1, p_2, p_3, \ldots$ of the symbols $Z_1, Z_2, Z_3, \ldots$, respectively and for the connectives $\vee, \wedge, \neg$ of the symbols $\cup, \cap, -$, respectively.

EXAMPLES:

α	α^*	Z_α
$p_1 \vee p_2$	$x \in Z_1 \vee x \in Z_2$	$Z_1 \cup Z_2$
$\neg(p_1 \wedge p_2)$	$\neg(x \in Z_1 \wedge x \in Z_2)$	$\overline{Z_1 \cap Z_2}$

L2.1a.
$$(\alpha \vee \beta)^* \equiv \alpha^* \vee \beta^*,$$
b.
$$(\alpha \wedge \beta)^* \equiv \alpha^* \wedge \beta^*,$$
c.
$$(\neg \alpha)^* \equiv \neg(\alpha^*).$$

These formulae follow directly from I.

L2.2a.
$$Z_{\alpha \vee \beta} = Z_\alpha \cup Z_\beta,$$
b.
$$Z_{\alpha \wedge \beta} = Z_\alpha \cap Z_\beta,$$
c.
$$Z_{\neg \alpha} = \overline{Z_\alpha}.$$

These formulae follow directly from II.

L2.3. *The equivalence*

(1)
$$\alpha^* \equiv x \in Z_\alpha$$

is a theorem of the algebra of sets.

Proof. The proof is constructed by an induction on the number of constant symbols occurring in α.

If no constant symbols occur in α then α will consist of a single variable, say p_j. In view of notations I and II both the formula α^* and the formula $x \in Z_\alpha$ have the form $x \in Z_j$. Hence formula (1) will become

$$x \in Z_j \equiv x \in Z_j$$

which will be true.

Let us assume now that k $(k > 0)$ constant symbols occur in α, and that formula (1) is true for those formulae in which fewer than k constant symbols occur.

The formula α has one of the following forms:

a. $\qquad\qquad\qquad\qquad \beta \vee \gamma,$

b. $\qquad\qquad\qquad\qquad \beta \wedge \gamma,$

c. $\qquad\qquad\qquad\qquad \neg\beta.$

In each of the formulae β and γ there are fewer than k constant symbols. Hence by the inductive assumption,

(2) $\qquad\qquad\qquad\qquad \beta^* \equiv x \in Z_\beta,$

(3) $\qquad\qquad\qquad\qquad \gamma^* \equiv x \in Z_\gamma.$

We shall now prove the lemma in case a.

(4) $\qquad \alpha^* \equiv (\beta \vee \gamma)^* \equiv \beta^* \vee \gamma^* \qquad\qquad \{\text{L2.1a}\}$

(5) $\qquad \alpha^* \equiv x \in Z_\beta \vee x \in Z_\gamma \qquad\qquad \{4, 2, 3\}$

$\quad \alpha^* \equiv x \in Z_\beta \cup Z_\gamma \equiv x \in Z_{\beta\vee\gamma} \equiv x \in Z_\alpha{}^{(1)} \quad \{5, \text{D1.4, L2.2a}\}$

Thus the formula is true in case a. The proofs for cases b. and c. are analogous.

T2.1. *If the formula*

a. $\alpha \equiv \beta$ *is a thesis of the sentential calculus, then* $Z_\alpha = Z_\beta,$

b. $\alpha \to \beta$,, ,, ,, ,, ,, ,,   ,,   ,, $Z_\alpha \subset Z_\beta,$

c. α ,, ,, ,, ,, ,,   ,,   ,, $Z_\alpha = 1.$

Proof. a. (1) $\alpha \equiv \beta$ $\qquad\qquad \{\text{a.}\}$

$\qquad\qquad$ (2) $\alpha^* \equiv \beta^*$ $\qquad\qquad \{1, \text{I}\}$

$\qquad\qquad$ (3) $x \in Z_\alpha \equiv x \in Z_\beta$ $\qquad \{\text{L2.3, 2}\}$

$\qquad\qquad\qquad Z_\alpha = Z_\beta$ $\qquad\qquad \{\text{D1.1, 3}\}$

Proof. b. (1) $\alpha \to \beta$ $\qquad\qquad \{\text{a.}\}$

$\qquad\qquad$ (2) $\alpha^* \to \beta^*$ $\qquad\qquad \{1, \text{I}\}$

$\qquad\qquad$ (3) $x \in Z_\alpha \to x \in Z_\beta$ $\qquad \{\text{L2.3, 2}\}$

$\qquad\qquad\qquad Z_\alpha \subset Z_\beta$ $\qquad\qquad \{\text{D1.2, 3}\}$

Proof. c. (1) α $\qquad\qquad\qquad \{\text{a.}\}$

$\qquad\qquad$ (2) α^* $\qquad\qquad\qquad \{1, \text{I}\}$

$\qquad\qquad$ (3) $x \in Z_\alpha$ $\qquad\qquad \{\text{L2.3, 2}\}$

$\qquad\qquad$ (4) $x \in Z_\alpha \equiv x \in 1$ $\quad \{p \to [q \to (p \equiv q)], 3, \text{A1.1}\}$

$\qquad\qquad\qquad Z_\alpha = 1$ $\qquad\qquad \{\text{D1.1, 4}\}$

The theorem proved above shows the close relationship between the sentential calculus and the algebra of sets. It also enables us

---

(1) The formula $\phi \equiv \psi \equiv \chi$ is the conjunction of the formulae $\phi \equiv \psi$ and $\psi \equiv \chi$.

to obtain directly from theses of the sentential calculus theorems
of the algebra of sets consisting of two expressions, incorporating set
operations, and connected by the symbol of equality or inclusion.
These theorems are those of the algebra of sets which are most
frequently applied. The theorem below quotes some formulae of
the algebra of sets, accompanied on the right by those theses of the
sentential calculus from which these formulae can be derived.

T2.2a. $X \cup Y = Y \cup X$ $\{p \vee q \equiv q \vee p\}$
b. $X \cap Y = Y \cap X$ $\{p \wedge q \equiv q \wedge p\}$
c. $(X \cup Y) \cup Z = X \cup (Y \cup Z)$ $\{(p \vee q) \vee r \equiv p \vee (q \vee r)\}$
d. $(X \cap Y) \cap Z = X \cap (Y \cap Z)$ $\{(p \wedge q) \wedge r \equiv p \wedge (q \wedge r)\}$
e. $X \cap (Y \cup Z) = (X \cap Y) \cup (X \cap Z)$ $\{p \wedge (q \vee r) \equiv p \wedge q \vee p \wedge r\}$
f. $X \cup Y \cap Z = (X \cup Y) \cap (X \cup Z)$ $\{p \vee q \wedge r \equiv (p \vee q) \wedge (p \vee r)\}$
g. $X \cup X = X$ $\{p \vee p \equiv p\}$
h. $X \cap X = X$ $\{p \wedge p \equiv p\}$
i. $\overline{(X \cup Y)} = \bar{X} \cap \bar{Y}$ $\{\neg(p \vee q) \equiv \neg p \wedge \neg q\}$
j. $\overline{(X \cap Y)} = \bar{X} \cup \bar{Y}^{(1)}$ $\{\neg(p \wedge q) \equiv \neg p \vee \neg q\}$
k. $X \subset X \cup Y$ $\{p \to p \vee q\}$
l. $X \cap Y \subset X$ $\{p \wedge q \to p\}$
m. $X \cup \bar{X} = 1$ $\{p \vee \neg p\}$

Exercises

1. Which theses of the sentential calculus enable the following formulae of the
algebra of sets to be deduced?

(a) $\bar{\bar{X}} = X$,
(b) $X \cap Y \cup X = X$,
(c) $(X \cup Y) \cap X = X$.

2. Write out the formulae of the algebra of sets which can be obtained from
the following theses of the sentential calculus:

(a) $\neg(\neg p \vee \neg q) \equiv p \wedge q$,
(b) $\neg(\neg p \wedge \neg q) \equiv p \vee q$,
(c) $p \wedge q \to p \vee q$.

3. Generalize the theorems and lemmas given in this Section so that they will
cover the subtraction of sets.

---

(1) Formulae i. and j. are called De Morgan's laws for sets.

4. The formula

$$p \not\equiv q$$

is defined as follows:

$$p \not\equiv q \underset{\mathrm{df}}{=} \neg(p \equiv q).$$

Complete the results obtained in this Section so that they will cover not only the logical constants $\vee, \wedge, \neg$, but also the relation $\not\equiv$, when the symbol of symmetric difference $\div$ (cf. Exercise 10 in the preceding Section) is correlated with this relation.

5. Demonstrate that if the formula $\neg \alpha$ is a thesis of the sentential calculus, then

$$Z_\alpha = \emptyset.$$

6. Prove the formula

$$X \cap \overline{X} = \emptyset.$$

3. Boolean Algebra[1]

Nearly all the theorems in the two preceding Sections contained only variables representing sets, and not variables representing individuals. Consequently such theorems did not contain the symbol "$\in$", while they did contain symbols of operations on sets and symbols of relations between sets.

But the concept of membership (symbolized by "$\in$") was needed both to define such operations and relations and to prove theorems concerning their properties. The question arises whether it is possible to construct a theory including all those theorems in the algebra of sets which contain only variables representing sets without using the concept of membership of a set. Such a theory was formulated by George Boole (cf. footnote on p. 158); it has become known as Boolean Algebra[2]. Its importance rests mainly on its applications. A brief description of the theory is given below.

Boolean Algebra has as its primitive terms the symbols: $\cup, \cap, -, B$. The first three stand as before for the addition, multiplication, and complementation of sets, while B stands for the family of all subsets of a definite non-empty set which is treated as the universal set. Hence the sum, product, and complement of sets belonging

---

[1] The definitions and theorems given in this Section will not be referred to later.

[2] The term "Boolean algebra" is used also in another, but closely related, meaning.

to the family **B** also belong to **B**. It is assumed that the variables stand for sets belonging to the family **B**. In setting out the theorems of Boolean Algebra we use logical symbols, in particular the symbol of equality.

We list below the axioms of Boolean Algebra, leaving out the conditions pertaining to the set **B** which have been formulated above in verbal form. The numbering of the theses will be explained later.

I. $$u \cup v = v \cup u,$$

II. $$u \cap (v \cup w) = u \cap v \cup u \cap w,$$

II*. $$u \cup v \cap w = (u \cup v) \cap (u \cup w),$$

III. $$u \cup v \cap \bar{v} = u,$$

IV. $$u \cup \bar{u} = v \cup \bar{v},$$

IV*. $$u \cap \bar{u} = v \cap \bar{v}.$$

Formulae I, II, and II* were proved in Sec. 2 as Theorems T2.2a, e, f. Formulae III, IV, and IV* can be proved in a similar way.

Note, first, that the following formulae, whose fairly difficult proofs are not given here, result from the axioms listed above:

I*. $$u \cap v = v \cap u,$$

III*. $$u \cap (v \cup \bar{v}) = u.$$

There is a clear connection between the formulae marked with the same Roman numbers (once without and once with an asterisk). Each such formula is obtained from the other by interchanging the symbols $\cup$ and $\cap$. We say that formulae satisfying this condition are *dual*. Thus every axiom of Boolean Algebra has a dual axiom or theorem. It follows from the above that every theorem of Boolean Algebra has this property. This can be illustrated by the following example:

V. $$u = u \cup u.$$

Proof. (1) $u = u \cup (u \cap \bar{u})$ {III}

(2) $u \cup (u \cap \bar{u}) = (u \cup u) \cap (u \cup \bar{u})$ {II*}

(3) $(u \cup u) \cap (u \cup \bar{u}) = u \cup u$ {III*}

$u = u \cup u$ {1–3}

V*. $$u = u \cap u.$$

Proof. (1*) $u = u \cap (u \cup \overline{u})$ {III*}

(2*) $u \cap (u \cup \overline{u}) = (u \cap u) \cup (u \cap \overline{u})$ {II}

(3*) $(u \cap u) \cup (u \cap \overline{u}) = u \cap u$ {III}

$u = u \cap u$ {1*–3*}

Note that the corresponding lines of these proofs are dual formulae.

The terms of the algebra of sets, introduced in Sec. 1, which are not primitive terms of Boolean Algebra, are: 1, Ø, $-$, $\subset$ (the term $\doteq$ is disregarded here, since it is replaced by the symbol $=$). These terms can be defined in Boolean algebra in the following manner:

VI. $1 = u \cup \overline{u},$

VI*. $\emptyset = u \cap \overline{u},$

VII. $u - v = u \cap \overline{v},$

VIII. $u \subset v \equiv u \cup v = v.$

Note that the formulae which occur in VI and VI* on the right-hand side of the equality sign are dual. Hence in order to obtain its dual formula from the formula α, which contains the terms Ø and 1 in addition to the primitive terms of Boolean Algebra, the symbol Ø in α must be replaced by 1, and the symbol 1 by Ø.

Formulae III and III* can now be replaced by simpler ones:

IX. $u \cup \emptyset = u,$

IX*. $u \cap 1 = u.$

We also give, by way of example, the proofs of two other theorems of Boolean algebra.

X. $u = (u \cap v) \cup (u \cap \overline{v}).$

Proof. (1) $u = u \cap 1$ {III*, IV*, VI}

(2) $u \cap 1 = u \cap (v \cup \overline{v})$ {VI}

(3) $u \cap (v \cup \overline{v}) = (u \cap v) \cup (u \cap \overline{v})$ {II}

$u = u \cap v \cup u \cap \overline{v}$ {1–3}

In line (2) of the proof we have made use of the rule of extensionality for identity.

The formula

X*. $u = (u \cup v) \cap (u \cup \overline{v})$

is dual with respect to X.

In the theorem below some of the terms are sentential connectives:

XI. $$u \cup v = 1 \wedge \bar{u} \cup \bar{v} = 1 \to u = \bar{v}.$$

Proof.

(1) $u \cup v = 1$ }
(2) $\bar{u} \cup \bar{v} = 1$ }   {a.}

(3) $u = (u \cup v) \cap (u \cup \bar{v})$ {X*}

(4) $(u \cup v) \cap (u \cup \bar{v}) = u \cup \bar{v}$ {IX*, 1, I*}

(5) $u \cup \bar{v} = (\bar{v} \cup u) \cap (\bar{v} \cup \bar{u})$ {IX*, 2, I}

(6) $(\bar{v} \cup u) \cap (\bar{v} \cup \bar{u}) = \bar{v}$ {X*}

$\qquad u = \bar{v}$ {3–6}

The theorem

XI*. $$u \cap v = \varnothing \wedge \bar{u} \cap \bar{v} = \varnothing \to u = \bar{v}$$

is dual with respect to XI.

Boolean algebra has an interpretation in many different theories, this being its chief theoretical value. We shall now explain what is meant by the statement that a theory has an interpretation in another theory.

Let $\pi_1, ..., \pi_k$ be all the primitive terms of a given theory T, and $\pi_1^*, ..., \pi_k^*$, primitive or defined terms of another theory, T*. If, when we replace in the axioms of theory T the π_1 by π_1^*, π_2 by π_2^*, etc., they become axioms or theorems of theory T*, we say that *theory T has an interpretation in theory* T*. The following statement is a less precise definition of this concept, but is perhaps easier to grasp.

If the axioms of a theory T remain true sentences when the primitive terms which occur in them are given a sense taken from some other theory T*, then theory T has an interpretation in theory T*.

Hence to interpret Boolean Algebra in a theory T*, we must first select from the concepts—primitive or defined—of that theory some set B^*, which has at least two elements, and the operations $\cup^*$, $\cap^*$, and $-^*$, the first two of which are performed on two arguments, and the third on one argument. These operations must have the property that when they are performed on elements of the set B^* they always yield elements of that set. Next we must check whether the axioms of Boolean algebra become axioms or theorems of theory

T* when the symbols ∪, ∩, ⁻, which occur in them, are replaced, respectively, by ∪*, ∩*, ⁻*.

It is obvious that if a theory T has an interpretation in a theory T*, then every theorem of T has its analogue among the theorems of T*. Consequently, the results obtained in theory T may be mechanically transferred to any theory in which T has an interpretation. In particular, every theory in which Boolean algebra can be interpreted has a part which does not differ formally from that algebra.

The simplest interpretation of Boolean algebra is obtained when the symbols ∪, ∩, ⁻ retain their meaning unchanged, and the symbol B is understood as any field of sets, i.e., as a family of sets consisting of at least two elements and having the property that the sum, the product, and the complement of sets which belong to that family also belong to it. It is obvious that the family of all the subsets of any non-empty set is a field of sets, but not conversely. The family of sets whose only elements are the null set and the set of all natural numbers, treated as the universal set, is an example of a field of sets which is not a family of all the subsets of any set.

The following interpretation of Boolean algebra in the probability calculus is extremely important: let B be the set of all possible events; by

$$a \cup b, \quad a \cap b, \quad \bar{a}$$

we mean, respectively, events which occur when at least one of the events a or b occurs, when both these events occur, and when the event a does not occur.

There is also an interpretation of Boolean algebra in the theory of electric circuits. That interpretation has considerable practical importance. Another interpretation of Boolean algebra will be referred to in later Sections.

Exercises

1. Give formulae dual with respect to the formulae
   (a) $\overline{(u \cup v)} = \bar{u} \cap \bar{v}$,
   (b) $u \cup \bar{u} = 1$,
   (c) $\overline{\varnothing} = 1$.
2. Demonstrate that if the formulae α^* and β^* are the duals of the formulae α and β respectively, then the formulae $\alpha^* \subset \beta^*$ and $\beta \subset \alpha$ are equivalent.

4. Infinite Operations

Let

(1) $$X_1, X_2, X_3, \ldots$$

be an arbitrary sequence of sets, and let

$$Y = \varphi(X)$$

be a function defined for every term of the sequence (1) such that the values of that function are sets.

D4.1. $$x \in \bigcup_{i=1}^{\infty} \varphi(X_i) \equiv \sum_{i \in \mathbf{N}} [x \in \varphi(X_i)]. \,^{(1)}$$

D4.2. $$x \in \bigcap_{i=1}^{\infty} \varphi(X_i) \equiv \prod_{i \in \mathbf{N}} [x \in \varphi(X_i)].$$

The set $\bigcup_{i=1}^{\infty} \varphi(X_i)$ is called the *sum* of the terms of the sequence

(2) $$\varphi(X_1), \varphi(X_2), \varphi(X_3), \ldots$$

and the set $\bigcap_{i=1}^{\infty} \varphi(X_i)$ is called the *product* of the terms of that sequence. Thus those and only those elements belong to the set $\bigcup_{i=1}^{\infty} \varphi(X_i)$ which belong to at least one of the sets (2), while those and only those elements belong to the set $\bigcap_{i=1}^{\infty} \varphi(X_i)$ which belong to each of the sets (2). In particular, the symbols $\bigcup_{i=1}^{\infty} X_i$ and $\bigcap_{i=1}^{\infty} X_i$ (in this case $\varphi(X) = X$) denote the sum and product of the sets (1), and the symbols $\bigcup_{i=1}^{\infty} \bar{X}_i$ and $\bigcap_{i=1}^{\infty} \bar{X}_i$ (in this case $\varphi(X) = \bar{X}$) denote the sum and product of the terms of the sequence

$$\bar{X}_1, \bar{X}_2, \bar{X}_3, \ldots$$

The following two formulae are generalizations of De Morgan's laws:

T4.1a. $$\overline{\bigcup_{i=1}^{\infty} X_i} = \bigcap_{i=1}^{\infty} \bar{X}_i,$$

b. $$\overline{\bigcap_{i=1}^{\infty} X_i} = \bigcup_{i=1}^{\infty} \bar{X}_i.$$

---

(1) The symbol **N** stands for the set of all natural numbers.

Proof. a. (1) $\quad x \in \overline{\bigcup_{i=1}^{\infty} X_i} \equiv \neg \left(x \in \bigcup_{i=1}^{\infty} X_i \right) = \neg \left(\sum_{i \in N} x \in X_i \right)$

$$\equiv \prod_{i \in N} \neg (x \in X_i) \equiv \prod_{i \in N} x \in \overline{X_i}$$

$$\equiv x \in \bigcap_{i=1}^{\infty} \overline{X_i} \qquad \{D1.7,\ D4.1,\ D1.7,\ D4.2\}$$

(2) $\quad \prod_x \left[x \in \overline{\bigcup_{i=1}^{\infty} X_i} \equiv x \in \bigcap_{i=1}^{\infty} \overline{X_i} \right] \qquad \{1\}$

$$\overline{\bigcup_{i=1}^{\infty} X_i} = \bigcap_{i=1}^{\infty} \overline{X_i} \qquad\qquad \{D1.1,\ 2\}$$

The proof of formula b is similar.

T4.2a. $\qquad\qquad \prod_{i \in N} X_i \subset X \rightarrow \bigcup_{i=1}^{\infty} X_i \subset X,$

b. $\qquad\qquad \prod_{i \in N} X \subset X_i \rightarrow X \subset \bigcap_{i=1}^{\infty} X_i.$

The proofs of these theorems, which are analogues of the first two theses given in Exercise 5 on p. 172, are left to the reader.

Let X be an arbitrary family of sets. It is assumed that the function $Y = \varphi(X)$ is defined for every element of the family of sets X, and that its values are sets. The following two definitions are generalizations of D4.1 and D4.2:

D4.3. $\qquad\qquad x \in \bigcup_{X \in X} \varphi(X) \equiv \sum_{X \in X} x \in \varphi(X),$

D4.4. $\qquad\qquad x \in \bigcap_{X \in X} \varphi(X) \equiv \prod_{X \in X} x \in \varphi(X).$

In particular, the symbols $\bigcup_{X \in X} X$ and $\bigcap_{X \in X} X$ (in this case $\varphi(X) = X$) denote the sum and product of all sets belonging to the family X.

Exercises

1. Prove the following formulae:

(a) $\quad \bigcup_{i=1}^{\infty} (X \cap Y_i) = X \cap \bigcup_{i=1}^{\infty} Y_i,$

(b) $\quad \bigcap_{i=1}^{\infty} (X \cap Y_i) = X \cap \bigcap_{i=1}^{\infty} Y_i,$

(c) $\bigcup\limits_{i=1}^{\infty} (X_i \cup Y_i) = \bigcup\limits_{i=1}^{\infty} X_i \cup \bigcup\limits_{i=1}^{\infty} Y_i,$

(d) $\bigcap\limits_{i=1}^{\infty} (X_i \cap Y_i) = \bigcap\limits_{i=1}^{\infty} X_i \cap \bigcap\limits_{i=1}^{\infty} Y_i.$

2. Prove the formulae[1]

(a) $\overline{\bigcup\limits_{X \in X} X} = \bigcap\limits_{X \in X} \overline{X},$

(b) $\overline{\bigcap\limits_{X \in X} X} = \bigcup\limits_{X \in X} \overline{X}.$

3. Prove the following theorems:

(a) $\prod\limits_{i \in N} (A_i \subset B_i) \to \bigcup\limits_{i=1}^{\infty} A_i \subset \bigcup\limits_{i=1}^{\infty} B_i,$

(b) $\prod\limits_{X \in X} \sum\limits_{Y \in Y} (X \subset Y) \to \bigcup\limits_{X \in X} X \subset \bigcup\limits_{Y \in Y} Y.$

4. Show that if X is an empty set then the set $\bigcup\limits_{X \in X} \varphi(X)$ is also empty, and the set $\bigcap\limits_{X \in X} \varphi(X)$ is a universal set regardless of how we define the function φ.

5. Prove that if $X = \{Z, U\}$, then

$$\bigcup\limits_{X \in X} \varphi(X) = \varphi(Z) \cup \varphi(U),$$

and

$$\bigcap\limits_{X \in X} \varphi(X) = \varphi(Z) \cap \varphi(U).$$

6. Formulate definitions of $\bigcup\limits_{X \in X} X$ and $\bigcap\limits_{X \in X} X$ analogous to the definitions of the sum and product of sets, as given in Exercise 4 in Sec. 1. By the definitions thus obtained prove the formulae given in the previous exercises and demonstrate the equivalence of the new definitions to D4.3 and D4.4.

5. The Cartesian Product of Sets

We introduce a new primitive concept, namely that of an *ordered pair*.[2] An ordered pair with its first element x and its second element y will be denoted by $\langle x, y \rangle$. Complex numbers are ordered pairs whose elements are real numbers. The fundamental property

---

[1] These formulae are also called De Morgan's laws for sets.

[2] The concept of an ordered pair need not be included in the primitive terms of set theory, but can be defined from the primitive terms already introduced. A definition of an ordered pair is given in Sec. 4 of the *Supplement*.

of ordered pairs is formulated in the following axiom:

A5.1 $\langle x, y \rangle = \langle z, t \rangle \equiv x = z \land y = t.$

Thus two ordered pairs are identical if and only if their first elements are identical and their second elements are also identical. Consequently, if $a \neq b$, then the pairs $\langle a, b \rangle$ and $\langle b, a \rangle$ differ. Instead of $\langle \langle x, y \rangle, z \rangle$ we write $\langle x, y, z \rangle$, and instead of $\langle \langle x, y, z \rangle, t \rangle$ we write $\langle x, y, z, t \rangle$. In general, an ordered system of objects, marked with the symbols $x_1, ..., x_n$, is denoted by $\langle x_1, ..., x_n \rangle$. Some, and even all, of these objects may be identical.

D5.1. $\langle x, y \rangle \in X \times Y \equiv x \in X \land y \in Y.$

A more exact form of this definition is

$$\alpha \in X \times Y \equiv \sum_{x \in X} \sum_{y \in Y} \alpha = \langle x, y \rangle.$$

The set $X \times Y$ is called the *Cartesian product* of the sets X and Y. It is clear that there is a Cartesian product of any two sets.

If by **I** we denote the set of all integers, and by **N**, as before, the set of all natural numbers, then the product **I** $\times$ **N** is the set of all ordered pairs whose first elements are integers and whose second elements are natural numbers. The introduction of this set serves to define the concept of a rational number.

The following theorem will be referred to in later Sections:

T5.1a. $(X \cup Y) \times Z = X \times Z \cup Y \times Z,$

   b. $X \cap Y = \emptyset \rightarrow (X \times Z) \cap (Y \times Z) = \emptyset,$

   c. $\{1, 2, ..., n\} \times X = \{1\} \times X \cup \{2\} \times X \cup ... \cup \{n\} \times X,$

   d. $X \subset Y \rightarrow X \times Z \subset Y \times Z,$

   e. $X \neq \emptyset \rightarrow \{X\} \times X \neq \emptyset,$

   f. $\{X\} \times X \neq \{Y\} \times Y \rightarrow (\{X\} \times X) \cap (\{Y\} \times Y) = \emptyset.$

We shall give only the proof of formula c, leaving the proofs of the remaining formulae to the reader.

Proof. c. To prove formula c it suffices to note that the ordered pair $\langle i, x \rangle$ is an element of the set $\{1, 2, ..., n\} \times X$ as well as of the set $\{1\} \times X \cup \{2\} \times X \cup ... \cup \{n\} \times X$ if and only if i is a natural number $\leqslant n$ and the element x belongs to the set X.

Let $\Phi(x)$ be an arbitrary condition in which the free variable x occurs; other free variables may also occur in this formula. By $\Phi(a)$

we denote the formula obtained from $\Phi(x)$ by the substitution of the variable a for the variable x. The symbols $\Phi(z, t)$ and $\Phi(a, b)$ have an analogous meaning.

We now introduce two definitions which enable us to define in a simple way the sets satisfying given conditions:

D5.2. $x \in \underset{a}{E}\, \Phi(a) \equiv \Phi(x).$ [1]

Thus the set $\underset{a}{E}\, \Phi(a)$ is the set of all objects satisfying condition $\Phi(x)$.

By means of D5.2 we can prove the equivalence of T1.5 and TVI (Part I, Chap. II, Sec. 6): from the right-hand side of equivalence T1.5 we obtain the right-hand side of equivalence TVI by substituting the expression "$\underset{a}{E}\, A(a)$" for "X", and making use of D5.2. And from the right-hand side of equivalence TVI we obtain the right-hand side of equivalence T1.5 by the substitution $A(t)/t \in X.$ [2]

D5.3. $\alpha \in \underset{a,b}{E}\, \Phi(a, b) \equiv \sum_z \sum_t [\alpha = \langle z, t \rangle \wedge \Phi(z, t)].$

The meaning of this concept is explained by

T5.2. $\langle x, y \rangle \in \underset{a,b}{E}\, \Phi(a, b) \equiv \Phi(x, y).$

Thus the set $\underset{a,b}{E}\, \Phi(a, b)$ is the set of all ordered pairs whose elements satisfy the condition $\Phi(x, y)$.

Proof. (1.1) $\langle x, y \rangle \in \underset{a,b}{E}\, \Phi(a, b)$ {ad.a.}

(1.2) $\sum_z \sum_t [\langle x, y \rangle = \langle z, t \rangle \wedge \Phi(z,t)]$ {D5.3, 1.1}

(1.3) $\langle x, y \rangle = \langle z_1, t_1 \rangle$ ⎫
(1.4) $\Phi(z_1, t_1)$ ⎬   {1.2}
(1.5) $x = z_1 \wedge y = t_1$ {A5.1, 1.3}
(1.6) $\Phi(x, y)$ {1.4, 1.5}
(2.1) $\Phi(x, y)$ {ad.a.}
(2.2) $\langle x, y \rangle = \langle x, y \rangle \wedge \Phi(x, y)$ {2.1}

---

[1] The letter E is the first letter of the French word *ensemble*, which means "set".

[2] We make use here of the rule of substitution for functional formulae, discussed in Part I, Chap. 2, Sec. 5.

(2.3) $\sum_z \sum_t [\langle x, y \rangle = \langle z, t \rangle \wedge \Phi(z, t)]$ {2.2}

(2.4) $\langle x, y \rangle \in \underset{a,b}{E} \Phi(a, b)$ {D5.3, 2.3}

thesis {1.1→1.6, 2.1→2.4}

By D5.2, the set of all positive real numbers may be represented by $\underset{a}{E}(a > 0 \wedge a \in \mathbf{R})$, and the set of all the roots of the polynomial $W(x)$ by $\underset{a}{E}(W(a) = 0)$. By T5.2, the set of all ordered pairs in which the first element is equal to the second may be represented by $\underset{a,b}{E}(a = b)$, and the set of all the solutions of the equation $x^2 + y^2 = 10$, by $\underset{a,b}{E}(a^2 + b^2 = 10)$.

Exercises

1. Show that
$$\langle x_1, y_1, z_1 \rangle = \langle x_2, y_2, z_2 \rangle \equiv x_1 = x_2 \wedge y_1 = y_2 \wedge z_1 = z_2.$$

2. Prove the equivalence
$$\langle x, y \rangle = \langle y, x \rangle \equiv x = y.$$

3. Prove the formulae:
   (a) $(X \cap Y) \times Z = (X \times Z) \cap (Y \times Z)$,
   (b) $(X-Y) \times Z = X \times Z - Y \times Z$.

4. Give an example which proves the falsehood of the formula
$$X \times Y = Y \times X.$$

5. Prove the formulae:
   (a) $\underset{a}{E}[\Phi(a) \vee \Psi(a)] = \underset{a}{E}\Phi(a) \cup \underset{a}{E}\Psi(a)$,
   (b) $\underset{a}{E}[\Phi(a) \wedge \Psi(a)] = \underset{a}{E}\Phi(a) \cap \underset{a}{E}\Psi(a)$,
   (c) $\prod_x [\Phi(x) \equiv \Psi(x)] \equiv \underset{a}{E}\Phi(a) = \underset{a}{E}\Psi(a)$,
   (d) $\prod_x [\Phi(x) \to \Psi(x)] \equiv \underset{a}{E}\Phi(a) \subset \underset{a}{E}\Psi(a)$.

6. Elements of the Theory of Relations

The theory of relations is an important branch of mathematical logic. However, we shall confine ourselves to discussing only those concepts of the theory which bear upon the foundations of set theory.

We shall first give the definition of the concept of a function, which is of fundamental importance in mathematics:

D6.1. $R \in \text{funct} \equiv \prod_x \prod_y \prod_z (xRy \wedge xRz \rightarrow y = z).$

Thus a relation is a function if and only if any object may bear that relation to at most one object. Hence, for instance, the following relations are functions:

$$xR_1 y \equiv y = |x|,$$

$$xR_2 y \equiv y = x^2.$$

T5.2 and D6.1 yield

CONCLUSION 6.1. *The relation R is a function if and only if no two ordered pairs having the same first element and different second elements belong to the set $\underset{a,b}{E}(aRb).$*

D6.2a. $x \in D_l(R) \equiv \sum_y xRy.$

   b. $y \in D_r(R) \equiv \sum_x xRy.$

The sets $D_l(R)$ and $D_r(R)$ are called the *left* and *right domain* of the relation R.[1] If as the universal set we take the set of all natural numbers, then the left domain of the relation "greater than" is the set of all natural numbers > 1, and the right domain is the set of all natural numbers. The left domain of the relation: x is the father of y, is a subset of the set of all human males, and the right domain is the set of all human beings.

T5.2 and D6.2a, b yield

CONCLUSION 6.2. *The set $D_l(R)$ is the set of all first elements, and the set $D_r(R)$ is the set of all second elements of the ordered pairs which are elements of the set $\underset{a,b}{E}(aRb).$*

D6.3. $R \in \text{funct} \wedge x \in D_l(R) \rightarrow [y = R(x) \equiv xRy].$

---

[1] The left domain of a relation is also often called the domain and symbolized $D(R)$, while the right domain is called the converse domain or range and symbolized $D(R)$.

The element $R(x)$ is called the *value of the function R* for the argument x. Note that if R is not a function or if $x \notin D_l(R)$, the symbol $R(x)$ has no definite meaning.

D6.2a, b and D6.3 easily yield

CONCLUSION 6.3a. $\quad x \in D_l(R) \equiv \sum\limits_{y \in D_r(R)} x R y,$

b. $\quad y \in D_r(R) \equiv \sum\limits_{x \in D_l(R)} x R y,$

c. $\quad R \in \text{funct} \wedge x \in D_l(R) \rightarrow R(x) \in D_r(R).$

D6.4. $\qquad\qquad y \in R(X) \equiv \sum\limits_{x \in X} x R y.$

The set $R(X)$ is called the R *image* of the set X, and if R is a function, the *set of values* of that function for the arguments belonging to the set X. If X is a set whose only element is zero, and the relation R is the relation "less than", then the set $R(X)$ is the set of all positive numbers. Note that the symbol $R(x)$ stands for an individual, while the symbol $R(X)$ stands for a set. The former has a meaning only if R is a function and $x \in D_l(R)$; the latter always has a meaning.

T6.1a. $\quad X \subset Y \rightarrow R(X) \subset R(Y),$

b. $\quad R(X \cup Y) = R(X) \cup R(Y),$

c. $\quad R(\bigcup\limits_{i=1}^{\infty} X_i) = \bigcup\limits_{i=1}^{\infty} R(X_i),$

d. $\quad \prod\limits_{i \in N} X_{i+1} = R(X_i) \rightarrow \bigcup\limits_{i=1}^{\infty} X_i = X_1 \cup R(\bigcup\limits_{i=1}^{\infty} X_i).$

The easy proofs of formulae a and b are left to the reader. Formula c will be proved by the demonstration of the equivalence:

$$y \in R(\bigcup\limits_{i=1}^{\infty} X_i) \equiv y \in \bigcup\limits_{i=1}^{\infty} R(X_i),$$

Proof. $y \in R(\bigcup\limits_{i=1}^{\infty} X_i) \equiv \sum\limits_{x \in \bigcup\limits_{i=1}^{\infty} X_i} x R y \equiv \sum\limits_{x} [\sum\limits_{i \in N} x \in X_i \wedge x R y]$

$\equiv \sum\limits_{x} \sum\limits_{i \in N} [x \in X_i \wedge x R y] \equiv \sum\limits_{i \in N} \sum\limits_{x} [x \in X_i \wedge x R y]$

$\equiv \sum\limits_{i \in N} y \in R(X_i) \equiv y \in \bigcup\limits_{i=1}^{\infty} R(X_i)$

{D6.4, D4.1, formula 2.(n), p. 124, formula 1.(e), p. 123}

To prove formula d, it suffices to demonstrate that with the assumption:

$$\prod_{i\in\mathbf{N}} X_{i+1} = R(X_i),$$

(a) *if* $y \in \bigcup_{i=1}^{\infty} X_i$, *then* $y \in X_1 \cup R(\bigcup_{i=1}^{\infty} X_i)$;

(b) *if* $y \in X_1 \cup R(\bigcup_{i=1}^{\infty} X_i)$, *then* $y \in \bigcup_{i=1}^{\infty} X_i$.

Proof. (a) (1) $\prod_{i\in\mathbf{N}} X_{i+1} = R(X_i)$ ⎫

                                        ⎬ {a.}

        (2) $y \in \bigcup_{i=1}^{\infty} X$ ⎭

        (3) $i_1 \in \mathbf{N}$ ⎫

        (4) $y \in X_{i_1}$ ⎭ {D4.1, 2}

        (5) $i_1 = 1 \lor i_1 > 1$ {3}

      (1.1) $i_1 = 1$ {ad.a.}

      (1.2) $y \in X_1$ {4, 1.1}

      (2.1) $i_1 > 1$ {ad.a.}

      (2.2) $i_1 - 1 \in \mathbf{N}$ {3, 2.1}

      (2.3) $X_{i_1} = R(X_{i_1-1})$ {1, 2.2}

      (2.4) $y \in R(X_{i_1-1})$ {4, 2.3}

      (2.5) $y \in \bigcup_{i=1}^{\infty} R(X_i)$ {D4.1, 2.2, 2.4}

      (2.6) $y \in R(\bigcup_{i=1}^{\infty} X)$ {T6.1c, 2.5}

        (6) $y \in X_1 \cup R(\bigcup_{i=1}^{\infty} X_i)$ {5, 1.1 → 1.2, 2.1 → 2.6}

(b) (1) $\prod_{i\in\mathbf{N}} X_{i+1} = R(X_i)$ ⎫

                                        ⎬

        (2) $y \in X_1 \cup R(\bigcup_{i=1}^{\infty} X_i)$ ⎭ {a.}

      (1.1) $y \in X_1$ {ad.a.}

      (1.2) $y \in \bigcup_{i=1}^{\infty} X_i$ {D4.1, 1.1}

      (2.1) $y \in R(\bigcup_{i=1}^{\infty} X_i)$ {ad.a.}

      (2.2) $y \in \bigcup_{i=1}^{\infty} R(X_i)$ {T6.1c, 2.1}

(2.3) $i_1 \in \mathbf{N}$ }
(2.4) $y \in R(X_{i_1})$ }   {D4.1, 2.2}

(2.5) $X_{i_1+1} = R(X_{i_1})$ {1, 2.3}

(2.6) $y \in X_{i_1+1}$ {2.4, 2.5}

(2.7) $i_1 + 1 \in \mathbf{N}$ {2.3}

(2.8) $y \in \bigcup\limits_{i=1}^{\infty} X_i$ {D4.1, 2.7}

$y \in \bigcup\limits_{i=1}^{\infty} X_i$ {2, 1.1 → 1.2, 2.1 → 2.8}

The concept of a sequence is a special case of the concept of a function:

An infinite sequence is a function whose left domain is the set of all natural numbers; the nth term of the sequence being the value of the function for the argument n. The infinite sequence with the terms $a_1, a_2, a_3, \ldots$ is represented by $\{a_n\}$.

A finite sequence of n terms is a function whose left domain is the set of natural numbers $\leqslant n$.

D6.5. $x R^{-1} y \equiv y R x.$

The relation R^{-1} is called the *converse* of the relation R. The converse of the relation $<$ is the relation $>$, and vice versa. The converse of the relation: is the husband of, is the relation: is the wife of.

CONCLUSION 6.4. $\langle x, y \rangle \in \underset{a,b}{E}\,(aR^{-1}b) \equiv \langle y, x \rangle \in \underset{a,b}{E}\,(aRb)$

                                           {T5.2, D6.5}

L6.1a. $D_l(R) = D_r(R^{-1}).$

b. $D_r(R) = D_l(R^{-1}).$

Proof. a. (1) $x \in D_l(R) \equiv \sum\limits_{y} x R y \equiv \sum\limits_{y} y R^{-1} x \equiv x \in D_r(R^{-1})$

                                  {D6.2a, D6.5, D6.2b}

         $D_l(R) = D_r(R^{-1})$ {D1.1, 1}

The proof of formula b is similar.

D6.6. $R \in 1{-}1 \equiv R, R^{-1} \in \text{funct.}$

The formula $R \in 1{-}1$ is read: R is a one-one relation. A relation is one-one if and only if it and its converse are both functions.

The following arithmetical relations are examples of one-one relations:

$$x R_1 y \equiv y = 2x \wedge x, y \in \mathbf{N},$$
$$x R_2 y \equiv y = x^2 \wedge x, y \in \mathbf{N},$$
$$x R_3 y \equiv y = x^3 \wedge x, y \in \mathbf{N}.$$

Conclusions 6.1 and 6.4 together with D6.6 yield

CONCLUSION 6.5. *R is a one-one relation if and only if the set* $\underset{a,b}{E}(aRb)$ *does not contain two different ordered pairs whose first or second elements are identical.*

L6.2. $R \in 1-1 \rightarrow R^{-1} \in 1-1.$

Proof. It follows from the fact that $R \in 1-1$ and from Conclusion 6.5 that no two different ordered pairs whose first or second elements are identical belong to the set $\underset{a,b}{E}(aRb)$. Hence from Conclusion 6.4 it follows that no two such pairs belong to the set $\underset{a,b}{E}(aR^{-1}b)$ either. Then, in view of Conclusion 6.5, the lemma holds.[1]

L6.3a. $R \in 1-1 \wedge x \in D_l(R^{-1}) \rightarrow x = R(R^{-1}(x)),$

    b. $R \in 1-1 \wedge x, y \in D_l(R) \wedge x \neq y \rightarrow R(x) \neq R(y),$

    c. $R^{-1} \in \text{funct} \rightarrow R(X-Y) = R(X)-R(Y).$

Proof. a. (1) $R \in 1-1$ ⎫
         (2) $x \in D_l(R^{-1})$ ⎭   {a.}
         (3) $R, R^{-1} \in \text{funct}$ {D6.6, 1}
         (4) $R^{-1}(x) = R^{-1}(x) \equiv x R^{-1} R^{-1}(x)$ {D6.3, 3, 2}
         (5) $x R^{-1} R^{-1}(x)$ {4}
         (6) $R^{-1}(x) R x$ {D6.5, 5}
         (7) $R^{-1}(x) \in D_l(R)$ {D6.2a, 6}
              $x = R[R^{-1}(x)]$ {D6.3, 3, 7, 6}

---

[1] The proof of L6. 2 is markedly less detailed than the proofs given previously. A proof of L6.2 written down exclusively by means of formulae is not difficult either, but such a proof does not fully explain the intuitive aspect of the reasoning involved. Subsequent proofs will often be given in words if this serves to clarify the proof.

Proof. b. (1) $R \in 1-1$ ⎫
      (2) $x, y \in D_l(R)$ ⎬   {a.}
      (3) $x \neq y$ ⎭

      (4) $R(x) = R(y)$ {a.i.p.}

      (5) $R, R^{-1} \in$ funct   {D6.6, 1}

      (6) $x RR(x) \land y RR(y)$ {D6.3, 5, 2}

      (7) $R(x) R^{-1}x \land R(y) R^{-1}y$ {D6.5, 6}

      (8) $x = y$ {D6.1, 5, 4, 7}

          contr.   {3, 8}

The proof of formula c is left to the reader.

T6.2. $R \in 1-1 \land X \subset D_l(R) \to R^{-1}[R(X)] = X.$

The easy though rather long proof of this theorem is left to the reader.

D6.7. $x R_X y \equiv x R y \land x \in X.$

The relation R_X is called a *relation limited to the set X*.
D5.3 and D6.7 yield

CONCLUSION 6.6. *The set* $\underset{a,b}{E}(a R_X b)$ *contains those and only those ordered pairs which belong to the set* $\underset{a,b}{E}(a R b)$ *and whose first elements belong to X.*

L6.4. $R \in 1-1 \to R_X \in 1-1.$

Proof. From the fact that $R \in 1-1$ and from Conclusion 6.5 it follows that no two different ordered pairs whose first or second elements are identical belong to the set $\underset{a,b}{E}(a R b)$. Hence from Conclusion 6.6 it follows that the set $\underset{a,b}{E}(a R_X b)$ also does not contain any two such pairs. From this fact and Conclusion 6.5 the lemma follows.

L6.5a. $X \subset D_l(R) \to D_l(R_X) = X,$

    b. $X \subset D_l(R) \to D_r(R_X) = R(X).$

Proof. a. (1) $X \subset D_l(R)$ {a.}

          (1.1) $x \in D_l(R_X)$ {ad.a.}

          (1.2) $x R_X y_1$ {D6.2a, 1.1}

          (1.3) $x \in X$ {D6.7, 1.2}

          (2.1) $x \in X$ {ad.a.}

(2.2) $x \in D_l(R)$ $\{1,\ 2.1\}$

(2.3) $x R y_2$ $\{D6.2a,\ 2.2\}$

(2.4) $x R_X y_2$ $\{D6.7,\ 2.3,\ 2.1\}$

(2.5) $x \in D_l(R_X)$ $\{D6.2a,\ 2.4\}$

$\quad\ \ D_l(R_X) = X$ $\{D1,\ 1.1 \rightarrow 1.3,\ 2.1 \rightarrow 2.5\}$

The proof of formula b is left to the reader.

D6.8. $$xR;Sy \equiv \sum_z (xRz \wedge zSy).$$

The relation $R;S$ is called the *relative product* of relations R and S. The relation of perpendicularity of straight lines in space is a relative product of the relation of the parallelism of straight lines and the relation of perpendicularity of straight lines in a plane. In other words the straight line l is perpendicular in space to the straight line m if and only if there is a straight line which is parallel to l and at the same time intersects m at right angles. The relation: x is the father-in-law of y, is a relative product of the relation: x is the father of z, and the relation: z is the wife of y.

T5.2 and D6.8 yield

CONCLUSION 6.7. *The ordered pair $\langle x, y \rangle$ belongs to the set $\underset{a,b}{E}(aR;$ Sb) if and only if for a certain z the pair $\langle x, z \rangle \in \underset{a,b}{E}(aRb)$ and at the same time the pair $\langle z, y \rangle \in \underset{a,b}{E}(aSb)$.*

L6.6. $R,S \in 1{-}1 \rightarrow R;S \in 1{-}1.$

Proof. Suppose that $R;S$ is not a one–one relation. Then by Conclusion 6.5 there exist x, y, z such that $y \neq z$ and

a. $\langle x, y \rangle,\ \langle x, z \rangle \in \underset{a,b}{E}(aR;Sb)$

or

b. $\langle y, x \rangle,\ \langle z, x \rangle \in \underset{a,b}{E}(aR;Sb)$

We analyse case a. It follows from Conclusion 6.7 that there exist t_1 and t_2 such that

(1) $\langle x, t_1 \rangle,\ \langle x, t_2 \rangle \in \underset{a,b}{E}(aRb)$

and

(2) $$\langle t_1, y \rangle, \langle t_2, z \rangle \in \underset{a,b}{E}(aSb).$$

If $t_1 = t_2$, then formula (2) contradicts the assumption that $S \in 1-1$. If $t_1 \neq t_2$, then formula (1) contradicts the assumption that $R \in 1-1$. Hence case a cannot occur. Similarly we can show that case b is impossible. Hence the assumption that $R;S \notin 1-1$ leads to a contradiction.

L6.7a. $\qquad D_r(R) = D_l(S) \rightarrow D_l(R;S) = D_l(R),$

b. $\qquad D_r(R) = D_l(S) \rightarrow D_r(R;S) = D_r(S).$

Proof. a. (1) $\quad D_r(R) = D_l(S)$ $\qquad$ {a.}

$\quad$ (1.1) $\quad x \in D_l(R;S)$ $\qquad$ {ad.a.}

$\quad$ (1.2) $\quad x R;Sy_1$ $\qquad$ {D6.2a, 1.1}

$\quad$ (1.3) $\quad x Rz_1$ $\qquad$ {D6.8, 1.2}

$\quad$ (1.4) $\quad x \in D_l(R)$ $\qquad$ {D6.2a, 1.3}

$\quad$ (2.1) $\quad x \in D_l(R)$ $\qquad$ {ad.a.}

$\quad$ (2.2) $\quad x Rz_2$ $\qquad$ {D6.2a, 2.1}

$\quad$ (2.3) $\quad z_2 \in D_r(R)$ $\qquad$ {D6.2b, 2.2}

$\quad$ (2.4) $\quad z_2 \in D_l(S)$ $\qquad$ {1, 2.3}

$\quad$ (2.5) $\quad z_2 Sy_2$ $\qquad$ {D6.2a, 2.4}

$\quad$ (2.6) $\quad x R;Sy_2$ $\qquad$ {D6.8, 2.2, 2.5}

$\quad$ (2.7) $\quad x \in D_l(R;S)$ $\qquad$ {D6.2a, 2.6}

$\qquad D_l(R;S) = D_l(R)$ $\quad$ {D1.1, 1.1 → 1.4, 2.1 → 2.7}

The proof of formula b is analogous.

The lemma proved above will be referred to in the next Section.

D6.9. A relation is said to be an *equivalence relation* if and only if it is reflexive, symmetric, and transitive.

The relation of equality, the similarity of polygons, the equivalence of plane figures, and having blood of the same group are all example of equivalence relations.

The following definition is an analogue of D1.1:

D6.10. $$R \doteq S \equiv \prod_x \prod_y (xRy \equiv xSy).$$

The formula $R \doteq S$ is read: the relation R is extensionally equal to the relation S.

An analogue of T1.2 is the following theorem, the straightforward proof of which is omitted:

T6.2a. $\qquad\qquad R \doteq R,$

   b. $\qquad\qquad R \doteq S \to S \doteq R,$

   c. $\qquad\qquad R \doteq S \wedge S \doteq T \to R \doteq T.$

An analogue of T1.6 is

T6.3. $\qquad\qquad\qquad R = S \to R \doteq S \qquad\qquad$ {EI, T6.2a}

An analogue of the axiom of extensionality for sets (A1.2) is the axiom of extensionality for relations:

A6.1. $\qquad\qquad\qquad R \doteq S \to R = S.$

T6.3 and A6.1 yield

T6.4 $\qquad\qquad\qquad R \doteq S \equiv R = S.$

Formulae of the type $R \doteq S$ may thus always be replaced by formulae of the type $R = S$. For this reason we shall henceforth use only the symbol $=$. One more lemma will be referred to in further Sections:

L6.8. $\qquad\qquad\qquad X = D_l(R) \to R_X = R.$

> *Proof.* (1)$\quad X = D_l(R)$ $\qquad\qquad$ {a.}
> $\qquad$(1.1)$\quad x R_X y$ $\qquad\qquad\qquad$ {ad. a.}
> $\qquad$(1.2)$\quad x R y$ $\qquad\qquad\qquad$ {D6.7, 1.1}
> $\qquad$(2.1)$\quad x R y$ $\qquad\qquad\qquad$ {ad. a.}
> $\qquad$(2.2)$\quad x \in D_l(R)$ $\qquad\qquad$ {D6.2a, 2.1}
> $\qquad$(2.3)$\quad x \in X$ $\qquad\qquad\qquad$ {2.2, 1}
> $\qquad$(2.4)$\quad x R_X y$ $\qquad\qquad\qquad$ {D6.7, 2.1, 2.3}
> $\qquad\qquad R_X = R$ $\qquad\qquad$ {D6.10, 1.1 → 1.2, 2.1 → 2.4}

D6.11. $\qquad\qquad x R \cup S y \equiv x R y \vee x S y.$

The relation $R \cup S$ is called the *sum* of the relations R and S.

L6.9a.$\quad D_l(R \cup S) = D_l(R) \cup D_l(S),$

   b.$\quad D_r(R \cup S) = D_r(R) \cup D_r(S),$

   c.$\quad (R \cup S)^{-1} = R^{-1} \cup S^{-1},$

   d.$\quad R, S \in \text{funct} \wedge (D_l(R)) \cap (D_l(S)) = \emptyset \to R \cup S \in \text{funct}.$

The easy proofs of formulae a–c are left to the reader. Formula d will be proved by demonstrating that, with the assumption L6.9d, *if* $xR \cup Sy \wedge xR \cup Sz$, *then* $y = z$.

Proof of L6.9d:

(1) $R \in$ funct

(2) $S \in$ funct

(3) $(D_l(R)) \cap (D_l(S)) = \emptyset$ $\{a.\}$

(4) $xR \cup Sy \wedge xR \cup Sz$

(5) $(xRy \vee xSy) \wedge (xRz \vee xSz)$ $\{D6.11, 4\}$

(6) $(xRy \wedge xRz) \vee (xSy \wedge xSz) \vee (xRy \wedge xSz)$
$\vee (xSy \wedge xRz)$ $\{T39. \text{ Part I, Chap. III, Sec. 3 } 1.2\}$

(7) $\neg(xSy \wedge xRz)$ $\{3\}$

(8) $\neg(xRy \wedge xSz)$ $\{3\}$

(9) $(xRy \wedge xRz) \vee (xSy \wedge xSz)$ $\{6, 7, 8\}$

(1.1) $xRy \wedge xRz$ $\{ad.a.\}$

(1.2) $y = z$ $\{D6.1, 1, 1.1\}$

(2.1) $xSy \wedge xSz$ $\{ad.a.\}$

(2.2) $y = z$ $\{D6.1, 2, 2.1\}$

  $y = z$ $\{9, 1.1 \rightarrow 1.2, 2.1 \rightarrow 2.2\}$

T6.5. $R, S \in 1-1 \wedge (D_l(R)) \cap (D_l(S)) = \emptyset \wedge (D_r(R)) \cap (D_r(S)) = \emptyset$

$\rightarrow R \cup S \in 1-1.$

Proof. (1) $R \in 1-1$

(2) $S \in 1-1$

(3) $(D_l(R)) \cap (D_l(S)) = \emptyset$ $\{a.\}$

(4) $(D_r(R)) \cap (D_r(S)) = \emptyset$

(5) $R \in$ funct $\{D6.6, 1\}$

(6) $S \in$ funct $\{D6.6, 2\}$

(7) $R \cup S \in$ funct $\{L6.9d, 5, 6, 3\}$

(8) $R^{-1} \in$ funct $\{D6.6, 1\}$

(9) $S^{-1} \in$ funct $\{D6.6, 2\}$

(10) $(D_l(R^{-1})) \cap (D_l(S^{-1})) = \emptyset$ $\{L6.1b, 4\}$

(11) $R^{-1} \cup S^{-1} \in$ funct $\{L6.9d, 8, 9, 10\}$

(12) $(R \cup S)^{-1} \in$ funct $\{L6.9c, 11\}$

  $R \cup S \in 1-1$ $\{D6.6, 7, 12\}$

In addition to the sum of relations, the calculus of relations also includes the product and the complement of relations, defined by the formulae:

$$xR \cap Sy \equiv xRy \wedge xSy,$$
$$x\overline{R}y \equiv \neg xRy.$$

It can be demonstrated that these three operations on relations satisfy the axioms of Boolean algebra specified in Sec. 3. In this way we obtain one more interpretation of that algebra.

Thus far reference has been made to relations with only two arguments. The calculus of relations also deals with relations having more arguments. The concept of a function can, for instance, be generalized to cover the concept of many variables. We shall give here only a definition of a function of two variables:

R is a function of two variables

$$\equiv \prod_{x} \prod_{y} \prod_{z} \prod_{u} [R(x, y, z) \wedge R(x, y, u) \rightarrow z = u].$$

Thus functions of two variables are relations involving three arguments. We shall also introduce the concept of a relation defined and performable in a given set, in which we also confine ourselves to relations with three arguments:

R is a relation defined and performable in the set X

$$\equiv \prod_{x \in X} \prod_{y \in X} \sum_{z \in X} R(x, y, z).$$

Relations of three arguments, which are defined and performable in the set X and which are functions, are called *operations of two arguments* in the set X. Thus, for instance, addition and multiplication are operations of two arguments in the set of natural numbers, subtraction is an operation in the set of integers, division is an operation in the set of rational numbers other than 0.

Operations of one argument, of three arguments, etc., are also objects of study in logic.

Exercises

1. Which of the following relations are functions:

$$xR_1 y \equiv y^2 = x \wedge x, y \in \mathbf{R},$$
$$xR_2 y \equiv y^2 = x \wedge x, y \in \mathbf{R} \wedge y \geqslant 0,$$
$$xS_1 y \equiv x < y \leqslant x+1 \wedge x, y \in \mathbf{N},$$
$$xS_2 y \equiv x < y \leqslant x+1 \wedge x, y \in \mathbf{R},$$
$$x \text{ is the son of } y, \quad x \text{ is the father of } y?$$

2. Prove the formulae

(a) $R \in 1-1 \land x \in D_l(R) \to x = R^{-1}(R(x))$,

(b) $R \in 1-1 \land x \in D_r(R) \to x = R(R^{-1}(x))$,

(c) $R \in \text{funct} \land x \in D_l(R) \to [x \in X \equiv R(x) \in R(X)]$,

(d) $R(X \cup Y) = R(X) \cup R(Y)$,

(e) $R(X \cap Y) \subset R(X) \cap R(Y)$,

(f) $R \in \text{funct} \equiv \prod_{x \in D_l(R)} \sum_1 x R y$,

(g) $R \in 1-1 \equiv \prod_{x \in D_l(R)} \sum_1 x R y \land \prod_{y \in D_r(R)} \sum_1 x R y$.

Show by means of an example that the formulae

$$R(X) \cap R(Y) \subset R(X \cap Y)$$

is not true.

3. Give the left and the right domains of the relations specified in Exercise 1. Which of them are one-one relations?

4. Is the formula $R_X \in 1-1 \to R \in 1-1$ true?

5. Of what two relations are the following relations relative products:

x is the son-in-law of y, x is the brother-in-law of y?

6. Find an example to show that the formulae

$$x R; S y \to x S; R y$$

is false.

7. Give examples of a relation which is

(a) reflexive and symmetric, but not transitive,

(b) reflexive and transitive, but not symmetric,

(c) symmetric and transitive, but not reflexive.

8. Prove the formulae

(a) $R = (R^{-1})^{-1}$,

(b) $(R;S); T = R;(S;T)$,

(c) $(R;S)^{-1} = S^{-1}; R^{-1}$,

(d) $R \in \text{sym} \equiv R = R^{-1}$.

9 Prove the formula

$$D_l(R) \subset X \to R(X) = D_r(R).$$

10. Show by an example that the following formula is false

$$R_X = R \to X = D_l(R).$$

7. Equinumerous Sets. Cardinal Numbers

The concepts to be defined in this Section are among the most important and the most essential in set theory.

D7.1.　$X \sim_R Y \equiv R \in 1-1 \land X = D_l(R) \land Y = D_r(R).$

The formula $X \sim_R Y$ is read: the relation R establishes that the sets X and Y are equinumerous.

Thus the relation

$$x R_1 y \equiv y = 2x \wedge x \in \mathbf{N}$$

establishes that the set of all natural numbers and the set of all even integers are equinumerous;

$$x R_2 y \equiv y = x^2 \wedge x \in \mathbf{N}$$

establishes that the set of all natural numbers and the set of all squares of natural numbers are equinumerous.

As a further example, let C_1 and C_2 be the circumferences of two concentric circles. Associate a point on C_2 with a point on C_1 if they both lie on the same straight half line which starts from the common centre of C_1 and C_2. This relation establishes that the sets of points on these circumferences are equinumerous. These examples show that sets can be equinumerous although one of them is a proper part of the other. Obviously, only infinite sets can have this property.

T5.2, D7.1 and Conclusions 6.5 and 6.3 yield

CONCLUSION 7.1. *The relation R establishes that the sets X and Y are equinumerous if and only if the set $\underset{a,b}{E}(aRb)$ contains no two different ordered pairs whose first or second elements are identical, and if the first elements of the ordered pairs belonging to the set $\underset{a,b}{E}(aRb)$ form the set X, and the second elements form the set Y.*

D7.2. $$X \sim Y \equiv \sum_R X \sim_R Y.$$

The formula $X \sim Y$ is read: the sets X and Y are equinumerous. Hence, two sets are *equinumerous* if and only if there exists a relation establishing that they are.

It is obvious that two finite sets are equinumerous if and only if they have the same number of elements.

For the discussion in this Section it is convenient to denote the identity relation by the symbol I. We shall therefore write xIy instead of $x = y$. This notation is fairly often used in logic. Note also that the formula xIx is a logical thesis.

L7.1. $$X = I(X).$$

Proof. (1.1) $y \in X$ {ad.a.}

    (1.2) $y \in X \wedge yIy$ {1.1}

    (1.3) $\sum_{x \in X} xIy$ {1.2}

    (1.4) $y \in I(X)$ {D6.4, 1.3}

    (2.1) $y \in I(X)$ {ad.a.}

    (2.2) $\sum_{x \in X} xIy$ {D6.4, 2.1}

    (2.3) $x_1 \in X$ ⎫

    (2.4) $x_1 Iy$ ⎭   {2.2}

    (2.5) $y \in X$ {EI: 2.4, 2.3}

          $X = I(X)$ {D1.1, 1.1 → 1.4, 2.1 → 2.5}

T7.1a. $X \sim X,$

   b. $X \sim Y \to Y \sim X,$

   c. $X \sim Y \wedge Y \sim Z \to X \sim Z.$

This theorem can be given the following verbal formulation: *The relation of equinumerosity of sets is reflexive, symmetric and transitive*, and hence is an equivalence relation.

Proof. a. (1) $I \in \text{funct}$ {D6.1}

     (2) $I^{-1} \in \text{funct}$ {D6.5, D6.1}

     (3) $I \in 1{-}1$ {D6.6, 1, 2}

     (4) $I_X \in 1{-}1$ {L6.4, 3}

     (5) $x \in D_l(I)$ {D6.2a}

     (6) $D_l(I) = 1$ {D1.1, 5, A1.1}

     (7) $X \subset D_l(I)$ {T1.1, 6}

     (8) $D_l(I_X) = D_r(I_X) = X$ {L6.5a, b, 7, L7.1}

     (9) $X \sim_{I_X} X$ {D7.1, 4, 8}

        $X \sim X$ {D7.2, 9}

   b. (1) $X \sim Y$ {a.}

     (2) $X \sim_{R_1} Y$ {D7.2, 1}

     (3) $R_1 \in 1{-}1$ ⎫

     (4) $X = D_l(R_1) \wedge Y = D_r(R_1)$ ⎭ {D7.1, 2}

     (5) $R_1^{-1} \in 1{-}1$ {L6.2, 3}

     (6) $D_l(R_1^{-1}) = Y \wedge D_r(R_1^{-1}) = X$ {L6.1a, b, 4}

     (7) $Y \sim_{R_1^{-1}} X$ {D7.1, 5, 6}

        $Y \sim X$ {D7.2, 7}

c. (1) $X \sim Y \wedge Y \sim Z$ {a.}

(2) $X \sim_{R_1} Y \wedge Y \sim_{S_1} Z$ {D7.2, 1}

(3) $R_1, S_1 \in 1-1$

(4) $X = D_l(R_1) \wedge Y = D_r(R_1)$ ⎫
     $\quad = D_l(S_1) \wedge Z = D_r(S_1)$ ⎬   {D7.1, 2}
                                              ⎭

(5) $R_1; S_1 \in 1-1$ {L6.6, 3}

(6) $X = D_l(R_1; S_1) \wedge Z = D_r(R_1; S_1)$ {L6.7a, b, 4}

(7) $X \sim_{R_1; S_1} Z$ {D7.1, 5, 6}

$X \sim Z$ {D7.2, 7}

The following two lemmas will prove useful in a later Section.

L7.2. $X \sim_R Y \to Y = R(X)$.

Proof. (1) $X \sim_R Y$ {a.}

(2) $X = D_l(R)$ ⎫
                        ⎬   {D7.1, 1}
(3) $Y = D_r(R)$ ⎭

(4) $D_r(R_X) = R(X)$ {L6.5b, 2}

(5) $R_X = R$ {L6.8, 2}

(6) $D_r(R_X) = D_r(R)$ {5}

$Y = R(X)$ {3, 6, 4}

L7.3. $R \in 1-1 \wedge X \subset D_l(R) \to X \sim_{R_X} R(X)$.

Proof. (1) $R \in 1-1$ ⎫
                              ⎬   {a.}
(2) $X \subset D_l(R)$ ⎭

(3) $R_X \in 1-1$ {L6.4, 1}

(4) $D_l(R_X) = X$ {L6.5a, 2}

(5) $D_r(R_X) = R(X)$ {L6.5b, 2}

$X \sim_{R_X} R(X)$ {D7:1, 3, 4, 5}

We introduce a new primitive concept, namely that of the *power of a set*.[1] It is one of the most fundamental concepts of set theory.

The power of the set X is represented by $\overline{\overline{X}}$. Georg Cantor, the founder of set theory, thought that we arrived at the concept of the power of a set by abstracting from the properties of its elements and of ordering within the set. The two lines in the symbol for

---

[1] The term "power of a set" is accepted in mathematical terminology, although "the number of elements of a set" would perhaps be clearer.

the power of a set are supposed to indicate this process of double abstraction.

A fresh axiom containing the newly introduced primitive term is added to the system of set theory:

A7.1. $$\bar{\bar{X}} = \bar{\bar{Y}} \equiv X \sim Y.$$

Remark 7.1. It is assumed that the power of the null set is the number zero, and that the power of a finite set is equal to the number of its elements.

Thus the powers of two sets are identical if and only if the sets are equinumerous. Note that A7.1 and the fact that the identity relation is reflexive, symmetric and transitive, immediately yield T7.1.[1]

D7.3. $$\mathrm{m} \in \mathrm{Nc} \equiv \sum_{X} \mathrm{m} = \bar{\bar{X}}.$$

The formula $\mathrm{m} \in \mathrm{Nc}$ is read: m is a cardinal number. A mathematical object is thus a cardinal number if and only if it is the power of a set. It follows from remark 7.1 and D7.3 that natural numbers and the number zero are cardinal numbers. In further Sections we shall demonstrate that there exists non-equinumerous infinite sets. Hence there exist different cardinal numbers which are neither zero nor natural numbers. Arbitrary cardinal numbers will be represented by small german letters: m, n, t.

D7.3 easily gives

T7.2a. $$\bar{\bar{X}} \in \mathrm{Nc},$$

b. $$\prod_{X} \sum_{\mathrm{m} \in \mathrm{Nc}} {}_1 \, \mathrm{m} = \bar{\bar{X}},$$

c. $$\prod_{\mathrm{m} \in \mathrm{Nc}} \sum_{X} \mathrm{m} = \bar{\bar{X}}.$$

This theorem has the following verbal formulation:
a. *the power of any set is a cardinal number;*
b. *for every set there exists exactly one cardinal number which is the power of that set;*
c. *for every cardinal number there is a set whose power is that cardinal number.*

---

[1] Cf. the remark after T1.7 on p. 162.

L7.4. *For every pair of sets A and B there exist sets A_1 and B_1 such that*

$$A_1 \sim A, B_1 \sim B, \text{ and } A_1 \cap B_1 = \varnothing.$$

Proof. The sets A_1 and B_1 are defined as follows:

$$\langle 1, a \rangle \in A_1 \equiv a \in A,$$

$$\langle 2, b \rangle \in B_1 \equiv b \in B.$$

It can easily be seen that sets A_1 and B_1 satisfy the conditions imposed by the lemma.

When we consider cardinal numbers, it is irrelevant which of the equinumerous sets are taken into account. By L7.4, we can always replace two given sets by disjoint sets.

L7.5a. $A \sim B \wedge A_1 \sim B_1 \wedge A \cap A_1 = B \cap B_1 = \varnothing \rightarrow$
$$A \cup A_1 \sim B \cup B_1,$$

b. $A \sim B \wedge A_1 \sim B_1 \rightarrow A \times A_1 \sim B \times B_1.$

Proof. a. Suppose that relation R establishes that sets A and B are equinumerous, and relation R_1 establishes that sets A_1 and B_1 are too. Then on the strength of T6.5, the relation $R \cup R_1$ establishes that sets $A \cup A_1$ and $B \cup B_1$ are equinumerous.

Note that the condition that the sets A and A_1, and B and B_1, should be disjoint is essential, as is shown by the following example:

$$A = \{1, 2, 3\}, \ A_1 = \{1, 2, 3, 4\}, \ B = \{1, 2, 3\}, \ B_1 = \{1, 4, 5, 6\}.$$

The proof of part b of the lemma is left to the reader.

L7.4 and L7.5 will be used in the next Section.

Note also that the power of the set of all natural numbers will be represented by $\aleph_0$,[1] and the power of the set of all real numbers by c. Thus the following formulae are satisfied:

$$\overline{\overline{N}} = \aleph_0, \qquad \overline{\overline{R}} = c.$$

Sets whose power is the number $\aleph_0$ are called *enumerable*, and sets whose power is the number c are called *sets of continuum power*.

---

[1] The symbol $\aleph$, introduced in this role by Cantor, is the first letter of the Hebrew alphabet and is called *aleph*. The symbol $\aleph_0$ is read: aleph nought.

D7.4. Set A is the *range* of the sequence $\{a_n\}$

$$\equiv \prod_a [a \in A \equiv \sum_{i \in N} a = a_i].$$

T7.3 $\overline{\overline{A}} = \aleph_0 \equiv \sum_{\{a_n\}} [\prod_i \prod_j (i \neq j \rightarrow a_i \neq a_j)$

$\wedge$ the set A is the range of the sequence $\{a_n\}].$

Thus the necessary and sufficient condition for a set A to be enumerable is that there exists an infinite sequence of non-repeating terms having the set A as its range.

Proof. Let

$$a_1, a_2, a_3, \ldots$$

be an infinite sequence of non-repeating terms having the set A as its range. With each term of this sequence let us associate its index so that the number 1 is associated with a_1, the number 2 with a_2, etc. It can easily be seen that this establishes the fact that the sets A and N are equinumerous. From this the theorem follows.

Note that for nearly all theorems concerning powers of sets there exist equivalent theorems in which that concept does not occur and which refer only to the concept of equinumerous sets. Hence the omission of the concept of the power of a set from the list of primitive terms of set theory (accompanied by the omission of A7.1) would not make the theory essentially poorer, but would merely complicate the formulation of many theorems.

Exercises

1. Prove that the following sets are equinumerous:
   (a) the set of all positive real numbers and the set of all negative real numbers;
   (b) the set of all natural numbers and the set of all natural numbers greater than a given number n;
   (c) the set of the real numbers in the interval $(0, 1)$ and the set of the real numbers in the interval $(k, k+1)$;
   (d) the set of all the points on a chord of a circle and the set of all the points on the arc intersected by that chord.
2. Prove that the following definition is equivalent to D7.2:

$$X \sim Y \equiv \sum_{R \subseteq 1-1} [X \subset D_l(R) \wedge R(X) = Y].$$

3. Find an example which disproves the statement:

$$X \sim Y \wedge X_1 \sim Y_1 \rightarrow X \cap X_1 \sim Y \cap Y_1.$$

4. Demonstrate in detail that the sets $\{x\}$ and $\{y\}$ are equinumerous.

8. The Arithmetic of Cardinal Numbers

We shall now define the addition, multiplication and raising to a power of cardinal numbers. We denote these operations in the same way as in the arithmetic of "ordinary" numbers, e.g., the arithmetic of natural or real numbers. We shall also retain the terminology of elementary arithmetics (e.g., the result of addition of cardinal numbers will be called the *sum*, and the cardinal numbers being added will be called *elements*).

D8.1. $$X \cap Y = \emptyset \rightarrow \bar{\bar{X}} + \bar{\bar{Y}} = \overline{\overline{X \cup Y}}.$$

This is an abbreviated form of the definition of addition of cardinal numbers. The exact form of the definition is as follows:

$$\mathfrak{m} + \mathfrak{n} = \mathfrak{k} \equiv \sum_X \sum_Y (\bar{\bar{X}} = \mathfrak{m} \wedge \bar{\bar{Y}} = \mathfrak{n} \wedge X \cap Y = \emptyset \wedge \mathfrak{k} = \overline{\overline{X \cup Y}}).$$

The definitions of other operations will be given in their abbreviated forms only.

D8.2. $$\bar{\bar{X}} \cdot \bar{\bar{Y}} = \overline{\overline{X \times Y}}.$$

We now give an auxiliary definition:

D8.3. $$R \in Y^X \equiv R \in \text{funct} \wedge D_l(R) = X \wedge R(X) \subset Y.$$

The formula $R \in Y^X$ is read: the function R is the mapping of the set X into the set Y.

D8.3 yields:

$$A \sim A_1 \wedge B \sim B_1 \rightarrow B^A \sim B_1^{A_1}.$$

The proof of this formula, which is analogous to L7.5a, is left to the reader.

D8.3 also gives

CONCLUSION 8.1. $$Z \subset Y \rightarrow Z^X \subset Y^X.$$

D8.4. $$\overline{\overline{X^Y}} = (\overline{\overline{X}}^{\bar{\bar{Y}}}).$$

The power of the set X raised to the power equal to the power of the set Y is thus equal to the power of the set of all mappings of the set Y onto the set X.

We shall show next that D8.1 is correct. To do so we must prove that for every two cardinal numbers $\mathfrak{m}$ and $\mathfrak{n}$ there exists exactly one cardinal number which is their sum.

By T7.2c there exist sets A and A_1 such that $\overline{\overline{A}} = \mathfrak{m}$ and $\overline{\overline{A_1}} = \mathfrak{n}$. By L7.4 we may assume that these sets are disjoint. From T7.2b there exists a cardinal number $\mathfrak{k} = \overline{\overline{A \cup A_1}}$.

Thus for every two cardinal numbers their sum exists.

Now let B and B_1 be any sets satisfying the conditions

$$\overline{\overline{B}} = \mathfrak{m}, \qquad \overline{\overline{B_1}} = \mathfrak{n}, \qquad B \cap B_1 = \varnothing,$$

and let $\mathfrak{k}_1 = \overline{\overline{B \cup B_1}}$. It follows immediately from L7.5a that $A \cup A_1 \sim B \cup B_1$. From this and A7.1 it follows that $\mathfrak{k} = \mathfrak{k}_1$. Hence there exists a unique sum of given cardinal numbers. Likewise we can show that the definitions of the remaining operations of cardinal numbers are correct.

We next show that operations on natural numbers are special cases of operations on cardinal numbers.

Let

$$X = \{x_1, x_2, \ldots, x_m\}, \qquad Y = \{y_1, y_2, \ldots, y_n\}^{(1)}$$

so that

$$\overline{\overline{X}} = m \quad \text{and} \quad \overline{\overline{Y}} = n.$$

We also assume that every element of the set X is different from every element of the set Y. It is clear that $\overline{\overline{X \cup Y}} = m+n$. Thus the addition of natural numbers is a special cases of addition of cardinal numbers.

We shall now show that $\overline{\overline{X \times Y}} = m \cdot n$. The Cartesian product $X \times Y$ includes every ordered pair $\langle x_i, y_j \rangle$, where $i = 1, 2, \ldots, m$, $j = 1, 2, \ldots, n$. It is obvious that there are $m \cdot n$ such pairs. Thus

---

(1) It is assumed that of the elements $x_1, x_2, \ldots, x_m$ none are identical, and also that of the elements $y_1, y_2, \ldots, y_n$ none are identical.

multiplication of natural numbers is a special case of multiplication of cardinal numbers. Note further that every mapping of the set Y into the set X uniquely determines the sequence of n terms

(1) $x_1, x_2, \ldots, x_n,$

which are elements of the set X.[1] It can easily be seen that there are m^n such sequences. Hence the raising of natural numbers to powers is a special case of the raising of cardinal numbers to powers.

We shall give as examples, some results from the arithmetic of cardinal numbers.

T8.1a. $\bar{\bar{X}} + \bar{\bar{Y}} = \bar{\bar{Y}} + \bar{\bar{X}},$

b. $\bar{\bar{X}} \cdot \bar{\bar{Y}} = \bar{\bar{Y}} \cdot \bar{\bar{X}}.$

This theorem can also be written in the following form:

a. $\mathfrak{m} + \mathfrak{n} = \mathfrak{n} + \mathfrak{m},$

b. $\mathfrak{m} \cdot \mathfrak{n} = \mathfrak{n} \cdot \mathfrak{m}.$

Proof. a. By L7.4 we may assume that the sets X and Y are disjoint. Hence the following formulae can be derived

(1) $X \cap Y = Y \cap X = \emptyset$

(2) $\bar{\bar{X}} + \bar{\bar{Y}} = \overline{\overline{X \cup Y}}$ $\Big\}$

(3) $\bar{\bar{Y}} + \bar{\bar{X}} = \overline{\overline{Y \cup X}}$ $\{\text{D8.1, 1}\}$

(4) $X \cup Y = Y \cup X$ $\{\text{T2.2a}\}$

(5) $\overline{\overline{X \cup Y}} = \overline{\overline{Y \cup X}}$ $\{4\}$

$\bar{\bar{X}} + \bar{\bar{Y}} = \bar{\bar{Y}} + \bar{\bar{X}}.$ $\{5, 2, 3\}$

Proof. b. By D8.2, the following formulae are satisfied:

(1) $\bar{\bar{X}} \cdot \bar{\bar{Y}} = \overline{\overline{X \times Y}}, \quad \bar{\bar{Y}} \cdot \bar{\bar{X}} = \overline{\overline{Y \times X}}.$

We now define a relation needed in the proof:

$\langle x, y \rangle \, R_1 \, \langle z, t \rangle \equiv x = t \wedge y = z \wedge x \in X \wedge y \in Y.$

It is clear that this relation establishes that the sets $X \times Y$ and $Y \times X$ are equinumerous. From this and D8.2 and A7.1 follows the equality

(2) $\overline{\overline{X \times Y}} = \overline{\overline{Y \times X}}.$

---

[1] Note that terms of sequence (1) with different indices may be identical.

And from (1) and (2) follows

$$\overline{\overline{X.\,Y}} = \overline{\overline{Y.\,X}}.$$

T8.2. $\qquad\qquad (\overline{\overline{X}}+\overline{\overline{Y}}).\,\overline{\overline{Z}} = \overline{\overline{X}}.\,\overline{\overline{Z}}+\overline{\overline{Y}}.\,\overline{\overline{Z}}.$

Proof. By L7.4, we can assume that

(1) $X \cap Y = \varnothing$

(2) $\overline{\overline{X}}+\overline{\overline{Y}} = \overline{\overline{X \cup Y}}$ $\qquad\qquad\qquad$ {D8.1, 1}

(3) $(\overline{\overline{X}}+\overline{\overline{Y}}).\,\overline{\overline{Z}} = \overline{\overline{X \cup Y}}.\,\overline{\overline{Z}} = \overline{\overline{(X \cup Y) \times Z}}$ $\qquad$ {2, D8.2}

(4) $(X \times Z) \cap (Y \times Z) = \varnothing$ $\qquad\qquad\qquad$ {T5.1b, 1}

(5) $\overline{\overline{X}}.\,\overline{\overline{Z}}+\overline{\overline{Y}}.\,\overline{\overline{Z}} = \overline{\overline{X \times Z}}+\overline{\overline{Y \times Z}} = \overline{\overline{X \times Z \cup Y \times Z}}$ $\quad$ {D8.2, D8.1, 4}

(6) $(X \cup Y) \times Z = X \times Z \cup Y \times Z$ $\qquad\qquad$ {T5.1a}

(7) $\overline{\overline{(X \cup Y) \times Z}} = \overline{\overline{X \times Z \cup Y \times Z}}$ $\qquad\qquad$ {6}

$\qquad (\overline{\overline{X}}+\overline{\overline{Y}}).\,\overline{\overline{Z}} = \overline{\overline{X}}.\,\overline{\overline{Z}}+\overline{\overline{Y}}.\,\overline{\overline{Z}}$ $\qquad\qquad$ {7, 3, 5}

T8.3. $\qquad\qquad n.\,\overline{\overline{X}} = \underbrace{\overline{\overline{X}}+\overline{\overline{X}}+ \ldots +\overline{\overline{X}}}_{n}$

Proof. Let $\mathbf{N}_1 = \{1, 2, \ldots, n\}$. Hence $\overline{\overline{\mathbf{N}}}_1 = n$. Let X_i (where $i = 1, 2, \ldots, n$) stand for the Cartesian product of the unit set $\{i\}$ and the set X. It can easily be seen that the following formulae are satisfied:

$$\overline{\overline{X}}_1 = \overline{\overline{X}}_2 = \ldots = \overline{\overline{X}}_n = \overline{\overline{X}},$$

$$i \neq j \to X_i \cap X_j = \varnothing.$$

From these formulae, and T5.1c, D8.1 and D8.2, follow the equations

$$n.\,\overline{\overline{X}} = \overline{\overline{\mathbf{N}_1 \times X}} = \overline{\overline{X_1 \cup X_2 \cup \ldots \cup X_n}} = \overline{\overline{X}}_1+\overline{\overline{X}}_2+ \ldots +\overline{\overline{X}}_n$$

$$= \underbrace{\overline{\overline{X}}+\overline{\overline{X}} + \ldots +\overline{\overline{X}}.}_{n}$$

Hence the theorem is true.

T8.4. $\qquad\qquad (\overline{\overline{X}})^n = \underbrace{\overline{\overline{X}}.\,\overline{\overline{X}}. \ldots .\overline{\overline{X}}.}_{n}$

Proof. Let $\mathbf{N}_1 = \{1, 2, \ldots, n\}$. Hence $\overline{\overline{\mathbf{N}}}_1 = n$. It follows from D8.3 that the set $X^{\mathbf{N}_1}$ is the set of all the sequences $x_1, x_2, \ldots, x_n$ whose terms are elements of the set X, and by D5.1 the set $\underbrace{X \times X \times \ldots \times X}_{n}$

is the set of all ordered n-tuples $\langle x_1, x_2, ..., x_n \rangle$, where x_1, $x_2, ..., x_n \in X$. It is obvious that the relation which associates with the sequence $x_1, x_2, ..., x_n$ the ordered n-tuple $\langle x_1, x_2, ..., x_n \rangle$ establishes that the sets X^{N_1} and $\underbrace{X \times X \times ... \times X}_{n}$ are equinumerous.

From this and D8.2 and D8.4 the required theorem follows.

T8.5.
$$\overline{\overline{X}}^{\overline{\overline{Y}}+\overline{\overline{Z}}} = \overline{\overline{X}}^{\overline{\overline{Y}}} \cdot \overline{\overline{X}}^{\overline{\overline{Z}}}.$$

The proof is given in outline.

Proof. By L7.4 we may assume that

(1)
$$Y \cap Z = \varnothing.$$

Let R be any function which maps the set $Y \cup Z$ into the set X. From this, formula (1), D6.7, and D8.3 follow the formulae

$$R_Y \in X^Y, \qquad R_Z \in X^Z.$$

Let the relation S associate with the function R the ordered pair $\langle R_Y, R_Z \rangle$. The relation S establishes that the sets $X^{Y \cup Z}$ and $X^Y \times X^Z$ are equinumerous. This easily leds to the required theorem.

The following theorem is given without proof.

T8.6.
$$(\overline{\overline{X}}^{\overline{\overline{Y}}})^{\overline{\overline{Z}}} = \overline{\overline{X}}^{\overline{\overline{Y}} \cdot \overline{\overline{Z}}}.$$

The theorems concerning cardinal numbers given in this Section are analogous to the corresponding theorems of the arithmetic of natural numbers. Note also that definitions of the converse operations of subtraction and division have not been given. The reasons why these operations are not defined in set theory will be explained in later Sections.

Exercises

1. Prove the following formulae:
   (a) $\mathfrak{m}+0 = \mathfrak{m}$,[1]
   (b) $\mathfrak{m}.1 = \mathfrak{m}$,
   (c) $\mathfrak{m}.0 = 0$,
   (d) $\mathfrak{m}^1 = \mathfrak{m}$,
   (e) $1^{\mathfrak{m}} = 1$,
   (f) $(\overline{\overline{X}} \cdot \overline{\overline{Y}}) \cdot \overline{\overline{Z}} = \overline{\overline{X}} \cdot (\overline{\overline{Y}} \cdot \overline{\overline{Z}})$.

---

[1] We recall that 0 is the power of the null set.

9. Inequalities

L9.1. $\quad X \sim X_1 \wedge Y \sim Y_1 \wedge \sum\limits_{Z \subset Y} (X \sim Z) \to \sum\limits_{U \subset Y_1} (X_1 \sim U).$

Proof.
(1) $\quad X \sim X_1$
(2) $\quad Y \sim Y_1$ $\{a.\}$
(3) $\quad \sum\limits_{Z \subset Y} X \sim Z$

(4) $\quad R_1 \in 1-1$ $\{D7.2, 2\}$
(5) $\quad Y \sim_{R_1} Y_1$
(6) $\quad Y_1 = R_1(Y)$ $\{L7.2, 5\}$
(7) $\quad Z_1 \subset Y$ $\{3\}$
(8) $\quad X \sim Z_1$
(9) $\quad R_1(Z_1) \subset R_1(Y)$ $\{T6.1, 7\}$
(10) $\quad R_1(Z_1) \subset Y_1$ $\{9, 6\}$
(11) $\quad Y = D_l(R_1)$ $\{D7.1, 5\}$
(12) $\quad Z_1 \subset D_l(R_1)$ $\{7, 11\}$
(13) $\quad Z_1 \sim_{R_1 Z_1} R_1(Z_1)$ $\{L7.3, 4, 12\}$
(14) $\quad Z_1 \sim R_1(Z_1)$ $\{D7.2, 4, 13\}$
(15) $\quad X_1 \sim R_1(Z_1)$ $\{T7.1b, c, 1, 8, 14\}$
$\quad \sum\limits_{U \subset Y_1} X_1 \sim U$ $\{10, 15\}$

We shall now define the relations of ordinary and proper inequality. These relations will be denoted by symbols taken from the arithmetic of "ordinary" numbers.

D9.1. $\qquad \overline{\overline{X}} \leqslant \overline{\overline{Y}} \equiv \sum\limits_{Z \subset Y} (X \sim Z).$

Thus the necessary and sufficient condition for the power of the set X to be less than or equal to the power of the set Y, is that the set X be equinumerous with a subset of the set Y. Note that it follows from L9.1 that if $\overline{\overline{X}} = \overline{\overline{X_1}}$, $\overline{\overline{Y}} = \overline{\overline{Y_1}}$, and $\overline{\overline{X}} \leqslant \overline{\overline{Y}}$, then $\overline{\overline{X_1}} \leqslant \overline{\overline{Y_1}}$.

The exact form of D9.1 is as follows:

$$\mathfrak{m} \leqslant \mathfrak{n} \equiv \sum\limits_{\overline{\overline{X}}=\mathfrak{m}} \sum\limits_{\overline{\overline{Y}}=\mathfrak{n}} \sum\limits_{Z \subset Y} X \sim Z.$$

From D9.1 and T7.1a we obtain

CONCLUSION 9.1.

$$X \subset Y \to \overline{\overline{X}} \leqslant \overline{\overline{Y}}.$$

Note also that the replacement, under the quantifier, of the formula $Z \subset Y$ by the formula $Z \underset{\neq}{\subset} Y$ would not permit us to replace the symbol $\leqslant$ by the symbol $<$, since, for instance, the set of all natural numbers greater than a definite number is a proper part of the set of all natural numbers, and yet the powers of these sets are equal.[1]

D9.2. $$\overline{\overline{X}} < \overline{\overline{Y}} \equiv \overline{\overline{X}} \leqslant \overline{\overline{Y}} \wedge \overline{\overline{X}} \neq \overline{\overline{Y}}.$$

It can easily be seen that the two relations defined above are generalizations of the corresponding relations in the arithmetic of natural numbers.

L9.2. $$\overline{\overline{X}} = \overline{\overline{Y}} \to \overline{\overline{X}} \leqslant \overline{\overline{Y}}.$$

Proof. (1) $\overline{\overline{X}} = \overline{\overline{Y}}$ {a.}

(2) $X \sim Y$ {A7.1, 1}

(3) $Y \subset Y$ {T1.4a}

(4) $\sum_{Z \subset Y} X \sim Z$ {2, 3}

$\overline{\overline{X}} \leqslant \overline{\overline{Y}}$ {D9.1, 4}

D9.2 and L9.2 can also be written in the following form:

$$\mathfrak{m} < \mathfrak{n} \equiv \mathfrak{m} \leqslant \mathfrak{n} \wedge \mathfrak{m} \neq \mathfrak{n},$$
$$\mathfrak{m} = \mathfrak{n} \to \mathfrak{m} \leqslant \mathfrak{n}.$$

On the strength of these formulae and the thesis of the sentential calculus

$$(r \equiv p \wedge \neg q) \to [(q \to p) \to (p \equiv q \vee r)],$$

we can state the theorem

T9.1. $$\mathfrak{m} \leqslant \mathfrak{n} \equiv \mathfrak{m} = \mathfrak{n} \vee \mathfrak{m} < \mathfrak{n}.$$

T9.2a. $$X_2 \subset X_1 \subset X \wedge X \sim X_2 \to X \sim X_1.$$

Proof. (1) $X_2 \subset X_1 \subset X$ $\Big\}$ {a.}

(2) $X \sim X_2$

(3) $S \in 1-1$

(4) $D_l(S) = X$ {D7.2, D7.1, L7.2, 2}

(5) $X_2 = S(X)$

(6) $S(X) \subset X_1$ {5, 1}

---
[1] Cf. Exercise 1b, Sec. 7.

We shall now define by induction the sequence of sets:

(7) $\quad Z_1 = X_1 - S(X)$

(8) $\quad \prod_{i \in N} Z_{i+1} = S(Z_i)$

(9) $\quad Z = \bigcup_{i=1}^{\infty} Z_i$ $\qquad\qquad$ {def.}

(10) $\quad X_1 = Z_1 \cup S(X)$ $\qquad$ {formula 8c, p. 173, 6, 7}

(1.1) $\quad Z_i \subset X_1$ $\qquad\qquad\qquad$ {a.p.}

(1.2) $\quad S(Z_i) \subset S(X_1)$ $\qquad\qquad$ {T6.1a, 1.1}

(1.3) $\quad S(X_1) \subset S(X)$ $\qquad\qquad$ {T6.1a, 1}

(1.4) $\quad Z_{i+1} \subset X_1$ $\qquad\qquad$ {8, 1.2, 1.3, 6}

(11) $\quad \prod_{i \in N} Z_i \subset X_1$ $\qquad\qquad$ {7, 1.1 → 1.4}

(12) $\quad Z \subset X_1$ $\qquad\qquad\qquad$ {T4.2a, 11, 9}

(13) $\quad Z \subset X$ $\qquad\qquad\qquad$ {12, 1}

(14) $\quad Z = Z_1 \cup S(Z)$ $\qquad\qquad$ {T6.1d, 8, 9}

(15) $\quad R = I_Z \cup S_{X-Z}$ $\qquad\qquad$ {Df.}

We shall now demonstrate that the relation R establishes equinumerosity of the sets X and X_1.

(16) $\quad D_l(I_Z) = Z$ $\qquad\qquad\qquad$ {formula 8, p. 201}

(17) $\quad D_l(S_{X-Z}) = X - Z$ $\qquad\qquad$ {L6.5a, 4}

(18) $\quad D_l(R) = Z \cup (X - Z) = X \quad$ {L6.9a, 15, 16, 17, 13,

$\qquad\qquad\qquad\qquad\qquad\qquad$ formula 8a, p. 173}

(19) $\quad D_r(I_Z) = Z$ $\qquad\qquad\qquad$ {formula 8, p. 201}

(20) $\quad D_r(S_{X-Z}) = S(X - Z)$ $\qquad$ {L6.5b, 4}

(21) $\quad S(X - Z) = S(X) - S(Z)$ $\quad$ {L6.3c, D6.6, 3}

(22) $\quad D_r(R) = Z_1 \cup S(Z) \cup [S(X) - S(Z)] = Z_1 \cup S(X) = X_1$

$\qquad\qquad\qquad\qquad\qquad\qquad$ {L6.9b, 15, 19, 14, 20,

$\qquad\qquad\qquad\qquad\qquad\qquad$ 21, 10, 12, formula 8j,

$\qquad\qquad\qquad\qquad\qquad\qquad\qquad\qquad$ p. 173}

(23) $\quad I_Z \in 1 - 1$ $\qquad\qquad\qquad$ {L6.4, $I \in 1 - 1$}

(24) $\quad S_{X-Z} \in 1 - 1$ $\qquad\qquad\qquad$ {L6.4, 3}

(25) $\quad (D_l(I_Z)) \cap (D_l(S_{X-Z})) = Z \cap (X - Z) = \varnothing$

$\qquad\qquad\qquad\qquad\qquad\qquad$ {16, 17, formula 1i,

$\qquad\qquad\qquad\qquad\qquad\qquad\qquad\qquad$ p. 171}

(26) $\quad Z_1 \cap S(X) = \varnothing$ $\qquad\qquad$ {formula 8e, p. 173}

$$(27) \quad \left(D_r(I_Z)\right) \cap \left(D_r(S_{X-Z})\right)$$
$$= \left(Z_1 \cup S(Z)\right) \cap \left(S(X) - S(Z)\right) = \varnothing$$

$\{19, 14, 20, 21, \text{formula } 8i, \text{ p. } 173\}$

(28) $\quad R \in 1 - 1$ $\{T6.5, 15, 23, 24, 25, 27\}$

(29) $\quad X \sim_R X_1$ $\{D7.1, 28, 18, 22\}$

       $X \sim X_1$ $\{D7.2, 29\}$

T9.2b. $\mathfrak{m} \leqslant \mathfrak{n} \wedge \mathfrak{n} \leqslant \mathfrak{m} \to \mathfrak{m} = \mathfrak{n}.$

Proof. (1) $\mathfrak{m} \leqslant \mathfrak{n}$ $\left.\right\}$ $\{a.\}$

      (2) $\mathfrak{n} \leqslant \mathfrak{m}$

      (3) $\overline{\overline{X}} = \mathfrak{m} \wedge \overline{\overline{Y}} = \mathfrak{n}$ $\{T7.2c\}$

      (4) $X_1 \subset X$ $\left.\right\}$ $\{D9.1, 2, 3\}$

      (5) $Y \sim X_1$

      (6) $\overline{\overline{Y}} = \overline{\overline{X}}_1$ $\{A7.1, 5\}$

      (7) $\overline{\overline{X}} \leqslant \overline{\overline{X}}_1$ $\{1, 3, 6\}$

      (8) $X_2 \subset X_1$ $\left.\right\}$ $\{D9.1, 7\}$

      (9) $X \sim X_2$

    (10) $X \sim X_1$ $\{T9.2a, 4, 9\}$

        $\mathfrak{m} = \mathfrak{n}$ $\{10, 3, 6\}$

Theorem T9.2b is called the *Cantor-Bernstein theorem*.

T9.3a. $\mathfrak{m} < \mathfrak{n} \to \mathfrak{m} \neq \mathfrak{n},$

    b. $\mathfrak{n} < \mathfrak{m} \to \mathfrak{m} \neq \mathfrak{n},$

    c. $\mathfrak{m} = \mathfrak{n} \to \neg\,(\mathfrak{m} < \mathfrak{n}),$

    d. $\mathfrak{m} = \mathfrak{n} \to \neg\,(\mathfrak{n} < \mathfrak{m}),$

    e. $\mathfrak{m} < \mathfrak{n} \to \neg\,(\mathfrak{n} < \mathfrak{m}).$

Proof. Formulae a and b follow from D9.2. From them, by contraposition we obtain formulae c and d.

The proof of formula e is as follows:

(1) $\mathfrak{m} < \mathfrak{n}$ $\{a.\}$

(2) $\mathfrak{n} < \mathfrak{m}$ $\{a.i.p.\}$

(3) $\mathfrak{m} \neq \mathfrak{n}$ $\left.\right\}$ $\{D9.2, 1\}$

(4) $\mathfrak{m} \leqslant \mathfrak{n}$

(5) $\mathfrak{m} = \mathfrak{n}$ $\{D9.2, T9.2b, 2, 4\}$

    contr. $\{3, 5\}$

T9.4. $\qquad\qquad\qquad$ $\mathfrak{m} \leqslant \mathfrak{n} \lor \mathfrak{n} \leqslant \mathfrak{m}.$

We omit the proof of the theorem and confine ourselves to indicating its fundamental idea. Let $\overline{\overline{X}} = \mathfrak{m}$ and $\overline{\overline{Y}} = \mathfrak{n}$. To prove T9.4 we have to show that there always exists either a subset U of the set Y and a relation R such that $X \sim_R U$, or a subset Z of the set X and a relation S such that $Y \sim_S Z$. But the assumptions so far adopted are not sufficient to demonstrate this. A new axiom, called the *axiom of choice* (to be formulated in Sec. 14) is required.

It follows from T9.4 and T9.1 that for any two cardinal numbers one of the following relations is satisfied:

$$\mathfrak{m} < \mathfrak{n}, \qquad \mathfrak{n} < \mathfrak{m}, \qquad \mathfrak{m} = \mathfrak{n}.$$

From T9.3 it follows that at most one of these relations holds. This fact can be stated briefly: the principle of trichotomy is satisfied for cardinal numbers.

From the fact that $\aleph_0 = \overline{\overline{\mathbf{N}}}$ and $\mathbf{c} = \overline{\overline{\mathbf{R}}}$ and from Conclusion 9.1 together with D9.2 it follows that

T9.5a. $\qquad\qquad\qquad$ $\aleph_0 < \mathbf{c},$

b. $\qquad$ *If n is a natural number, then* $n < \aleph_0.$

The following lemma will be used in later Sections:

L9.3a. $\qquad\qquad$ $\mathfrak{m} < \mathfrak{n} \to \mathfrak{m}.\mathfrak{k} \leqslant \mathfrak{n}.\mathfrak{k},$

b. $\qquad\qquad$ $\mathfrak{m} < \mathfrak{n} \to \mathfrak{m}^{\mathfrak{k}} \leqslant \mathfrak{n}^{\mathfrak{k}}.$

Proof. a. (1) $\quad \mathfrak{m} < \mathfrak{n}$ $\qquad\qquad\qquad$ {a.}

(2) $\quad \overline{\overline{X_1}} = \mathfrak{m} \land \overline{\overline{Y_1}} = \mathfrak{n} \land \overline{\overline{T_1}} = \mathfrak{k}$ $\qquad$ {T7.2c}

(3) $\quad Z_1 \subset Y_1 \land X_1 \sim Z_1$ $\qquad$ {D9.2, D9.1, 1, 2}

(4) $\quad \overline{\overline{Z_1}} = \mathfrak{m}$ $\qquad\qquad\qquad$ {A7.1, 3, 2}

(5) $\quad Z_1 \times T_1 \subset Y_1 \times T_1$ $\qquad\qquad$ {T5.1d, 3}

(6) $\quad \overline{\overline{Z_1 \times T_1}} \leqslant \overline{\overline{Y_1 \times T_1}}$ $\qquad$ {Conclusion 9.1, 5}

$\qquad\quad \mathfrak{m}.\mathfrak{k} \leqslant \mathfrak{n}.\mathfrak{k}$ $\qquad\qquad\qquad$ {D8.2, 2, 4}

In the proof of formula b we make use of Conclusion 8.1.

Exercises

1 Prove the formulae:

(a) $\overline{\overline{X}} < \overline{\overline{Y}} \to \overline{\overline{X+Z}} \leqslant \overline{\overline{Y+Z}},$

(b) $\neg (\mathfrak{n} < \mathfrak{n}),$

(c) $\mathfrak{m} < \mathfrak{n} \wedge \mathfrak{n} < \mathfrak{k} \rightarrow \mathfrak{m} < \mathfrak{k}$,

(d) $\mathfrak{m} \leqslant \mathfrak{n} \wedge \mathfrak{n} \leqslant \mathfrak{k} \rightarrow \mathfrak{m} \leqslant \mathfrak{k}$.

Show by examples that in the consequent of implication (a) the ordinary inequality may not be replaced by a proper one.

2. Prove the following theorem:

$$\aleph_0 < \overline{\overline{X}} \equiv \sum_{Y \subset X} \overline{\overline{Y}} = \aleph_0.$$

10. Power Sets

D10.1. $X \in 2^Y \equiv X \subset Y.$

The set 2^Y is called a *power set*. Thus the elements of the power set 2^Y are all the subsets of the set Y.

L10.1. $\overline{\overline{M}} \leqslant \overline{\overline{(2^M)}}.$

Proof. Let M_1 be the family of all unit sets $\{m\}$ whose only element m belongs to M, and of such sets only. Hence

$$\{m\} \in M_1 \equiv m \in M.$$

It is evident that each of the sets $\{m\}$ is a subset of the set M. Consequently

(1) $M_1 \subset 2^M.$

Denote by R_1 the relation which associates the unit set $\{m\}$ with every element m of the set M. It is clear that this relation shows that sets M and M_1 are equinumerous. The lemma follows from this fact, formula (1) and Conclusion 9.1.

Let R be any function which associates sets of individuals with individuals. We now define the set $Z(R)$, dependent on the function R:

D10.2. $R \in \text{funct} \rightarrow [x \in Z(R) \equiv x \in D_l(R) \wedge x \notin R(x)].$

The element x of the left domain of relation R belongs to the set $Z(R)$ if and only if it is not an element of the set which is associated with it by function R. This definition will be used in the proofs of further lemmas.

L10.2. $R \in \text{funct} \rightarrow Z(R) \notin D_r(R).$

Proof. (1) $R \in \text{funct}$ {a.}

(2) $Z(R) \in D_r(R)$ {a.i.p.}

It follows from formula (2) that there exists an element x_1 of the left domain of relation R, with which the set $Z(R)$ is associated. Hence the following formulae are satisfied:

$$(3) \quad x_1 \in D_l(R)$$
$$(4) \quad Z(R) = R(x_1)$$
$$(5) \quad x_1 \in Z(R) \equiv x_1 \notin R(x_1) \qquad \{\text{D10.2, 1, 3}\}$$
$$(6) \quad x_1 \in Z(R) \equiv x_1 \notin Z(R) \qquad \{5, 4\}$$

It is obvious that formula (6) gives a contradiction.

L10.3. $$\overline{\overline{M}} \neq \overline{\overline{(2^M)}}.$$

Proof.

(1)	$\overline{\overline{M}} = \overline{\overline{(2^M)}}$	$\{\text{a.i.p.}\}$
(2)	$M \sim_{R_1} 2^M$	$\{\text{A7.1, D7.2, 1}\}$
(3)	$R_1 \in \text{funct}$	
(4)	$D_l(R_1) = M$	$\{\text{D7.1, 2}\}$
(5)	$D_r(R_1) = 2^M$	
(6)	$Z(R_1) \subset D_l(R_1)$	$\{\text{D10.2, 3}\}$
(7)	$Z(R_1) \subset M$	$\{6, 4\}$
(8)	$Z(R_1) \in 2^M$	$\{\text{D10.1, 7}\}$
(9)	$Z(R_1) \in D_r(R_1)$	$(8, 5)$
(10)	$Z(R_1) \notin D_r(R_1)$	$\{\text{L10.2, 3}\}$
	contr.	$\{9, 10\}$

T10.1. $$\overline{\overline{M}} < \overline{\overline{(2^M)}} \qquad \{\text{D9.2, L10.1, L10.3}\}$$

The power of any set is less than the power of the set of all its subsets.

The following definition will be needed in the proof of the next theorem:

D10.3. Let M be any set, and let $A \subset M$. The function $\varphi_A(m)$ is called the *characteristic function* of the set A if and only if, for every M, this function satisfies the condition

$$\varphi_A(m) = \begin{cases} 1, & \text{when} \quad m \in A, \\ 0, & \text{when} \quad m \notin A. \end{cases}$$

T10.2. $$2^{\overline{\overline{M}}} = \overline{\overline{(2^M)}}$$

Proof. It follows from D8.3 and D10.3 that every characteristic function of a set $A \subset M$ is a mapping of the set M into the set $\{0, 1\}$

and conversely, every mapping of the set M into the set $\{0, 1\}$ is a characteristic function of a subset of the set M. Let us associate with every mapping of the set M into the set $\{0, 1\}$ the subset of the set M which has the mapping as its characteristic function. It can easily be seen that this correspondence establishes that the set $\{0, 1\}^M$ of all mappings of the set M into the set $\{0, 1\}$ is equinumerous with the set 2^M of all subsets of the set M. Thus the following formula is satisfied:

(1) $\quad \{0, 1\}^M \sim 2^M$

(2) $\quad \overline{\overline{\{0, 1\}^M}} = \overline{\overline{(2^M)}}$ $\hspace{3cm}$ $\{\text{A7.1, 1}\}$

(3) $\quad \overline{\overline{\{0, 1\}^M}} = \overline{\overline{\{0, 1\}}}^{\overline{\overline{M}}} = 2^{\overline{\overline{M}}}$ $\hspace{2cm}$ $\{\text{D8.4}\}$

$\qquad 2^{\overline{\overline{M}}} = \overline{\overline{(2^M)}}.$ $\hspace{4cm}$ $\{2, 3\}$

T10.3. $\hspace{3cm}$ $\overline{\overline{M}} < 2^{\overline{\overline{M}}}$ $\hspace{2cm}$ $\{\text{T10.1, T10.2}\}$

This theorem can also be written

$$\mathfrak{m} < 2^{\mathfrak{m}}.$$

T10.4. *There exist infinitely many (different) cardinal numbers none of which is a natural number.*

Proof. To demonstrate the truth of this theorem it suffices to note that by T10.3 every term of the sequence

$$\aleph_0, \; 2^{\aleph_0}, \; 2^{\left(2^{\aleph_0}\right)}, \; \ldots$$

is greater than the preceding term.

Exercises

1. Prove the following theorem by induction:

$$n \in \mathbf{N} \wedge n = \overline{\overline{M}} \to 2^n = \overline{\overline{(2^M)}}.$$

2. Let the sequence

$$\mathfrak{m}_1, \mathfrak{m}_2, \ldots$$

be any sequence of cardinal numbers such that, for every i, $\mathfrak{m}_i < \mathfrak{m}_{i+1}$. Let $M_1, M_2, \ldots$ be a sequence of sets which for every i satisfies the condition

$\overline{\overline{M_i}} = \mathfrak{m}_i$. Prove that the power of the set $\bigcup\limits_{i=1}^{\infty} M_i$ is greater than any of the numbers $\mathfrak{m}_1, \mathfrak{m}_2, \ldots$.

11. Infinite Sets in Dedekind's Sense

We usually interpret the concept of infinite sets in such a way that those and only those non-empty sets whose powers are not natural numbers are infinite sets. We shall now give a definition of an infinite set, in which the concept of a natural number does not occur.[1] This definition is due to the German mathematician Richard Dedekind (1831–1916).

D11.1. X is an *infinite set in Dedekind's sense* $\equiv \sum\limits_{\substack{Y \subseteq X \\ \neq}} Y \sim X$.

Thus for a set to be infinite in Dedekind's sense it is a necessary and sufficient condition that it be equinumerous with a proper part of itself.

D11.2. The powers of the infinite sets in Dedekind's sense are called *transfinite cardinal numbers*.

It is evident that the power of any set infinite in Dedekind's sense is not zero nor any natural number. The question arises whether, conversely, any set whose power is not zero nor any natural number is infinite in Dedekind's sense. This theorem seems quite obvious, but the proof requires a new axiom, called the *axiom of choice*, to be discussed in Section 14.

In the proof of the following lemma we shall refer to four theorems from the algebra of sets (cf. Exercise 8, c, d, e, f in Sec. 1):

I. $Y \subset X \wedge Z = X - Y \rightarrow X = Z \cup Y$,

II. $X \underset{\neq}{\subseteq} Y \wedge Y \cap Z = \emptyset \rightarrow Z \cup X \underset{\neq}{\subseteq} Z \cup Y$,

III. $Z = X - Y \rightarrow Y \cap Z = \emptyset$,

IV. $X \subset Y \wedge Y \cap Z = \emptyset \rightarrow X \cap Z = \emptyset$.

L11.1. $P \subset X \wedge \overline{\overline{P}} = \aleph_0 \rightarrow \sum\limits_{\substack{Y \subseteq X \\ \neq}} X \sim Y$.

Proof. (1) $P \subset X$ ⎫
                                              $\{a.\}$
         (2) $\overline{\overline{P}} = \aleph_0$ ⎭

---

[1] The importance of the fact that the concept of an infinite set can be defined without reference to the concept of a natural number is brought out in Sec. 6 of the *Supplement*.

It follows from (2) and T7.3 that there exists an infinite sequence

(a) $$p_1, p_2, p_3, \ldots$$

which ranges over the set P and in which no term is repeated. Denote by P_1 the range of the sequence

(b) $$p_2, p_4, p_6, \ldots$$

It is clear that

(3) $P_1 \underset{\neq}{\subseteq} P$.

It is also easily seen that the relation which associates with every term of sequence (a) a term of sequence (b) with an index which is twice as great establishes that the sets P and P_1 are equinumerous. Hence

(4) $P \sim P_1$

(5) $Z = X - P$ ⎱
(6) $X_1 = Z \cup P_1$ ⎰ {def.}

(7) $X = Z \cup P$ {I, 1, 5}

(8) $P \cap Z = \emptyset$ {III, 5}

(9) $P_1 \cap Z = \emptyset$ {IV, 3, 8}

(10) $Z \cup P \sim Z \cup P_1$ {L7.5a, T7.1a, 4, 8, 9}

(11) $X \sim X_1$ {10, 6, 7}

(12) $X_1 \underset{\neq}{\subseteq} X$ {II, 3, 8, 6, 7}

$\displaystyle\sum_{Y \underset{\neq}{\subseteq} X} X \sim Y$ {11, 12}

L11.2. $\displaystyle\sum_{Y \underset{\neq}{\subseteq} X} (X \sim Y) \to \sum_{P \subset X} (\overline{\overline{P}} = \aleph_0).$

Proof. (1) $\displaystyle\sum_{Y \underset{\neq}{\subseteq} X} X \sim Y$ {a.}

(2) $Y_1 \underset{\neq}{\subseteq} X$ ⎱
(3) $X \sim Y_1$ ⎰   {1}

(4) $X \sim_{R_1} Y_1$ {D7.2, 3}

(5) $R_1 \in 1-1 \;\wedge\; X = D_l(R_1) \;\wedge\; Y_1 = D_r(R_1)$ {D7.1, 4}

(6) $R_1 \in \text{funct}$ {D6.6, 5}

Note that for every $x \in X$ the symbol $R_1(x)$ is meaningful (by D6.3, 5, 6).

Denote by p_1 any term such that

(7) $p_1 \in X \;\wedge\; p_1 \notin Y_1$.

The existence of such a term is established by formula (2).

(8) $x \in D_l(R_1) \to R_1(x) \in D_r(R_1)$ {Conclusion 6.3c, 6}

(9) $x \in X \to R_1(x) \in Y_1$ {8, 5}

The sequence

(a) $p_1, p_2, p_3, \ldots$

is now defined as follows:

the term p_1 is already defined {(cf. 7)}

(10) $k > 1 \to p_k = R_1(p_{k-1})$

From formulae (7), (9), (10) and (2) it follows by induction that

(11) $p_i \in X, \quad i = 1, 2, 3, \ldots$

(12) $r > s \wedge s = 1 \to p_r \neq p_s$ {7, 9, 10, 11}

(2.1) $r > s \wedge s > 1$ }
(2.2) $p_{r-1} \neq p_{s-1}$ }   {ad. a.}

(2.3) $R_1(p_{r-1}) \neq R_1(p_{s-1})$ {L6.3b, 11, 5, 2.2}

(2.4) $p_r \neq p_s$.   {2.1, 2.3, 10}

From (12) and the fact that formulae (2.1) and (2.2) imply formula (2.4) we obtain by induction:

(13) $r > s \to p_r \neq p_s$.

Hence no term is repeated in sequence (a). Denote the range of that sequence by P. By T7.3 we obtain

(14) $\overline{\overline{P}} = \aleph_0$

(15) $P \subset X$ {11}

$\sum_{P \subset X} \overline{\overline{P}} = \aleph_0$.   {14, 15}

T11.1. *The set X is infinite in Dedekind's sense* $\equiv \sum_{P \subset X} \overline{\overline{P}} = \aleph_0$

{L11.1, L11.2, D11.1}

Thus for a set to be infinite in Dedekind's sense it is a necessary and sufficient condition for it to have an enumerable part.

From T11.1, D11.2 and Conclusion 9.1 follows

CONCLUSION 11.1. *The cardinal number* $\mathfrak{m}$ *is transfinite* $\equiv \aleph_0 \leqslant \mathfrak{m}$.

From this and T9.5a follows

CONCLUSION 11.2. *The cardinal numbers $\aleph_0$ and c are transfinite.*

L11.3. $$\bar{\bar{X}} = \bar{\bar{Y}} = \aleph_0 \rightarrow \overline{\overline{X \cup Y}} = \aleph_0.$$

Proof. To prove this lemma it is is sufficient to assume that the set X is the set of all odd natural numbers, and the set Y is the set of all even natural numbers.

L11.4. $$\bar{\bar{X}} = \aleph_0 \wedge \bar{\bar{Y}} = n \rightarrow \overline{\overline{X \cup Y}} = \aleph_0.^{(1)}$$

To prove this lemma it is sufficient to assume that the elements of the set Y are the numbers $1, 2, ..., n$, and the elements of the set X are the numbers $n+1, n+2, ...$

T11.2. *If X is an infinite set in Dedekind's sense and $\bar{\bar{Y}} = n^{(1)}$ or $\bar{\bar{Y}} = \aleph_0$, then*

(1) $$\overline{\overline{X \cup Y}} = \bar{\bar{X}}.$$

Proof. It follows from T11.1 that there exists an enumerable part P of the set X. From L11.3 and L11.4 we have that

(2) $$P \sim P \cup Y.$$

The formula

(3) $$X - P \sim X - P$$

is, of course, satisfied. We may assume that

$$X \cap Y = \emptyset.$$

It follows from formulae (2), (3) and (4) and L7.5a that

$$\overline{(X-P) \cup P} = \overline{(X-P) \cup (P \cup Y)},$$

from which we obtain

$$\bar{\bar{X}} = \overline{\overline{X \cup Y}},$$

i.e. formula (1) is true.

T11.3. *If $\bar{\bar{Y}} = n^{(1)}$ or $\bar{\bar{Y}} = \aleph_0$ and $X - Y$ is an infinite set in Dedekind's sense, then*

(1) $$\overline{\overline{X - Y}} = \bar{\bar{X}}.$$

---

(1) n stands here for any natural number.

Proof. From T11.2 and the assumptions adopted it follows that

(2) $$\overline{\overline{(X-Y) \cup Y}} = \overline{\overline{X-Y}}.$$

It is obvious that

$$X \subset (X-Y) \cup Y.$$

From this and conclusion 9.1 we obtain

$$\overline{\overline{X}} \leqslant \overline{\overline{(X-Y) \cup Y}}.$$

From this formula and formula (2) it follows that

(3) $$\overline{\overline{X}} \leqslant \overline{\overline{X-Y}}.$$

By referring once more to Conclusion 9.1 we obtain

$$\overline{\overline{X-Y}} \leqslant \overline{\overline{X}}.$$

Hence by formula (3), and T9.2, equation (1) follows.

The last two theorems may be formulated verbally as follows:
The power of a set which is infinite in Dedekind's sense does not change when any finite or enumerable number of elements is joined to it.

The power of a set does not change when we remove from it any finite or enumerable number of elements, provided that the set obtained in this way is infinite in Dedekind's sense.

It is obvious that in such a case the set from which we remove a finite or enumerable number of elements is infinite in Dedekind's sense.

12. The Cardinal Numbers $\aleph_0$ and c

As was mentioned above, the cardinal numbers $\aleph_0$ and c are defined by the formulae

$$\aleph_0 = \overline{\overline{N}},$$

$$c = \overline{\overline{R}},$$

where N is the set of all natural numbers, and R is the set of all real numbers.

It should also be remembered that the numbers $\aleph_0$ and c are transfinite (Conclusion 11.2).

The following theorem, given by Cantor, is chronologically one of the earliest theorems of set theory. In the proof, which scarcely differs from Cantor's original proof, we shall make use of results from the theory of numbers

T12.1. $\aleph_0 \neq \mathbf{c}$.

Proof. (1) $\aleph_0 = \mathbf{c}$ {a.i.p.}

From this and T7.3 it follows that there exists a sequence

(a) $x_1, x_2, x_3, \ldots$

in which no term is repeated and which ranges over the set **R**.

It follows from theorems of arithmetic that every real number has exactly one proper expansion[1] into an infinite decimal. Obviously from a certain stage onwards all the digits in this expansion may be 0.

We write out the proper expansions of the numbers of sequence (a):

$$x_1 = c_1, y_{11} \, y_{12} \, y_{13} \cdots y_{1n} \cdots$$
$$x_2 = c_2, y_{21} \, y_{22} \, y_{23} \cdots y_{2n} \cdots$$
$$x_3 = c_3, y_{31} \, y_{32} \, y_{33} \cdots y_{3n} \cdots$$
$$\cdot \quad \cdot \quad \cdot \quad \cdot \quad \cdot \quad \cdot \quad \cdot \quad \cdot \quad \cdot \quad \cdot \quad \cdot$$
$$x_n = c_n, y_{n1} \, y_{n2} \, y_{n3} \cdots y_{nn} \cdots$$
$$\cdot \quad \cdot \quad \cdot \quad \cdot \quad \cdot \quad \cdot \quad \cdot \quad \cdot \quad \cdot \quad \cdot \quad \cdot$$

The symbols

$$c_1, c_2, \ldots$$

here stand for the integral parts of the numbers $x_1, x_2, \ldots$. The symbol y_{ij} stands for the jth digit after the decimal point in the decimal expansion of the number x_i.

We define the sequence

$$z_n = \begin{cases} 1, & \text{when } y_{nn} \neq 1, \\ 2, & \text{when } y_{nn} = 1. \end{cases}$$

It is clear that

(2) $0, z_1 \, z_2 \, z_3 \ldots$

is the proper expansion of a real number as an infinite decimal.

It can easily be seen that for every natural number n, the nth digit of expansion (2) differs from the nth digit of the expansion of the

---

[1] An improper expansion is an expansion in which from a certain stage onwards, only the digit 9 occurs. A proper expansion is any expansion which is not improper.

number x_n. This conclusion contradict the previous conclusion that the set $\mathbf{R}$ is the range of sequence (a). Thus assumption (1) leads to a contradiction.

From D9.2, T9.5a, and T12.1, the inequality $\aleph_0 < \mathbf{c}$ follows.

T12.2. $$ X \subset \mathbf{R} \wedge \overline{\overline{X}} = \aleph_0 \rightarrow \overline{\overline{\mathbf{R}-X}} = \mathbf{c}. $$

Thus the power of the set of all real numbers does not change when an enumerable set of elements is removed from it.

Proof. (1) $X \subset \mathbf{R}$ ⎫

(2) $\overline{\overline{X}} = \aleph_0$ ⎭   {a.}

It follows from T7.3, (1) and (2) that there exists a sequence

(a) $$ x_1, x_2, x_3, \ldots $$

of distinct real numbers whose range is the set X. As in the proof of T12.1, we conclude that there exists a real number z_1 which differs from every term of sequence (a). Hence

$$ z_1 \in \mathbf{R}-X. $$

The real number z_2 is defined as a number different from every number in the sequence

$$ z_1, x_1, x_2, \ldots $$

By proceeding in this way we obtain the infinite sequence

$$ z_1, z_2, z_3, \ldots $$

of distinct real numbers, each of which is an element of the set $\mathbf{R}-X$. Thus, by T7.3 and T11.1, $\mathbf{R}-X$ is infinite in Dedekind's sense. Hence from formula (2) and T11.3 it follows that

$$ \overline{\overline{\mathbf{R}-X}} = \overline{\overline{\mathbf{R}}} = \mathbf{c}. $$

This establishes the theorem.

L12.1. $$ \underset{x}{E}\,(-1 < x < 1) \sim \mathbf{R}.^{[1]} $$

---

[1] The symbol $\underset{x}{E}(-1 < x < 1)$ indicates an open interval with end-points -1 and 1, that is the set of all real numbers between -1 and 1. This set is usually represented $(-1, 1)$.

That the sets $E_x(-1 < x < 1)$ and $\mathbf{R}$ are equinumerous is determined by the relation

$$x = \frac{y}{1+|y|} \wedge y \in \mathbf{R}.$$

L12.2. $a < b \to E_x (a < x < b) \sim E_y(-1 < y < 1).$

Proof. The relation used to establish this result is

$$x = \frac{b-a}{2} y + \frac{b+a}{2} \wedge -1 < y < 1.$$

T12.3. $a < b \to \overline{\overline{E_x (a < x < b)}} = \mathbf{c}$ {L12.1, L12.2}

Thus every non-empty interval of real numbers is a set of the continuum power.

In the following formula n stands for any natural number.

T12.4a.
   b.
   c.
   d.
   e.

$$\left.\begin{array}{c} \aleph_0+n = \aleph_0 \\ \aleph_0+\aleph_0 = \aleph_0 \\ \mathbf{c}+n = \mathbf{c} \\ \mathbf{c}+\aleph_0 = \mathbf{c} \\ \mathbf{c}+\mathbf{c} = \mathbf{c} \end{array}\right\} \quad \{\text{conclusion } 11.2, \text{ T11.2}\}$$

Proof. e. (1)   Let a, b, c be real numbers which satisfy the condition $a < b < c$

(2) $E_x (a < x < b) \cup \{b\} \cup E_x (b < x < c)$

$= E_x(a < x < c)$ {1}

(3) $\overline{\overline{E_x (a < x < b) \cup \{b\} \cup E_x (b < x < c)}}$

$= \overline{\overline{E_x (a < x < b)}} + \overline{\overline{\{b\}}} + \overline{\overline{E_x (b < x < c)}}$

{D8.1, 2}

(4) $\mathbf{c} = \mathbf{c}+1+\mathbf{c}$ {T12.3, 1, 3}

$\mathbf{c}+\mathbf{c} = \mathbf{c}$ {T12.4c, 4}

From T12.3 and T12.4c follows

Conclusion 12.1a. $a < b \to \overline{\overline{E(a \leqslant x < b)}} = \mathbf{c}.$
$\qquad\qquad\qquad\quad\; x$

b. $a < b \to \overline{\overline{\underset{x}{E}(a < x \leqslant b)}} = \mathbf{c}.$

c. $a < b \to \overline{\overline{\underset{x}{E}(a \leqslant x \leqslant b)}} = \mathbf{c}.$

T12.5. *Let*

$$Z_1, Z_2, Z_3, \ldots$$

be any infinite sequence of pairwise disjoint sets.
If

$$\overline{\overline{Z_i}} = \aleph_0, \quad i = 1, 2, \ldots$$

then

$$\overline{\overline{\bigcup_{i=1}^{\infty} Z_i}} = \aleph_0.$$

Proof. Let

$$z_1^{(i)}, z_2^{(i)}, z_3^{(i)}, \ldots$$

be an infinite sequence in which no term is repeated and whose range is the set Z_i. It can easily be seen that no term of the sequence

$$z_1^{(1)}, z_1^{(2)}, z_2^{(1)}, z_1^{(3)}, z_2^{(2)}, z_3^{(1)}, z_1^{(4)}, \ldots$$

is repeated and that the range of this sequence is the set $\bigcup_{i=1}^{\infty} Z_i$.

From T7.3 the theorem then follows.

L12.3. *If the sum of an enumerable number of finite sets is infinite, then it is enumerable.*

The proof of this lemma, which is analogous to the proof of T12.5, is left to the reader.

L12.4. *The set of all finite sequences whose terms are numbers*

(1) $0, 1, \ldots, n$

is enumerable.

Proof. Let $Z_i \, (i = 1, 2, \ldots)$ be the set of all sequences of i terms, where the terms are chosen from (1). It is clear that for every i the set Z_i is finite, whilst the sum of all these sets will be infinite. Hence from L12.3 the lemma follows.

T12.6. $\aleph_0 \cdot \aleph_0 = \aleph_0.$

Proof. Let P stand for the set of all ordered pairs $\langle m, n \rangle$ of natural numbers, and let P_i $(i = 2, 3, \ldots)$ stand for the set of all such pairs which also satisfy the condition $m+n = i$. It is clear that

$$P = \bigcup_{i=2}^{\infty} P_i$$

and that the set P is infinite. It can also easily be seen that the sets $P_2, P_3, \ldots$ are finite. Hence from L12.3 it follows that

(1) $\overline{\overline{P}} = \aleph_0.$

On the other hand, it follows from the definition of set P and D8.2 that

(2) $\overline{\overline{P}} = \aleph_0 \cdot \aleph_0,$

$$\aleph_0 \cdot \aleph_0 = \aleph_0. \qquad\qquad \{1, 2\}$$

T12.7. *If n is a natural number* $\geqslant 2$, *then* $n^{\aleph_0} = \mathbf{c}$.

The proof will be given for the case $n = 10$. The general proof does not differ essentially from the proof in this particular case.
Proof. Let $\mathbf{N}_1$ stand for the set of non-negative integers $\leqslant 9$, so that $\overline{\overline{\mathbf{N}_1}} = 10$. Hence from D8.4 it follows that

(1) $\overline{\overline{(\mathbf{N}_1^{\overline{\overline{\mathbf{N}}}})}} = 10^{\aleph_0}.$

From D8.3 the mapping of the set $\mathbf{N}$ into the set $\mathbf{N}_1$ uniquely determines an infinite sequence whose terms are the numbers $0, 1, \ldots, 9$. Denote by C the set of all such sequences.
Thus from (1) it follows that

(2) $\overline{\overline{C}} = 10^{\aleph_0}.$

Denote by C_1 the set of all sequences belonging to C, in which from a certain stage onwards, each term is equal to 9. It follows from L12.4 that

(3) $\overline{\overline{C_1}} = \aleph_0.$

Clearly the formula

(4) $C_1 \subset C$

is also satisfied.

It follows from results in number theory that to every sequence belonging to $C-C_1$ there corresponds exactly one number from the interval $E(0 \leqslant x < 1)$. [1] Hence from Conclusion 12.1a it follows that

(5) $$\overline{\overline{C-C_1}} = \mathbf{c}.$$

Formulae (2)–(5) and T12.4d give the following equations:

$$10^{\aleph_0} = \overline{\overline{C}} = \overline{\overline{(C-C_1) \cup C_1}} = \overline{\overline{C-C_1}} + \overline{\overline{C_1}} = \mathbf{c} + \aleph_0 = \mathbf{c}.$$

This completes the proof of the theorem for the case $n = 10$.

A special case of T12.7 is

T12.8. $2^{\aleph_0} = \mathbf{c}.$

By this theorem the set of all infinite sequences whose terms are the numbers 0 or 1 is of the continuum power.

Note that this theorem and T10.1 immediately give T12.1. But the proof of T12.1, as given above, is significant because of its use of what is called the diagonal method. This method has been used to prove many important mathematical theorems.

In the following formulae n stands for any natural number.

T12.9a. $$n \cdot \aleph_0 = \aleph_0,$$
b. $$\aleph_0^n = \aleph_0,$$
c. $$n \cdot \mathbf{c} = \mathbf{c},$$
d. $$\mathbf{c} \cdot \mathbf{c} = \mathbf{c},$$
e. $$\mathbf{c}^n = \mathbf{c},$$
f. $$\mathbf{c}^{\aleph_0} = \mathbf{c},$$
g. $$\aleph_0 \cdot \mathbf{c} = \mathbf{c},$$
h. $$\aleph_0^{\aleph_0} = \mathbf{c}.$$

---

[1] This number is the infinite decimal fraction having 0 as its integral part and to the right of the decimal point the figures which are, respectively, the terms of the sequence associated with the fraction. Beginning with a certain digit, not all the digits may be 9 (but they may be 0). In a generalized proof of this theorem we would have to refer to the analogous theorem in number theory concerning the expansion of real numbers into infinite fractions for any base $\geqslant 2$. This point is the only one in which the proof of T12.7, as given above, differs from the generalized proof.

Proof. a. $n \cdot \aleph_0 = \underbrace{\aleph_0 + \aleph_0 + \ldots + \aleph_0}_{n} = \aleph_0$ {T8.3, T12.4b}

Proof. b. $\aleph_0^n = \underbrace{\aleph_0 \cdot \aleph_0 \cdot \ldots \cdot \aleph_0}_{n} = \aleph_0$ {T8.4, T12.6}

Proof. c. $n \cdot c = \underbrace{c + c + \ldots + c}_{n} = c$ {T8.3, T12.4e}

Proof. d. $c \cdot c = 2^{\aleph_0} \cdot 2^{\aleph_0} = 2^{\aleph_0 + \aleph_0} = 2^{\aleph_0} = c$ {T12.8, T8.5, T12.4b}

Proof. e. $c^n = \underbrace{c \cdot c \cdot \ldots \cdot c}_{n} = c$ {T8.4, T12.9d}

Proof. f. $c^{\aleph_0} = (2^{\aleph_0})^{\aleph_0} = 2^{\aleph_0 \cdot \aleph_0} = 2^{\aleph_0} = c$ {T12.8, T8.6, T12.6}

Proof. g.   (1) $n < \aleph_0 < c$ {T9.5b, D9.2, T12.1}

(2) $n \cdot c \leqslant \aleph_0 \cdot c \leqslant c \cdot c$ {L9.3a, 1}

(3) $c \leqslant \aleph_0 \cdot c \leqslant c$ {T12.9c, d, 2}

$\aleph_0 \cdot c = c$ {T9.2, 3}

In the analogous proof of formula h, L9.3b is used.

It follows from Theorem T12.9d that the set of ordered pairs of real numbers is equinumerous with the set of real numbers. It is known, however, that the set of points of a straight line is equinumerous with the set of real numbers, and the set of points of a plane is equi-numerous with the set of ordered pairs of real numbers (which are coordinates of those points). Hence the conclusion that the set of points of a plane is equinumerous with the set of points of a straight line, so that there is a one-to-one mapping of the set of points of a plane onto the set of points of a straight line. Cantor's proof of Theorem T12.9d, which seemed paradoxical in view of this conse-quence, led mathematicians to investigate the mappings of a plane onto a straight line. Several years after T12.9d had been proved it was shown that there is no one-to-one continuous mapping of the set of points of a plane onto the set of points of a straight line.

From T12.9h follows

CONCLUSION 12.1. *The set of all infinite sequences whose terms are natural numbers is of the continuum power.*

Likewise, from T12.9f follows

CONCLUSION 12.2. *The set of all infinite sequences whose terms are real numbers is of the continuum power.*

Note that the following equations result from T12,4a, b:

$$\aleph_0 + 1 = \aleph_0 + 2 = \aleph_0 + 3 = \ldots = \aleph_0 + \aleph_0,$$

while the following equations result from T12.4c, d, e:

$$c + 1 = c + 2 = c + 3 = \ldots = c + \aleph_0 = c + c.$$

Thus sums of cardinal numbers may be equal when the first elements are equal and the second are unequal. Hence the operation converse to the addition of cardinal numbers is not uniquely determined. Hence the operation of substraction is not introduced in the study of cardinal numbers.

From T12.9a and T12.6 follow the equations

$$1 \cdot \aleph_0 = 2 \cdot \aleph_0 = 3 \cdot \aleph_0 = \ldots = \aleph_0 \cdot \aleph_0,$$

while from T12.9c, d, g follow the equations

$$1 \cdot c = 2 \cdot c = 3 \cdot c = \ldots = \aleph_0 \cdot c = c \cdot c.$$

Thus the operation converse to the multiplication of cardinal numbers is not uniquely determined either.

Note also that from T12.7 and T12.9f, h it follows that

$$2^{\aleph_0} = 3^{\aleph_0} = 4^{\aleph_0} = \ldots = \aleph_0^{\aleph_0} = c^{\aleph_0}.$$

Thus we see that the operation converse to raising cardinal numbers to powers (extraction of roots) is also not uniquely determined.

Note finally that the following equations result from T12.9b:

$$\aleph_0^1 = \aleph_0^2 = \aleph_0^3 = \ldots,$$

while the following equations result from T12.9e, f:

$$c^1 = c^2 = c^3 = \ldots = c^{\aleph_0}.$$

Thus the second operation converse to raising to powers (namely taking logarithms) is likewise not uniquely determined.

T12.10. *The set of all finite sequences whose terms are integers is enumerable.*

Proof. Let $Z_{i,j}$ $(i = 1, 2, 3, \ldots; j = 0, 1, 2, \ldots)$ stand for the set of all sequences of i elements each, in which the absolute value of the greatest term equals j. It can be seen that for any i and j the set $Z_{i,j}$ is finite. It follows from L12.3 that the set $Z_i = \bigcup_{j=0}^{\infty} Z_{i,j}$ is enumerable. From T12.5 the theorem follows.

T12.11. *The set of all negative integers, the set of all integers, and the set of all rational numbers are enumerable sets. The set of all complex numbers is of the continuum power.*

Proof. With every natural number let us associate a negative number equal to the former in absolute value. It is clear that this association establishes that the set of all negative integers and the set of all natural numbers are equinumerous. Hence the set of all negative integers is enumerable.

The set of all integers can be regarded as the sum of the set of all natural numbers, the set of all negative integers, and the set whose only element is the number 0. This and T12.4a, b show that the set of all integers is enumerable.

Then by D8.2, T12.6 and the fact that the set of all integers is enumerable it follows that the set of all ordered pairs whose first elements are integers and second elements are natural numbers, is also enumerable.

Every rational number may be written as an irreducible fraction having an integer as numerator and a natural number as denominator[1]. Hence it follows that to every rational number there corresponds exactly one ordered pair having an integer as first element and a natural number as second element, such numbers being relatively prime. The set of all such pairs is a subset of the set of all ordered pairs having integers as first elements and natural numbers as second elements—that is, of an enumerable set. Thus the power of the set of all rational numbers $\leqslant \aleph_0$. On the other hand, $\aleph_0 \leqslant$ the power of the set of all rational numbers, because the set N is a subset of the set of all rational numbers. Hence from T9.2 it follows that the set of all rational numbers is enumerable.

That the set of all complex numbers is of the continuum power we conclude immediately from the fact that complex numbers can be represented as ordered pairs of real numbers and from D8.2 and T12.9d.

Exercise

Prove the formula

$$2^c = \aleph_0^c = c^c.$$

---

[1] The number 0 may be represented as $\dfrac{0}{1}$.

13. Cantor's Proof of the Existence of Transcendental Numbers

In higher algebra, those and only those complex numbers which are roots of polynomials with integral coefficients are called *algebraic numbers*. Complex numbers which are not algebraic numbers are called *transcendental numbers*.

The existence of transcendental numbers was first proved in 1851 by the French mathematician Liouville; in 1873 Cantor demonstrated that the set of all transcendental real numbers is of the continuum power. We shall give this proof as an example of the solution of a classical mathematical problem by set-theoretical methods.

L13.1. *The set of all algebraic real numbers is enumerable.*

Proof. It follows easily from T12.10 that the set of polynomials with integral coefficients is enumerable. Hence from the fact that every polynomial has a finite number of real roots, L12.3, and the fact that the set of all algebraic real numbers is infinite (since every rational number is algebraic) the lemma easily follows.

T13.1. *The set of all transcendental real numbers is of the continuum power.*

Proof. This theorem is a direct consequence of T12.2 and L13.1.

14. The Axiom of Choice

In 1904 the German mathematician Ernst Zermelo formulated the following axiom, called the *axiom of choice* or *Zermelo's axiom*:

A14.1. $Z \neq \emptyset \wedge \prod_{M \in Z} (M \neq \emptyset) \wedge \prod_{P \in Z} \prod_{Q \in Z} (P \neq Q \rightarrow P \cap Q = \emptyset)$
$$\rightarrow \sum_Y \prod_{M \in Z} \sum_x {}_1 (x \in M \cap Y).$$

Here is a verbal formulation of this axiom:

We consider the family of sets Z with the properties that

(1) *it is non-empty,*

(2) *every set belonging to the family is also non-empty,*

(3) *any two (different) sets belonging to the family are disjoint.*

By Zermelo's axiom, it follows from these assumptions that there exists at least one set Y which has exactly one element in common with every set $M \in Z$.

As an example of a proof in which A14.1 is used we give the proof
of the following theorem, mentioned previously in Sec. 11:

T14.1. $X \neq \emptyset \wedge \prod_{n \in \mathbf{N}} (\overline{\overline{X}} \neq n) \rightarrow \sum_{Y \subset X} (\overline{\overline{Y}} = \aleph_0).$

By this theorem, *any non-empty set whose power differs from every
natural number includes an enumerable subset, and hence, by* T11.1, *is
an infinite set in Dedekind's sense.*

Proof. Let m be any natural number. Let $C^{(m)}$ stand for the set
of all those sequences of m terms whose terms are the various elements of the set X. We shall show by induction that for any natural
number k the set $C^{(k)}$ is non-empty.

The set X is non-empty by assumption. Hence there is an element
x_1 which belongs to that set. The sequence whose only term is x_1
belongs to the set $C^{(1)}$. Hence $C^{(1)}$ is non-empty.

Let k be any natural number. Assume that the property of the
sets $C^{(m)}$, now being proved, is satisfied for k. Hence there exists
the sequence

(1) $c_1, c_2, \ldots, c_k$

whose terms are the various elements of the set X. Should the set X
not contain an element different from every term of sequence (1)
then the power of set X would be equal to the number k. This conclusion contradicts the second assumption of the theorem to be proved.
Hence there exists an element of set X which is different from every
term of sequence (1). Denote that element by c_{k+1}. The sequence

$$c_1, c_2, \ldots, c_k, c_{k+1}$$

belongs, of course, to the set $C^{(k+1)}$. Hence the set is non-empty.
It is also clear that for $i \neq j$: $C^{(i)} \cap C^{(j)} = \emptyset$.

Thus the sets

(2) $C^{(1)}, C^{(2)}, C^{(3)}, \ldots$

satisfy the assumptions of A14.1. Hence there exists a set C which
has exactly one element in common with each of the sets (2). Let the
sequence $\{c_i^{(m)}\}$ be the only element common to the sets C and $C^{(m)}$.

In the sequence

$$c_1^{(1)}, c_1^{(2)}, c_2^{(1)}, c_1^{(3)}, c_2^{(2)}, c_3^{(1)}, c_1^{(4)}, \ldots$$

we cancel every term which equals a preceding one and denote by $\{y_n\}$ the sequence thus obtained. Suppose for the time being that this sequence is finite and has r terms. It is obvious that in a sequence $\{c_i^{(r+1)}\}$ there must be at least one term different from every term of the sequence $\{y_n\}$. However, this conclusion contradicts the definition of the sequence $\{y_n\}$. Hence that sequence must be infinite. It is also clear that it is the distinct elements of the set X which are the terms of that sequence.

Thus Theorem T14.1 is true.

Note that without A14.1, from the assumptions of T14.1 we could only draw the conclusion that for every natural number n there exists a sequence of n elements whose terms are different elements of the set X. We could not, however, demonstrate the existence of an infinite sequence having that property.

The axiom of choice is probably the most controversial axiom in mathematics. This is due to the fact that, on the one hand, the axiom is used to prove theorems whose truth is obvious and which nevertheless could not be proved without the axiom (for instance, T14.1), and on the other hand, the consequences of the axiom include paradoxical results which violate our intuition. Even among the most eminent mathematicians there are those who favour and those who oppose the adoption of the axiom of choice.

As early as 1902 Beppo Levi drew attention to the fact that the proof of the theorem stating that the power of the sum of a family of disjoint and non-empty sets is greater than or equal to the power of that family requires the possibility of choosing one element from each set belonging to the family. Likewise, the introduction of a definition of the sum of an infinite sequence of cardinal numbers (which is a generalized form of a definition of the sum of two cardinal numbers) assumes the adoption of the axiom. This was first noted by Wacław Sierpiński in a paper on the axiom of choice in 1918.

The same is true of the definition of the product of an infinite sequence of cardinal numbers, and of other operations on cardinal numbers, on order types and ordinal numbers (to be discussed in Chapter II), if the definitions of these operations are to be generalized to cover an infinite number of elements. These examples show that even when introducing fairly elementary concepts of set theory we have to use the axiom of choice.

In other branches of mathematics as well, for instance in analysis, algebra, topology, the axiom of choice, or theorems that are equivalent to it, are needed in the proofs of theorems whose truth is not questioned.

We shall give one of the many examples. In analysis two definitions of the continuity of a function $f(x)$ at the point x_0 are known. In one of them the concept is defined by the condition:

(a) $$\prod_{\varepsilon>0} \sum_{\delta>0} [|x-x_0| < \delta \rightarrow |f(x)-f(x_0)| < \varepsilon],$$

and in the other by the condition:

(b) $$\lim_{k\to\infty} x_k = x_0 \rightarrow \lim_{k\to\infty} f(x_k) = f(x_0).$$

Now in the proof of the equivalence of these two definitions, namely in the proof that (a) follows from (b), we have to refer to the axiom of choice.

Opposition to the axiom of choice falls under two main headings:

1. The axiom is used to prove theorems on the existence of objects of a certain kind, although we are not in a position to give an example of such an object (non-effective proofs of existence).

2. The axiom is used to prove theorems (in particular, certain theorems in geometry) which seem intuitively unsound; the best known of these is the Banach-Tarski theorem on the paradoxical decomposition of the sphere.[1]

Many theorems are known to be equivalent to the axiom of choice, provided that the remaining axioms of set theory remain unchanged. One such theorem is the law of trichotomy for cardinal numbers:

$$\mathfrak{m} < \mathfrak{n} \ \bigvee \ \mathfrak{m} = \mathfrak{n} \ \bigvee \ \mathfrak{m} > \mathfrak{n}.$$

The problem of independence of the axiom of choice from the remaining axioms of set theory was for a long time unsolved. Quite recently, in 1964, Paul J. Cohen proved that this axiom is independent of the other axioms of set theory.

---

[1] Cf. K. Kuratowski & A. Mostowski, *Teoria mnogości* (Set Theory), *Supplement*.

A14.1 can be proved only in the particular case when the family Z, mentioned in A14.1, is finite. We then prove that theorem first for the case $Z = \{X_1\}$:

By the assumptions of A14.1 we have

(1)　$\sum_x x \in X_1$

(2)　$a_1 \in X_1$　　　　　　　　　　　　　　　　　　$\{1\}$

(3)　$a_1 \in X_1 \cap \{a_1\}$　　　　　　　　　　　　　$\{2\}$

(4)　$\sum_x{}_1(x \in X_1 \cap \{a_1\})$　　　　　　　　　$\{3\}$

　　　$\sum_Y \sum_x{}_1(x \in X_1 \cap Y).$　　　　　　　$\{4\}$

Next by induction we prove the theorem for the case $Z = \{X_1, X_2, ..., X_n\}$. If, however, the family Z is enumerable, then we cannot prove A14.1 even when the elements of the family Z are sets of two elements each.

We shall mention here one more conclusion of the axiom of choice. For this purpose the following definition will be needed:

D14.1.　　　　　$\alpha \in \mathfrak{M}_Z \equiv \sum_{M \in Z} (\alpha = \{M\} \times M).$

This definition leads immediately to

CONCLUSION 14.1.　$\{M\} \times M \in \mathfrak{M}_Z \equiv M \in Z.$

Thus the set $\mathfrak{M}_Z$ is the set of Cartesian products of the set whose only element is any set belonging to the family Z, and that set itself.

T14.2.　$Z \neq \emptyset \wedge \prod_{M \in Z} (M \neq \emptyset) \to \sum_R \prod_{M \in Z} [R(M) \in M].$[1]

Hence if every set belonging to a non-empty family of sets is non-empty, then there is a function which to every set of the family in question associates an element of that set.

Proof. (1)　$Z \neq \emptyset$

　　　　(2)　$\prod_{M \in Z} (M \neq \emptyset)$　　　　　　$\{a.\}$

　　　　(3)　$\mathfrak{M}_Z \quad \emptyset$　　　　　　　　$\{Cncl.\ 14.1;\ 1\}$

---

[1] The symbol $R(M)$ stands here for an element associated by the function R with the set $R(M)$ (cf. D6.3), and not for the set of values of that function (cf. D6.4).

(4) $\mathfrak{X} \in \mathfrak{M}_Z \to \mathfrak{X} \neq \emptyset$ {Cncl. 14.1; 2; T5.1e}

(5) $\mathfrak{X}, \mathfrak{J} \in \mathfrak{M}_Z \wedge \mathfrak{X} \neq \mathfrak{J} \to \mathfrak{X} \cap \mathfrak{J} = \emptyset$ {T5.1f; Cncl. 14.1}

(6) $\sum_{\mathfrak{J}} \prod_{\mathfrak{X} \in \mathfrak{M}_Z} \sum_{\alpha}{}_1 (\alpha \in \mathfrak{X} \cap \mathfrak{J})$ {A14.1; 3; 4; 5}

(7) $\prod_{\mathfrak{X} \in \mathfrak{M}_Z} \sum_{\alpha}{}_1 (\alpha \in \mathfrak{X} \cap \mathfrak{J}_1)$ {6}

(8) $M R_1 m \equiv M \in Z \wedge \langle M, m \rangle \in (\{M\} \times M) \cap \mathfrak{J}_1$ {def.}

The relation R_1 thus associates to every set M of the family Z the second element of the ordered pair $\langle M, m \rangle$, belonging to the product of the set $\{M\} \times M$ and $\mathfrak{J}_1$. We shall first demonstrate that the relation R_1 is a function:

(1.1) $M R_1 m \wedge M R_1 m_1$ {ad.a.}

(1.2) $M \in Z$ {8; 1.1}

(1.3) $\langle M, m \rangle, \langle M, m_1 \rangle \in (\{M\} \times M) \cap \mathfrak{J}_1$ {8; 1.1}

(1.4) $\{M\} \times M \in \mathfrak{M}_Z$ {Cncl. 14.1; 1.2}

(1.5) $\sum_{\alpha} \alpha \in (\{M\} \times M) \cap \mathfrak{J}_1$ {7; 1.4}

(1.6) $\langle M, m \rangle = \langle M, m_1 \rangle$ {1.3; 1.5}

(1.7) $m = m_1$ {A5.1; 1.6}

(9) $R_1 \in \text{funct}$ {D6.1; 1.1 → 1.7}

We shall now demonstrate that the function R_1 associates to every set of the family Z an element of that set.

(2.1) $M \in Z$ {ad.a.}

(2.2) $\{M\} \times M \in \mathfrak{M}_Z$ {Cncl. 14.1; 2.1}

(2.3) $\alpha_1 \in (\{M\} \times M) \cap \mathfrak{J}_1$ {7; 2.2}

(2.4) $\alpha_1 = \{M\} \times M$ {2.3}

(2.5) $\alpha_1 = \langle M_1, m_2 \rangle$

(2.6) $M_1 \in \{M\}$ {D5.1; 2.4}

(2.7) $m_2 \in M$

(2.8) $M_1 = M$ {D1.3a; 2.6}

(2.9) $\langle M, m_2 \rangle \in (\{M\} \times M) \cap \mathfrak{J}_1$ {2.3; 2.5; 2.8}

(2.10) $M R_1 m_2$ {8; 2.1; 2.9}

(2.11) $M \in D_l(R_1)$ {D6.2a; 2.10}

(2.12) $m_2 = R_1(M)$ {D6.3; 9; 2.11; 2.10}

(2.13) $R_1(M) \in M$ {2.7; 2.12}

(10) $\displaystyle\prod_{M\in Z} [R_1(M)\in M]$ $\{2.1 \to 2.13\}$

$\displaystyle\sum_{R\in\mathrm{funct}} \prod_{M\in Z} [R(M)\in M]$ $\{9; 10\}$

The theorem proved above is called the *general principle of choice*. It is interesting because, among other things, the axiom of choice follows from it. To demonstrate this it suffices to verify that the set Y_1, defined thus:

$$y \in Y_1 \equiv y = R_1(M),$$

satisfies Thesis A14.1 whenever R_1 is a function the existence of which is stated by T14.2. Thus the axiom of choice and the general principle of choice are equivalent.

Exercise

1. By using the axiom of choice, prove the following theorem: If Z is a non-empty family of non-empty and disjoint sets, then

$$\overline{\overline{Z}} < \overline{\overline{\bigcup_{Z\in Z} Z}}.$$

CHAPTER 2

ORDERED SETS

1. Isomorphism

The consideration of ordered sets will be preceded by a brief explanation of the concept of isomorphism, one of the most important concepts in logic. This concept is applicable to various branches of mathematics.

We begin with an auxiliary definition:

D1.1. $$C(R) = D_l(R) \cap D_r(R).$$

The set thus defined is called the *field* of the relation R.

This definition gives immediately

T1.1. $$x \in C(R) \equiv x \in D_l(R) \lor x \in D_r(R).$$

To simplify the notation of the next definition we assume that

$$X = C(S) \quad \text{and} \quad Y = C(T).$$

The symbols

$$\langle X, S \rangle \quad \text{and} \quad \langle Y, T \rangle$$

thus stand for ordered pairs whose second elements are relations and whose first elements are the fields of those relations. This convention regarding symbols of the type $\langle X, R \rangle$ will also be used later in this chapter.

D1.2. $$\langle X, S \rangle \text{ is}_R \langle Y, T \rangle \equiv R \in 1{-}1 \land D_l(R) = X \land D_r(R) = Y$$
$$\land \prod_x \prod_y \prod_z \prod_u [x R z \land y R u \to (x S y \equiv z T u)].$$

The formula defined above is read as follows: the relation R establishes an isomorphism between sets X and Y for relations S and T. Hence for a relation to establish an isomorphism between sets X and Y for relations S and T it is necessary and sufficient that the following conditions be satisfied:

(a) relation R is perfect (i.e., is a one-to-one relation);

(b) sets X and Y are, respectively, the fields of relations S and T;

(c) the left domain of relation R is equal to the field of relation S, and its right domain to the field of relation T;

(d) relation S holds between two objects if and only if relation T holds between two other objects associated with the former by relation R.

If it is clear from the context for which relations two given sets are isomorphic, or if this is irrelevant, we usually refer to the isomorphism between the sets without referring to the relations concerned.

It is clear that if a relation R establishes an isomorphism between two sets, its also establishes that they are equinumerous.

EXAMPLE I. Let $\mathbf{R}^+$ be the set of all positive real numbers, and $\mathbf{R}^-$, the set of all negative real numbers. It can easily be verified that the relation S, defined by the equivalence

$$xSy \equiv x \in \mathbf{R}^+ \wedge y \in \mathbf{R}^- \wedge |x| = |y|,$$

establishes an isomorphism between the sets $\mathbf{R}^+$ and $\mathbf{R}^-$ for the relations "greater than" and "less than".

EXAMPLE II. Let X be the set of expressions in a language, and Y the set of expressions in some other language. It is assumed that these sets are translatable, i.e., that for any element of one set there exists an element of the other set, with the same significance as the former. It is also assumed that neither set X nor set Y includes elements with the same meaning. Let the relation S hold between elements of set X if and only if these elements belong to the same part of speech, and let the relation T, holding between elements of set Y, have the analogous meaning. It is clear that relation R, which associates with the elements of set X the elements of set Y, with the same significance, establishes an isomorphism between X and Y.

D1.3. $\langle X, S \rangle$ is $\langle Y, T \rangle \equiv \sum_{R} (\langle X, S \rangle \text{ is}_R \langle Y, T \rangle).$

The formula defined above is read: sets X and Y are isomorphic for relations S and T.

D1.2 and D1.3 refer to relations S and T, each of which has two arguments. These definitions can easily be generalized to cover relations with more arguments. Nor is it difficult to define isomorphism between two sets for a greater number of relations. In particular, definitions thus generalized may cover sets in which the operations of, for instance, addition and multiplication, are defined.[1]

Let Φ stand for any formula which in addition to the symbols of the sentential calculus and quantifiers includes only the symbols:

(1) $$x, y, ..., u, X, S.$$

We assume that the variables $x, y, ..., u$ range over a defined set X which is the field of relation S, and that in the formula Φ quantifiers may bind only these variables, and not the variable S. We also assume that the variable X may occur in formula Φ only in the contexts $\prod_{x \in X}, \sum_{x \in X}, \prod_{y \subset X}, ...$

Let Φ^* stand for the formula obtained from Φ by replacing symbols (1) by the corresponding symbols

(2) $$x^*, y^*, ..., u^*, X^*, S^*.$$

Concerning these symbols we make assumptions analogous to those pertaining to symbols (1).

These notations and assumptions are used in the formulation and proof of the following lemma:

L1.1. $\langle X, S \rangle$ is$_R$ $\langle X^*, S^* \rangle \wedge xRx^* \wedge yRy^* \wedge ... \wedge uRu^*$
$$\rightarrow (\Phi \equiv \Phi^*).$$

Proof. (1) $\langle X, S \rangle$ is$_R$ $\langle X^*, S^* \rangle$ $\left.\begin{array}{l} \\ \\ \end{array}\right\}$ {a.}
     (2) $xRy^* \wedge yRy^* \wedge ... \wedge uRu^*$

The proof of the lemma is in three parts.

a. We assume first that no quantifiers occur in formula Φ. This means no quantifiers will occur in Φ^*. The following equivalences follow from assumptions (1) and (2) and D1.2:

$$xSx \equiv x^*S^*x^*$$
$$xSy \equiv x^*S^*y^*$$
$$\cdot \quad \cdot \quad \cdot \quad \cdot \quad \cdot \quad \cdot$$
$$xSu \equiv x^*S^*u^*$$
$$ySx \equiv y^*S^*x^*$$
$$\cdot \quad \cdot \quad \cdot \quad \cdot \quad \cdot \quad \cdot$$
$$uSu \equiv u^*S^*u^*.$$

---

[1] We recall that operations involving two arguments are special cases of relations between three elements.

The consequent of the lemma follows easily from these equivalences, the assumptions concerning formula Φ and Φ^*, and the rule of extensionality for the formulae of the sentential calculus.[1]

b. We assume that only existential quantifiers occur in formula Φ. In this case the proof of the lemma is by induction with respect to the number of such quantifiers. If this number is zero, then part a of the proof is applicable.

Assume that the lemma is true for the formulae $\Psi(x)$ and $\Psi^*(x^*)$.[2]

Let χ stand for the conjunction of all the assumptions of the lemma to be proved, except for the assumption stating that xRx^*. It is evident that this lemma, when it pertains to the formulae $\Psi(x)$ and $\Psi^*(x^*)$, is equivalent to the formula

$$\chi \to \left[xRx^* \to \left(\Psi(x) \equiv \Psi^*(x^*)\right)\right].$$

Thus, insofar as the assumptions of the lemma are fulfilled, the formula

$$xRx^* \to \left(\Psi(x) \equiv \Psi^*(x^*)\right)$$

is also true. Since the variables x and x^* do not occur in formula χ, this gives the formula

$$(3) \qquad \prod_{x \in X} \prod_{x^* \in X^*} \left[xRx^* \to \left(\Psi(x) \equiv \Psi^*(x^*)\right)\right].$$

We shall now prove the implication $\sum_{x \in X} \Psi(x) \to \sum_{x^* \in X^*} \Psi^*(x^*)$

(1.1)	$\sum_{x \in X} \Psi(x)$	{ad.a.}
(1.2)	$x_1 \in X$ $\Big\}$	{1.1}
(1.3)	$\Psi(x_1)$	
(1.4)	$D_l(R) = X \wedge D_r(R) = X^*$	{D1.2, 1}
(1.5)	$x_1 \in D_l(R)$	{1.2, 1.4}
(1.6)	$\sum_{x^* \in D_r(R)} x_1 R x^*$	{concl. 6.3a, Chap. I, 1.5}
(1.7)	$x_1^* \in D_r(R)$ $\Big\}$	{1.6}
(1.8)	$x_1 R x_1^*$	
(1.9)	$x_1^* \in X^*$	{1.7, 1.4}

---

[1] See Sec. 3 in Part I.

[2] These formulae, in addition to the variables x and x^*, may also include other variables.

(1.10) $\Psi^*(x_1^*)$ $\{3,\ 1.2,\ 1.9,\ 1.8,\ 1.3\}$

(1.11) $\displaystyle\sum_{x^*\in X^*}\Psi^*(x^*)$ $\{1.9,\ 1.10\}$

(4) $\displaystyle\sum_{x\in X}\Psi(x)\to\sum_{x^*\in X^*}\Psi^*(x^*)$ $\{1.1\to 1.11\}$

Similarly we can prove the implication

$$\sum_{x^*\in X^*}\Psi^*(x^*)\to\sum_{x\in X}\Psi(x).$$

Hence the equivalence

$$\sum_{x\in X}\Psi(x)\equiv\sum_{x^*\in X^*}\Psi^*(x^*)$$

is true.

Hence by the rule of extensionality for formulae of the sentential calculus the lemma is true in this case.

c. Let Φ and Φ^* be formulae involving existential and universal quantifiers. Their equivalence follows from part b of the proof and the fact that a universal quantifier may be replaced by a combination of two symbols of negation with an existential quantifier inserted between them. Hence the lemma is true.

Note that L1.1 may be preceded by the quantifiers $\prod_{x\in X}$ and $\prod_{x^*\in X^*}$. If in formula Φ the variable x is bound (in which case the variable x^* in formula Φ^* is also bound), then by T13* and T16*, the formula xRx^* can be replaced in the antecedent of L1.1 by the formula

$$\sum_{x\in X}\sum_{x^*\in X^*}xRx^*,$$

which is true, provided that the sets X and X^* are not empty. Hence it follows that in the assumptions of L1.1 we can eliminate every formula of the type xRx^*, provided that the variables x and x^* are bound in formulae Φ and Φ^*.

In particular

CONCLUSION 1.1. *If no free variables occur in formula Φ and Φ^*, then* L1.1 *takes the form*

$$\langle X, S\rangle\ \mathrm{is}_R\ \langle Y, T\rangle\to(\Phi\equiv\Phi^*).$$

In the following theorem we retain the assumptions concerning formulae Φ and Φ^*, as well as the assumption that the sets X and X^* are non-empty.

T1.2. *If*

$$\langle X, S \rangle \text{ is}_{R_1} \langle Y, T \rangle$$

and formula Φ is true, then formula Φ^* is also true.

Proof. If free variables occur in formulae Φ and Φ^*, then we precede them with universal quantifiers binding all these variables. The formulae thus obtained are represented by Φ_1 and Φ_1^*, respectively. That formula Φ_1 is true follows from the assumption that formula Φ is true.

Assume also that the relation which establishes an isomorphism between sets X and Y is the relation R_1. Hence

$$\langle X, S \rangle \text{ is}_{R_1} \langle Y, T \rangle.$$

The truth of formula Φ_1^* follows from the formula above, the truth of formula Φ_1, and Conclusion 1.1. By omitting the universal quantifiers placed at the beginning of formula Φ_1^* we again obtain formula Φ^*. This establishes the theorem.

The theorem proved above is called the *fundamental theorem on isomorphism*. It may be generalized in various ways. For instance, it may cover formulae referring to not just one, but many relations. These relations may have arbitrary numbers of arguments. The fundamental theorem on isomorphism was first stated by the Polish logicians Adolf Lindenbaum and Alfred Tarski.

It follows from the theorem that two theories, one of which concerns the elements of set X, which is a field of certain relations, and the other concerns the elements of set Y which is a field of the same number of relations, do not differ formally from one another if the sets X and Y are isomorphic for the given relations. Thus, for instance, if a set Y, in which two operations of two arguments and one operation of one argument are defined, is isomorphic for those operations with Boolean algebra, then a theory concerning the elements of set Y does not differ formally from Boolean algebra.

Exercises

1. Generalize D1.2 to cover the case in which each of the sets X and Y is a field of two relations.
2. Generalize D1.2 for the case in which S and T are relations of three arguments

3. We assume that two operations of two arguments, represented by "$+$" and ". ", are defined and performable in the set X, and that two operations, represented by "$\oplus$" and "$\odot$", are defined and performable in the set Y. It is also assumed that the sets X and Y are isomorphic for these operations. Show without referring to the fundamental theorem on isomorphism that:

   a. if the operations "$+$" and ". " are commutative and associative, so are the operations "$\oplus$" and "$\odot$";

   b. if the former have moduli, so have the latter;

   c. if the former have converse operations, so have the latter;

   d. if the operation "$+$" is distributive over the operation ". ", then the operation "$\oplus$" is distributive over the operation "$\odot$".

4. Show that the definition

$$\langle X, S \rangle \text{ is}_R \langle Y, T \rangle \equiv X \sim_R Y \wedge \prod_x \prod_y [x S y \equiv R(x)\, TR(y)]$$

is equivalent to D1.2.

2. Similar Sets. Order Types

In many respects the theory of ordered sets is similar to general set theory, which was discussed in the preceding chapter. For example, the subject-matter of the present Section is analogous to that of Sec. 7, Chap. I.

We first give some auxiliary definitions.

D2.1a. $R \in \mathrm{con}(A) \equiv \prod\limits_{x \in A} \prod\limits_{y \in A} [x \neq y \rightarrow (x R y \vee y R x)],$

   b. $R \in \mathrm{trans}(A) \equiv \prod\limits_{x \in A} \prod\limits_{y \in A} \prod\limits_{z \in A} (x R y \wedge y R z \rightarrow x R z),$

   c. $R \in \mathrm{asym}(A) \equiv \prod\limits_{x \in A} \prod\limits_{y \in A} [x R y \rightarrow \neg (y R x)].$

The formulae defined above are read as follows:

   a. the relation R is *connected* in the set A,

   b. the relation R is *transitive* in the set A,

   c. the relation R is *asymmetric* in the set A.

Thus each of the symbols: $\mathrm{con}(A)$, $\mathrm{trans}(A)$, $\mathrm{asym}(A)$, stands for a set of relations dependent on the set A.

The following definition is of fundamental importance to the theory of ordered sets.

D2.2.

$R \in \mathrm{ord}(A) \equiv A = C(R) \wedge R \in \mathrm{con}(A) \cap \mathrm{trans}(A) \cap \mathrm{asym}(A).$

This formula is read: the relation R *orders* the set A. For a relation to order a set A it is necessary and sufficient that the set A be the field of that relation and that the relation be connected, transitive and asymmetric in that set.

Typical relations ordering the set of real numbers are the relations "less than" and "greater than". Also typical are the following two relations:

$$t_1 \text{ is earlier than } t_2, \quad t_1 \text{ is later than } t_2,$$

which order any set of times.

D2.3. $\langle A, R \rangle$ is an *ordered* set $\equiv R \in \mathrm{ord}(A)$.

Thus an ordered set is a pair having as its first element an arbitrary set, and as its second element a relation ordering that set. To explain the intuitive meaning of D2.3 we adopt the following convention about the elements x and y of the set A. If they satisfy the condition "xRy" then we shall call x the *earlier* element, and y the *later* element. We may now formulate D2.3 in the following inexact but intuitive way:

A set is ordered if and only if:

a. *for any two (different) elements of the set it can be determined which element precedes the other (connectivity),*

b. *if the element x precedes y, and y precedes z, then x precedes z (transitivity),*

c. *an element x cannot both precede and follow another element y (asymmetry).*

The pairs: $\langle \mathbf{N}, < \rangle$, $\langle \mathbf{N}, > \rangle$,[1] $\langle \mathbf{R}, < \rangle$, $\langle \mathbf{R}, > \rangle$, are examples of ordered sets. It follows from A5.1 in the preceding Chapter that the ordered sets $\langle \mathbf{N}, < \rangle$ and $\langle \mathbf{N}, > \rangle$ are not the same since the relations "less than" and "greater than" are, of course, different.[2] The same applies to the ordered sets $\langle \mathbf{R}, < \rangle$ and $\langle \mathbf{R}, > \rangle$.

We shall now introduce certain conventions and simplifications of notation. We shall call the set A the *range* of the ordered set $\langle A, R \rangle$. Note that we say an object is an element of an ordered set if and only

---

[1] By the convention introduced in Sec. 1 concerning symbols of the type $\langle X, R \rangle$, the relations "$<$" and "$>$" are here confined to the set N (i.e., the field of these relations is the set N).

[2] Cf. D6.10 in Chap. 1.

if it is an element of the range of that ordered set. To indicate that a belongs to $\langle A, R \rangle$ we write: $a \in \langle A, R \rangle$. Hence the following equivalence is true:

$$a \in \langle A, R \rangle \equiv a \in A.$$

The symbol $\in$ is used here in two different senses: the first states that an object is a member of an ordered set, the second that it is a member of an unordered set. It will always be clear from the context which is intended in a particular case.

When considering ordered sets it is often stated which relation orders a given set. In other cases this may be of no significance; when this is so, we shall replace the symbol $\langle A, R \rangle$ by the simpler symbol A^0. Thus the symbols $A^0, B^0, C^0, \ldots$ stand for ordered sets whose ranges are the sets $A, B, C, \ldots$. It is clear that if the symbol A^0 occurs several times in a formula or proof, then each time the symbol stands for a set ordered by the same relation.

To indicate that the elements x and y of the set $\langle A, R \rangle$ satisfy the condition "xRy" we shall write:

(1) $x \prec_A y$.

This relation is read: the element x *precedes* (*is earlier than*) the element y in the set A^0. Hence the following equivalence is true:

$$x \prec_A y \equiv x, y \in A \wedge xRy.$$

Note also that the index A may sometimes be omitted from formula (1).

Using the symbolism introduced above we describe an ordered set as one in which arbitrary elements x and y satisfy the conditions

a. $x \neq y \rightarrow (x \prec y \vee y \prec x),$
b. $x \prec y \wedge y \prec z \rightarrow x \prec z,$
c. $x \prec y \rightarrow \neg(y \prec x).$

The following definition is an analogue of D7.1 in Chap. I:

D2.4. $A^0 \simeq_R B^0 \equiv A \sim_R B \wedge \prod_x \prod_y [x \prec_A y \equiv R(x) \prec_B R(y)].$

The formula $A^0 \simeq_R B^0$ is read: the relation R establishes the similarity of the ordered sets A^0 and B^0. Hence for a relation R to establish the similarity of the sets A^0 and B^0, A and B have to be equinumerous sets under the relation R, i.e. R has to be one-to-one with $D_l(R) = A$ and $D_r(R) = B$.

The second condition which relation R must satisfy can be stated as follows:

those elements of set B^0 which are associated by the relation R with elements x and y of A^0 have the same order relation in B^0 as x and y have in A^0.

D2.4 may also be written in the following form:

$$S \in \mathrm{ord}(A) \wedge T \in \mathrm{ord}(B) \to [\langle A, S \rangle \simeq_R \langle B, T \rangle \equiv \langle A, S \rangle \mathrm{is}_R \langle B, T \rangle].$$

Let C_1 stand for the set of negative integers. It can easily be seen that the relation R_1, defined by the formula

$$n R_1 c \equiv n = |c| \wedge n \in \mathbf{N} \wedge c \in C_1,$$

establishes the similarity of the sets $\langle \mathbf{N}, < \rangle$ and $\langle C_1, > \rangle$.

D2.5. $$A^0 \simeq B^0 \equiv \sum_R (A^0 \simeq_R B^0).$$

The relation thus defined is read: the ordered sets A^0 and B^0 are similar. This relation is an analogue of the relation of sets being equinumerous (cf. D7.2 in Chap. I).

It is clear that the concept of similarity of ordered sets is a special case of isomorphism.[1]

D2.5 and D2.4 give

CONCLUSION 2.2.
$$A^0 \simeq B^0 \to A \sim B.$$

Thus the ranges of similar sets are equinumerous. The converse is not true, since, e.g., the sets $\langle \mathbf{N}, < \rangle$ and $\langle \mathbf{N}, > \rangle$ are not similar.

T2.1a. $$A^0 \simeq A^0,$$

b. $$A^0 \simeq B^0 \to B^0 \simeq A^0,$$

c. $$A^0 \simeq B^0 \wedge B^0 \simeq C^0 \to A^0 \simeq C^0.$$

This shows that the relation of similarity of ordered sets is an equivalence relation.

Proof. The proof of T2.1 is analogous to that of T7.1 in Chap. I. We confine ourselves here to the following brief comments:

---

[1] Cf. Exercise 4 in the preceding Section.

The proof of formula a makes use of the identity relation restricted to the set A.

If a relation R establishes the similarity of sets A^0 and B^0, then the relation R^{-1} establishes the similarity of sets B^0 and A^0.

If a relation R establishes the similarity of sets A^0 and B^0, and a relation S the similarity of sets B^0 and C^0, then the relative product of R and S establishes the similarity of sets A^0 and C^0.

We introduce a new primitive concept, analogous to that of the power of a set. It is the concept of the *order type* of an ordered set. The order type of the set A^0 is denoted by $\overline{A^0}$. According to Cantor, who developed not only general set theory, but also the theory of ordered sets, we arrive at the concept of the order type of a given set by abstraction from the properties of the objects which make up the set. The line in the symbol for an order type is intended to stand for this single abstraction.

The meaning of the concept now introduced is explained by the following axiom, analogous to A7.1 in Chap. I:

A2.1. $$\overline{A^0} = \overline{B^0} \equiv A^0 \simeq B^0.$$

Thus the order types of two ordered sets are identical if and only if the sets are similar.

Note that T2.1 follows immediately from A2.1.[1]

D2.6. $$\alpha \in \mathrm{Ot} \equiv \sum_{A^0} \alpha = \overline{A^0}.$$

The formula $\alpha \in \mathrm{Ot}$ is read: α is an order type. Order types will be denoted by lower-case Greek letters.

D2.6 easily yields

T2.2a. $$\overline{A^0} \in \mathrm{Ot},$$

b. $$\prod_{A^0} \sum_{\alpha \in \mathrm{Ot}} {}_1 \alpha = \overline{A^0},$$

c. $$\prod_{\alpha \in \mathrm{Ot}} \sum_{A^0} \alpha = \overline{A^0}.$$

D2.6 and T2.2 are analogues of D7.3 and T7.2 in Chap. I. A7.1 in Chap. I, Conclusion 2.2 and A2.1 lead to

CONCLUSION 2.3. $$\overline{A^0} = \overline{B^0} \rightarrow \overline{\overline{A}} = \overline{\overline{B}}.$$

---

[1] Cf. the remark that follows T1.7 (on p. 162).

The converse is not true, as can be seen from the example A^0 = $\langle \mathbf{N}, < \rangle$, and $B^0 = \langle \mathbf{N}, > \rangle$.

In the concluding part of Sec. 7 of the preceding Chapter we noted that many theorems on powers of sets may be formulated without reference to the concept of the power of a set, but merely by means of the concept of equinumerous sets. Likewise, theorems on order types may be replaced by theorems employing the concept of similarity of sets. Hence if the concept of order type were eliminated from the list of the primitive concepts of set theory (with a simultaneous elimination of A2.1) the theory would not be essentially poorer, just as it is not made essentially poorer by the elimination of the concept of the power of a set.

The following lemma, analogous to 7.4 in the preceding Chapter, will be used frequently:

L2.1. *For any ordered sets A^0 and B^0 there exist ordered sets A_1^0 and B_1^0 such that*

$$A_1^0 \simeq A^0, \quad B_1^0 \simeq B^0, \quad \text{and} \quad A_1 \cap B_1 = \varnothing.$$

Proof. We define the sets A_1 and B_1 thus:

$$\langle 1, a \rangle \in A_1 \equiv a \in A,$$
$$\langle 2, b \rangle \in B_1 \equiv b \in B.$$

We also assume that

$$\langle 1, a_1 \rangle \prec_{A_1} \langle 1, a_2 \rangle \equiv a_1 \prec_A a_2,$$
$$\langle 2, b_1 \rangle \prec_{B_1} \langle 2, b_2 \rangle \equiv b_1 \prec_B b_2.$$

It can easily be seen that the ordered sets A_1^0 and B_1^0 defined in this way satisfy the required conditions.

In discussing order types it is immaterial which of the similar sets are taken into account. By L2.1 any two given sets can always be replaced by disjoint sets.

Exercises

1. Give an example of a relation R for which $C(R) = D_l(R) = D_r(R)$.
2. Show that $C(R) = C(R^{-1})$.
3. Give an example of a relation which is connected (transitive, asymmetric) in the set $\mathbf{N}$, but is non-connected (non-transitive, non-asymmetric) in the set $\mathbf{R}$.

4. Show that the following relations order the set $\mathbf{R}$:

    (a) $x R_1 y \equiv x^3 < y^3$;

    (b) $x R_2 y \equiv x^2 < y^2 \wedge x . y > 0 \vee x < 0 \wedge y > 0$;

    (c) $x R_3 y \equiv |x| < |y| \vee |x| = |y| \wedge x < 0 \wedge y > 0$.

5. Specify a relation which orders the set of all complex numbers.

6. What relation orders the set of words in dictionaries? Taking this relation as a model, give the relation which orders the sets of all finite sequences whose elements are natural numbers $\leqslant 9$.

7. Check whether the ordered sets $\langle \mathbf{R}, R_2 \rangle$ and $\langle \mathbf{R}, R_3 \rangle$ are similar when the relations R_2 and R_3 are defined as in Exercise 4.

8. Show that any relation is reflexive, asymmetric and transitive in the null set.

9. Show that the relation $\subseteq$ does not order the family of all subsets of a set A which has at least two elements.

10. Prove the formula

$$R \in \text{ord}(A) \to R^{-1} \in \text{ord}(A).$$

3. The Arithmetic of Order Types

The following two definitions define the sum and product of ordered sets:

D3.1. $A \cap B = \emptyset \to \{A^0 \cup B^0 = C^0 \equiv C = A \cup B \wedge$

$$\prod_x \prod_y [x \prec_C y \equiv (x \prec_A y \vee x \prec_B y \vee x \in A \wedge y \in B)]\}.$$

D3.2. $A^0 \cap B^0 = C^0 \equiv C = A \times B \wedge \prod_a \prod_b \prod_x \prod_y [\langle a, b \rangle \prec_C \langle x, y \rangle$

$$\equiv (b \prec_B y \vee b = y \wedge a \prec_A x)].$$

It can easily be seen that the condition

$$x \prec_A y \vee x \prec_B y \vee x \in A \wedge y \in B$$

defines a relation R_1 which orders the set $A \cup B$ provided that $A \cap B = 0$.

Likewise the condition

$$b \prec_B y \vee b = y \wedge a \prec_A x$$

defines a relation S_1 which orders the set $A \times B$. Thus the sum and product of ordered sets are ordered sets.

Note that relation R_1 orders the set $A \cup B$ so that all the elements of the set A^0 in unchanged order precede all the elements of the set B^0, also in unchanged order.[1]

---

[1] See footnote on p. 253.

To explain the intuitive meaning of relation S_1, consider the following example:

Let the sets A^0 and B^0 be the sets of all real numbers ordered by magnitude. Hence the following equations are satisfied:

$$A^0 = B^0 = \langle \mathbf{R}, < \rangle.$$

The elements of the set $A \times B$ may therefore be treated as points of a plane in which a system of co-ordinates has been chosen. It can easily be seen that these points are so ordered that if two points lie on different horizontal straight lines then that point is earlier which lies on the lower straight line. If on the other hand the two points

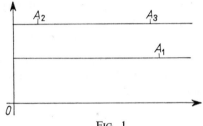

FIG. 1

lie on the same horizontal straight line then that point is earlier which lies more to the left (see Fig. 1).[1] Hence the points marked on Fig. 1 are ordered as follows:

$$A_1 \prec A_2 \prec A_3.$$

The following lemma is analogous to L7.5a, b in Chap. I.

L3.1a. $A^0 \simeq B^0 \wedge A_1^0 \simeq B_1^0 \wedge A \cap A_1 = B \cap B_1 = \varnothing$
$$\rightarrow A^0 \cup A_1^0 \simeq B^0 \cup B_1^0.$$

   b. $A^0 \simeq B^0 \wedge A_1^0 \simeq B_1^0 \rightarrow A^0 \cap A_1^0 \simeq B^0 \cap B_1^0.$

Proof. We shall prove only part b of the lemma, since the proof of part a is similar to the proof of L7.5a in Chap. I.

Let us assume that relation R establishes the similarity of the sets A^0 and B^0, and relation R_1, the similarity of the sets A_1^0 and B_1^0. It can readily be seen that the relation S, defined by the formula

$$\langle x, x_1 \rangle \, S \, \langle y, y_1 \rangle \equiv xRy \wedge x_1 \, R_1 \, y_1,$$

establishes the similarity of sets $A^0 \cap A_1^0$ and $B^0 \cap B_1^0$.

---

[1] The informal description given here is intended only to explain the intuitive meaning of the ordering relation under consideration.

We next define the sum and the product of order types.

D3.3. $\qquad A \cap B = \emptyset \rightarrow \overline{A^0} + \overline{B^0} = \overline{A^0 \cup B^0}.$

D3.4. $\qquad\qquad \overline{A^0} . \overline{B^0} = \overline{A^0 \cap B^0}.$

Note that D3.3 and D3.4 can also be written in the following more precise form:

$$\alpha + \beta = \gamma \equiv \sum_{\overline{A^0} = \alpha} \sum_{\overline{B^0} = \beta} (A \cap B = \emptyset \wedge \gamma = \overline{A^0 \cup B^0}).$$

$$\alpha . \beta = \gamma \equiv \sum_{\overline{A^0} = \alpha} \sum_{\overline{B^0} = \beta} (\gamma = \overline{A^0 \cap B^0}).$$

We shall show that D3.3 is correct. To do so we must prove that there exists exactly one order type which is the sum of any two order types α and β.

By T2.2c, there exist ordered sets A^0 and B^0 such that $\overline{A^0} = \alpha$ and $\overline{B^0} = \beta$. Let A_1^0 and B_1^0 be any ordered sets which satisfy the conditions $A_1^0 \simeq A^0$, $B_1^0 \simeq B^0$. By L2.1 we may assume that $A \cap B = A_1 \cap B_1 = \emptyset$. From L3.1a it follows that

$$A^0 \cup B^0 \simeq A_1^0 \cup B_1^0.$$

By A2.1 and D3.3 we obtain:

$$\overline{A^0} + \overline{B^0} = \overline{A_1^0} + \overline{B_1^0}.$$

Hence only one sum of two order types can exist. That such a sum always exists we infer from L2.1, D3.3, T2.2b, and the fact that the sum of two ordered sets is an ordered set.

Similarly D3.4 can be justified.

T3.1a. $\qquad\qquad (\alpha + \beta) + \gamma = \alpha + (\beta + \gamma).$

b. $\qquad\qquad (\alpha . \beta) . \gamma = \alpha . (\beta . \gamma).$

That is to say addition and multiplication of order types are associative. However, these operations, as we shall see later, are not commutative.

Proof. We shall prove only formula a.

Let A^0, B^0, C^0 be ordered sets which satisfy the formulae

(1) $\qquad\qquad \overline{A^0} = \alpha, \quad \overline{B^0} = \beta, \quad \overline{C^0} = \gamma.$

The existence of such sets is ensured by T2.2c. By L2.1 we can assume that

(2) $\qquad\qquad A \cap B = A \cap C = B \cap C = \emptyset.$

We introduce the following symbols:

(3)
$$D^0 = A^0 \cup B^0, \quad E^0 = B^0 \cup C^0,$$
$$X^0 = (A^0 \cup B^0) \cup C^0, \quad Y^0 = A^0 \cup (B^0 \cup C^0).$$

The following equations result from D3.1, (2), and (3):

$$X = (A \cup B) \cup C, \quad Y = A \cup (B \cup C).$$

Hence
(4)
$$X = Y.$$

D3.1, (2), and (3) also yield the equivalences

$$x \prec_X y \equiv x \prec_D y \lor x \prec_c y \lor x \in D \land y \in C$$
$$\equiv x \prec_A y \lor x \prec_B y \lor x \in A \land y \in B \lor x \prec_c y$$
$$\lor x \in A \cup B \land y \in C \equiv x \prec_A y \lor x \prec_B y \lor x \prec_c y$$
$$\lor x \in A \land y \in B \lor x \in A \land y \in C \lor x \in B \land y \in C,$$

$$x \prec_Y y \equiv x \prec_A y \lor x \prec_E y \lor x \in A \land y \in E$$
$$\equiv x \prec_A y \lor x \prec_B y \lor x \prec_c y$$
$$\lor x \in B \land y \in C \lor x \in A \land y \in B \lor x \in A \land y \in C.$$

Hence

(5)
$$x \prec_X y = x \prec_Y y.$$

It follows from D2.4 and formulae (4) and (5) that the identity relation restricted to set X establishes the similarity of the sets X^0 and Y^0. Hence from D2.5, and A2.1

(6)
$$\overline{X^0} = \overline{Y^0}.$$

On the other hand, D3.3, (2), and (3) give the equations

$$\overline{X^0} = (\overline{A^0} + \overline{B^0}) + \overline{C^0}, \quad \overline{Y^0} = \overline{A^0} + (\overline{B^0} + \overline{C^0}).$$

Formula a follows from these equations and formulae (6) and (1).
We state the following theorem:

T3.2.
$$\gamma \cdot (\alpha + \beta) = \gamma \cdot \alpha + \gamma \cdot \beta.$$

Thus the multiplication of order types is distributive over addition. The proof of T3.2 is omitted.

4. Cuts. Dense and Continuous Sets

The concept of a cut of all rational numbers was first used by Dedekind to construct the arithmetic of real numbers.[1] The concept defined below is a generalization of Dedekind's concept

D4.1. $\langle A^0, B^0 \rangle$ is a *cut* of the ordered set C^0

$$\equiv A \neq \varnothing \wedge B \neq \varnothing \wedge A \cap B = \varnothing \wedge A^0 \cup B^0 = C^0.$$

Thus cuts are ordered pairs of non-empty ordered sets with disjoint ranges. The set A^0 is called the *lower class* of the cut, and B^0 the *upper class*. It follows from D4.1 and D3.1 that

$$x \in A^0 \wedge y \in B^0 \to x \prec_c y.$$

Thus a cut is a division of all elements of a given ordered set into two classes such that every element of the first class precedes every element of the second class.

D4.2a. x is the *first element* of an ordered set A^0

$$\equiv x \in A \wedge \prod_{y \in A} (x \neq y \to x \prec_A y).$$

b. x is the *last element* of an ordered set A^0

$$\equiv x \in A \wedge \prod_{y \in A} (x \neq y \to y \prec_A x).$$

The first element of the set $\langle \mathbf{N}, < \rangle$ is the number 1; this set has no last element.

D4.3a. A cut $\langle A^0, B^0 \rangle$ is a *jump* if and only if there exists a last element of the set A^0 and a first element of the set B^0.

b. A cut $\langle A^0, B^0 \rangle$ is a *gap* if and only if there exists no last element of the set A^0 and no first element of the set B^0.

It is clear that each cut of the set $\langle \mathbf{N}, < \rangle$ is a jump. It can also be shown, on the other hand, that the cut $\langle A^0, B^0 \rangle$ of the set of all positive rational numbers, where

$$x \in A \equiv x^2 < 2, \quad x \in B \equiv x^2 > 2,$$

is a gap. There are, of course, cuts which are neither jumps nor gaps. For instance, there is the cut $\langle A^0, B^0 \rangle$ of the set $\langle \mathbf{R}, < \rangle$, in which the set A includes all non-positive numbers, and the set B all positive

[1] Dedekind's definition of an infinite set was explained in the previous Chapter.

numbers. The last element of the set A^0 is the number 0, while the set B^0 has no first element.

D4.4a. An ordered set is *dense* if and only if none of its cuts is a jump.

  b. An ordered set is *continuous* if and only if none of its cuts is a jump or a gap.

The set $\langle \mathbf{N}, < \rangle$ is, obviously, neither dense nor continuous. The set of all rational numbers ordered by magnitude is dense but not continuous, since, as is shown by the example given above, it has cuts which are gaps. On the other hand, the set $\langle \mathbf{R}, < \rangle$ is continuous.

We now introduce the important concept of *inclusion* of ordered sets.

D4.5. $A^0 \subset B^0 \equiv A \subset B \wedge \prod\limits_{x \in A} \prod\limits_{y \in A} (x \prec_A y \equiv x \prec_B y).$

Thus an ordered set A^0 is included in an ordered set B^0 if and only if the range of A^0 is included (in the sense defined in the algebra of sets) in the range of B^0 and the elements of A^0 are ordered within it in the same way as in B^0. Thus the symbol "$\subset$" has two meanings. There is, however, no danger of confusion, since the context will always indicate which meaning is intended. Note also that if $A^0 \subset B^0$, then the set A^0 is called a *part* or a *subset* of set B^0.

D4.5, D3.1, and D4.1 give immediately

CONCLUSION 4.1a. $A^0 \cup B^0 = C^0 \to A^0 \subset C^0 \wedge B^0 \subset C^0.$

  b. $\langle A^0, B^0 \rangle$ *is a cut of* $C^0 \to A^0 \subset C^0 \wedge B^0 \subset C^0.$

T4.1. $A^0 \subset B^0 \wedge B^0 \subset C^0 \to A^0 \subset C^0.$

Thus the relation of inclusion of ordered sets is transitive as is also the relation of inclusion of non-ordered sets.

Proof. (1) $A^0 \subset B^0 \wedge B^0 \subset C^0$ {a.}

(2) $A \subset B \wedge B \subset C$ {D4.5, 1}

(3) $A \subset C$ {2}

(1.1) $x, y \in A$ {ad.a.}

(1.2) $x, y \in B \wedge x, y \in C$ {1.1, 2}

(1.3) $x \prec_A y \equiv x \prec_B y$ }
(1.4) $x \prec_B y \equiv x \prec_C y$ }   {D4.5, 1, 1.1 $\to$ 1.2}

(1.5) $x \prec_A y \equiv x \prec_C y$ {1.3, 1.4}

$A^0 \subset C^0$ {D4.5, 3, 1.1 $\to$ 1.5}

T4.2a. *If $A^0 \simeq B^0$ and there exists a first element of the set A^0, then there also exists a first element of the set B^0.*

   b. *If $A^0 \simeq B^0$ and there exists a last element of the set A^0, then there also exists a last element of the set B^0.*

   c. *If $A^0 \simeq B^0$ and the set A^0 is dense, then the set B^0 is also dense.*

   d. *If $A^0 \simeq B^0$ and the set A^0 is continuous, then the set B^0 is also continuous.*

Proof. Although this theorem can be deduced from the fundamental theorem on isomorphism, given in Sec. 1, the proof of the theorem on isomorphism is difficult. We therefore give easier proofs of T4.2a, b, c, d, which are not based on the fundamental theorem on isomorphism. We shall first prove part a of the theorem.

Assume that a perfect (one-one) relation R establishes the similarity of sets A^0 and B^0, and that a_1 is the first element of A^0. We shall show that $b_1 = R(a_1)$ is the first element of B^0. For if there existed an element b_2 of set B^0 such that $b_2 \prec_B b_1$, then there would also exist an element a_2 of set A^0 satisfying the condition $b_2 = R(a_2)$. By D2.4, the following equivalence would then be true:

$$a_2 \prec_A a_1 \equiv R(a_2) \prec_B R(a_1).$$

Thus, contrary to the assumption, the element a_1 would not be the first element of A^0.

The proof of part b is similar. We shall now prove part c of the theorem.

Assume again that a perfect relation R establishes the similarity of sets A^0 and B^0 and that set A^0 is dense. Assume also, contrary to what we intend to prove, that a certain cut $\langle X^0, Y^0 \rangle$ of set B^0 is a jump. We denote by Z^0 and U^0 those subsets of set A^0 which satisfy the conditions

$$R(Z) = X, \quad R(U) = Y.$$

It can easily be seen that $Z^0 \simeq_R X^0$, $U^0 \simeq_R Y^0$, and that the ordered pair $\langle Z^0, U^0 \rangle$ is a cut of set A^0. Hence from the assumption that the cut $\langle X^0, Y^0 \rangle$ is a jump, and from T4.2b it follows that the set Z^0 has a last element, and the set U^0 a first element. Thus, the cut $\langle Z^0, U^0 \rangle$ is a jump. But this conclusion contradicts the assumption that set A^0 is dense.

The proof of part d is along similar lines.

T4.3.   *An ordered set A^0 is dense*

$$\equiv \prod_x \prod_y \sum_z (x \prec_A y \to x \prec_A z \prec_A y).^{(1)}$$

Proof. Assume first that the set A^0 is dense. Let x and y be a pair of its elements satisfying the condition

(1) $x \prec_A y.$

Let the subsets X^0 and Y^0 of set A^0 satisfy the conditions

(2) $t \in X^0 \equiv t \prec_A x \lor t = x;$

(3) $t \in Y^0 \equiv y \prec_A t \lor t = y.$

If there did not exist any element z satisfying the condition

(4) $x \prec_A z \prec_A y,$

then the ordered pair $\langle X^0, Y^0 \rangle$ would be a cut of set A^0 and that cut would be a jump. But this conclusion contradicts the assumption that set A^0 is dense. Hence it follows that for any elements x, y of set A^0 which satisfy formula (1), there exists an element z which, satisfies formula (4).

Assume now that set A^0 is not dense. Then there exists a cut $\langle X^0, Y^0 \rangle$ of that set which is a jump. It can easily be seen that in this case there exist elements x and y of set A^0 which satisfy formulae (1), (2), and (3). If an element z of set A^0 satisfied condition (4), then that element would belong neither to set X^0 nor to set Y^0, and thus the ordered pair $\langle X^0, Y^0 \rangle$ would, contrary to assumption not be a cut of set A^0. The proof of the theorem is thus complete.

Exercises

1. Find a relation R for which the ordered set $\langle N, R \rangle$ has neither a first nor a last element.

2. Find a relation R for which some cut of the set $\langle R, R \rangle$ will be a jump, and an other cut will be a gap.

3. A set X^0 is said to be *bounded above* in set A^0 if and only if

$$X^0 \subset A^0 \land \sum_a \prod_{x \subset X} x \prec_A a.$$

---

(1) The formula $x \prec_A z \prec_A y$ stands for the conjunction of the formulae: $x \prec_A z$ and $z \prec_A y.$

An element a of set A^0, such that

$$\prod_{x \in X} (x \prec_A a \lor x = a) \land \prod_y [\prod_{x \in X} (x \prec_A y) \to (a \prec_A y \lor a = y)]$$

is called the *upper bound* of set $X^0 \subset A^0$.

(a) Define by analogy a set bounded below and the lower bound.

(b) Prove that every subset of a continuous set which is bounded above has an upper bound, and that every subset which is bounded below has a lower bound.

(c) Show that ordered sets which are dense and non-continuous do not possess this property.

4. Prove the formulae:

    (a) $A^0 \subset A^0$,

    (b) $A^0 \subset B^0 \land B^0 \subset A^0 \to A^0 = B^0$.

5. Let X be any non-empty subset of the family of all subsets (in the sense of D4.5) of an ordered set X^0. The following formula is a generalization of the concept of the sum of ordered sets:

$$Z^0 = \bigcup_{Y \in X} Y \equiv Z = \bigcup_{Y \in X} Y \land \prod_{x \in Z} \prod_{y \in Z} (x \prec_Z y \equiv x \prec_X y).$$

Show that if none of the elements belonging to the family X has a first (last) element, then the set $\bigcup_{Y \in X} Y$ also has no first (last) element.

6. (a) Show that if any finite number of elements is removed from a dense set, the set remains dense.

    (b) Show that continuous sets have no analogous property.

5. Finite Ordered Sets. Order Types ω, η, λ. Converse Types

An ordered set A^0 is said *to be of power* $\mathfrak{m}$ if and only if the range of the set is a set of power $\mathfrak{m}$, i.e. if $\overline{\overline{A}} = \mathfrak{m}$. Similarly, the set A^0 is called *enumerable, finite,* or *empty* according as the set A is, respectively, enumerable, finite, or empty. Note also that ordered sets are said to be *equinumerous* when their ranges are equinumerous.

L5.1.   *Every finite ordered set has a first element.*

Proof. Let a natural number n be the power of an ordered set A^0. If $n = 1$, then the only element of set A^0 is its first element. We assume that the lemma is true for a natural number k, and that $A = k+1$. Let a be an arbitrary element of set A^0. We denote by A_1^0 the subset of A^0 whose range is the difference between set A and the unit set $\{a\}$. It is clear that $\overline{\overline{A}}_1 = k$. By the inductive assump-

tion there exists an element a_1 which is the first element of set A_1^0.
The first element of set A^0 is the earlier of elements a and a_1. The
inductive proof of the lemma is thus complete.

T5.1.   *If finite ordered sets A^0 and B^0 are equinumerous, then the
sets are similar.*

Proof. We assume that

$$\overline{\overline{A}} = \overline{\overline{B}} = n, \quad n \in \mathbf{N}.$$

We define by induction a finite sequence

(1) $a_1, a_2, ..., a_n.$

Let the first element of sequence (1) be the first element of set A^0.
The existence of such an element is guaranteed by L5.1. Let us
assume that we have defined the elements

(2) $a_1, a_2, ..., a_k, \quad k < n,$

of sequence (1). We denote by A_1 the difference between set A and
the range of sequence (2). Let a_{k+1} be the first element of the subset
A^0 of set A_1^0.[1] In this way sequence (1) is completely determined.
We likewise define the sequence

(3) $b_1, b_2, ..., b_n,$

whose terms are elements of set B.

Let a relation R hold between terms of sequences (1) and (3)
if and only if the indices of these terms are equal. It can easily be
shown that

I.
$$\begin{cases} R \in 1-1, \\ D_l(R) = A \wedge D_r(R) = B, \\ \prod_x \prod_y [x \prec_A y \equiv R(x) \prec_B R(y)]. \end{cases}$$

Thus relation R establishes the similarity of sets A^0 and B^0.

T5.1 leads to

CONCLUSION 5.1. *Every ordered set of n elements is similar to the set
of all natural numbers $\leqslant n$, ordered by magnitude.*

From T5.1 it follows also that *finite and equinumerous ordered
sets have the same order type*. These types will be denoted by the

---

[1] The existence of this element is guaranteed by L5.1.

symbols used to denote natural numbers. For instance, the order type of a set of five elements is represented by "5". In accordance with the conventions adopted in the preceding Chapter, this symbol also represents the power of a set of five elements. This ambiguity does not lead to a contradiction, since the properties of order types and cardinal numbers are the same for finite sets. In particular, the rules governing operations on order types and on cardinal numbers of finite sets are the same. Hence the arithmetic of natural numbers can be constructed as the arithmetic of cardinal numbers of finite sets or as the arithmetic of order types of finite sets.

To illustrate these remarks we give

T5.2.   *If m and n are order types of finite sets, then*

(1) $$m+n = n+m,$$

(2) $$m.n = n.m.^{(1)}$$

Proof. Let

(3) $$\overline{A^0} = m, \quad \overline{B^0} = n, \quad A \cap B = \varnothing.$$

It is clear that the sets $A^0 \cup B^0$ and $B^0 \cup A^0$ are finite and equinumerous. The same applies to the sets $A^0 \cap B^0$ and $B^0 \cap A^0$. Hence, by T5.1,

$$\overline{A^0 \cup B^0} = \overline{B^0 \cup A^0},$$

$$\overline{A^0 \cap B^0} = \overline{B^0 \cap A^0}.$$

This result together with D3.3, D3.4, and (3) leads to formulae (1) and (2).

We shall now analyse in greater detail the properties of sets which satisfy the following three conditions:

a. *Ordered sets which have a first element*;
b. *Ordered sets which have no last element*;
c. *Ordered sets such that every cut is a jump.*

T5.3.   *If the ordered sets A^0 and B^0 satisfy conditions a–c, then these sets are similar.*

Proof. We define by induction an infinite sequence

(1) $$a_1, a_2, a_3, \ldots$$

---

(1) As mentioned previously, the addition and multiplication of arbitrary order types are not commutative.

The term a_1 is the first element of set A^0; that such a term exists follows from the fact that set A^0 satisfies condition a. Let us assume that the terms $a_1, a_2, \ldots, a_k$ of sequence (1) are already defined. The following equations define two subsets of the ordered set A^0:

$$x \in A_1^0 \equiv x \prec a_k \lor x = a_k,$$
$$x \in A_2^0 \equiv a_k \prec x.$$

Since the set A^0 satisfies condition b, it follows that set A_2^0 is non-empty. It can easily be verified that the ordered pair $\langle A_1^0, A_2^0 \rangle$ is a cut of set A^0. Since this set satisfies condition c, it follows that set A_2^0 has a first element. Let it be the $(k+1)$-th term of sequence (1). Similarly we define an infinite sequence

$$(2) \qquad\qquad b_1, b_2, b_3, \ldots$$

whose terms are elements of set B^0.

We also define the relation R:

$$(3) \qquad\qquad aRb \equiv \sum_{i \in N} (a = a_i \land b = b_i).$$

We shall show first that

$$(4) \qquad\qquad D_l(R) = A.$$

It follows from (3) that the set $D_l(R)$ is equal to the range of sequence (1). It is also clear that every term of that sequence is an element of set A^0. Hence to prove formula (4) it suffices to show that every element of set A^0 is a term of sequence (1). Let us assume, contrary to what we intend to prove, that there exists an element a of set A^0 which is not equal to any term of sequence (1). This means that the element a satisfies one of the three conditions:

α. $\quad a \prec_A a_1$,

β. $\quad \displaystyle\sum_{k \in N} (a_k \prec_A a \prec_A a_{k+1})$,

γ. $\quad \displaystyle\prod_{i \in N} a_i \prec_A a$.

But case α cannot occur since a_1 is the first element of set A^0. If case β holds, then, as can easily be seen, the element a would belong to neither A_1^0 nor A_2^0, defined at the beginning of the proof. This is impossible since the pair $\langle A_1^0, A_2^0 \rangle$ is a cut of set A^0. Suppose now that case γ holds. The following equivalences define two subsets of the ordered set A^0:

$$x \in A_3^0 \equiv \sum_{i \in \mathbb{N}} x \prec_A a_i$$

$$x \in A_4^0 \equiv \prod_{i \in \mathbb{N}} a_i \prec_A x.$$

It is clear that the pair $\langle A_3^0, A_4^0 \rangle$ is a cut of set A^0 and that set A_3^0 has no last element. But this conclusion contradicts the assumption that set A^0 satisfies condition c. Hence case γ cannot hold either. Hence formula (4) is true.

The proof that

$$D_r(R) = B$$

is analogous.

The proof that relation R satisfies the remaining formulae I, given in the proof of T5.1, is left to the reader. From the fact that relation R satisfies these formulae, and from D2.4 it follows that $A^0 \simeq_R B^0$. The theorem then follows from D2.5.

D5.1. $\omega = \overline{A^0} \equiv$ the set A^0 satisfies conditions a–c.

It is clear that the set $\langle \mathbb{N}, < \rangle$ satisfies conditions a–c. Hence

CONCLUSION 5.2.

$$\omega = \overline{\langle \mathbb{N}, < \rangle}.$$

This formula may serve as a definition of the order type ω.

Thus every ordered set which satisfies conditions a–c is enumerable. Hence the cardinal number $\aleph_0$ can be defined independently of the concept of a natural number. But if we were to construct set theory without assuming the natural number concept we would have to adopt an axiom guaranteeing the existence of at least one enumerable set.

We shall now be concerned with ordered sets which

a′. have neither a first nor a last element;

b′. are enumerable;

c′. are dense.

T5.4. *If the ordered sets A^0 and B^0 satisfy conditions a′–c′, then they are similar.*

Proof. From the assumption that sets A^0 and B^0 satisfy condition b′ and from T7.3 in Chap. I it follows that there exist infinite sequences

(1) $\qquad\qquad\qquad a_1, a_2, a_3, \ldots$

(2) $\qquad\qquad\qquad b_1, b_2, b_3, \ldots$

which satisfy the conditions:

(i) *the sets A and B are, respectively, the ranges of the sequences* (1) *and* (2);

(ii) *if* $i \neq j$, *then* $a_i \neq a_j$ *and* $b_i \neq b_j$.

We define by induction an infinite sequence

(3) $\qquad\qquad\qquad b_{i_1}, b_{i_2}, b_{i_3}, \ldots$

Let $b_{i_1} = b_1$. Assume that we have already defined the terms b_{i_1}, $b_{i_2}, \ldots, b_{i_k}$. Assume also that the formula

(4) $\qquad\qquad\qquad a_m \prec_A a_n \equiv b_{i_m} \prec_B b_{i_n}$

holds for any natural numbers m and n less than or equal to k.

It is clear that the term a_{k+1} of sequence (1) must satisfy one of the conditions:

α. $\quad \prod_{j \leqslant k} a_{k+1} \prec_A a_j$,

β. $\quad a_j \prec_A a_{k+1} \prec_A a_{j+1}, \quad 1 \leqslant j < k,$

γ. $\quad \prod_{j \leqslant k} a_j \prec_A a_{k+1}.$

In each of these cases let the term $b_{i_{k+1}}$ of sequence (3) equal b_l which is the first term of sequence (2) and satisfies the appropriate condition given below:

α′. $\quad \prod_{j \leqslant k} b_l \prec_B b_{i_j}$,

β′. $\quad b_{i_j} \prec_B b_l \prec_B b_{i_{j+1}}$,

γ′. $\quad \prod_{j \leqslant k} b_{ij} \prec_B b_l.$

The existence of the term b_l results,

in cases α′ and γ′, from the fact that set B^0 satisfies condition a′;

in case β′, from the fact that set B^0 satisfies condition c′ and from formula (4).

It is clear that formula (4) remains true when m and n can also take the value $k+1$. Hence the formula is true for all terms of sequences (1) and (3).

The following equivalence defines relation R:

$$aRb \equiv \sum_{k \in \mathbf{N}} (a = a_k \wedge b = b_{i_k}).$$

We shall show that every term of sequence (2), and hence every element of set B, is a term of sequence (3). We assume for the time being that this is not the case; let b_l be the first term of sequence (2) which is not included in sequence (3). Hence each of the terms $b_1, \ldots$ $\ldots, b_{l-1}$ is equal to a term of sequence (3). Denote by $b_{i_{r_t}}$ the term of that sequence which satisfies the formula

$$b_{i_{r_t}} = b_t \qquad \text{where} \qquad t = 1, \ldots, l-1.$$

Let k be the largest of the numbers $r_1, \ldots, r_{l-1}$. We further assume that b_l satisfies condition α' and that a_s is the first term in sequence (1), which precedes all of the terms

(5) $$a_1, \ldots, a_k$$

in set A^0. Hence each of the terms $a_{k+1}, \ldots, a_{s-1}$ follows at least one of the terms (5). Thus the term a_s precedes any of the terms $a_1, \ldots, a_{s-1}$. Since formula (4) is satisfied for all the terms of sequences (1) and (3), it follows that each of the terms $b_{i_{k+1}}, \ldots, b_{i_{s-1}}$ follows at least one of the terms $b_{i_1}, \ldots, b_{i_k}$ in set B^0. Thus the term b_l precedes any of the terms $b_{i_1}, \ldots, b_{i_{s-1}}$, and b_l is the first term of sequence (2) which has this property, since each of the terms $b_1, \ldots, b_{l-1}$ equals one of the terms $b_{i_1}, \ldots, b_{i_{s-1}}$. From this fact and the definition of sequence (3) it follows that $b_l = b_{i_s}$. The analogous proofs demonstrating that b_l is one of the terms of sequence (3) when that term satisfies condition β' or γ' are left to the reader. It then follows that $D_r(R) = B$. Proving the remaining formulae I, given in the proof of T5.1, presents no difficulty.

Hence the relation R establishes the similarity of sets A^0 and B^0 This completes the proof of T5.4.

The theorem proved above enables the following definition to be stated:

D5.2. $\qquad \eta = \overline{A^0} \equiv$ the set A^0 satisfies conditions a'–c'.

It is clear that the set $\langle \mathbf{M}, < \rangle$ satisfies conditions a'–c'.

Conclusion 5.3.
$$\eta = \overline{\langle \mathbf{M}, < \rangle}.$$

This formula may serve as a definition of the order type η.
We shall now consider sets which satisfy the conditions:

a''. An ordered set X^0 has neither a first nor a last element;

b''. An ordered set X^0 is continuous;

c''. $\sum_{W} \{ W^0 \subset X^0 \wedge \overline{\overline{W}} = \aleph_0 \wedge \prod_{x} \prod_{y} [x \prec_x y \to \sum_{w \in W} (x \prec_x w \prec_x y)] \}.$

In words, condition c'' may be stated:

There exists an enumerable subset W^0 of set X^0 such that for any two different elements of X^0 there exists an intermediate element of the subset W.

T5.5. *If the ordered sets A^0 and B^0 satisfy conditions a''–c'', then they are similar.*

Proof.[1] Let W_1^0 and W_2^0 be enumerable subsets of sets A^0 and B^0, respectively, satisfying the conditions

(1)
$$\prod_{x} \prod_{y} [x \prec_A y \to \sum_{w \in W_1} (x \prec_A w \prec_A y)],$$
$$\prod_{x} \prod_{y} [x \prec_B y \to \sum_{w \in W_2} (x \prec_B w \prec_B y)].$$

The existence of such subsets is guaranteed by the assumption that sets A^0 and B^0 satisfy condition c''. It is evident that sets W_1^0 and W_2^0 satisfy conditions a'–c', and hence, by T5.4, they are similar. Let the relation S establish the similarity of sets W_1^0 and W_2^0.

Let x be an element of set A^0. Denote by A_1^0 the subset of set A^0 which includes all the elements of the set which precede the element x, and by A_2^0 the subset of set A^0 which includes all the remaining elements of the set. It is obvious that the ordered pair $\langle A_1^0, A_2^0 \rangle$ is a cut of set A^0 and that x is the first element of set A_2^0. Denote by C^0 the product of sets W_1 and A_1^0, and by B_1^0 the subset of set B^0 which includes every one of its elements a for which an element of the set $S(C)$ follows a in set B^0. Denote by B_2^0 the subset of set B^0 which includes all its elements not belonging to B_1^0. It is clear that the ordered pair $\langle B_1^0, B_2^0 \rangle$ is a cut of set B^0 and that set B_1^0 has no

---

[1] The proof is not given in full.

last element. It follows from the fact that set A_1^0 has no last element and from formula (1) that the set C^0, which is a subset of A^0, has no last element. Hence set $S(C)$, which is a subset of B^0, also has no last element. From this it follows easily that set B_1^0 has no last element.

From this and the assumption that set B^0 satisfies condition b'', it follows that there exists a first element in set B_2^0. Denote that element by y and associate it with the element x. In this way exactly one element of set B^0 is associated with every element of set A^0. If this association, which is, of course, a relation, denoted by R, we obtain the formula

(2) $$D_l(R) = A,$$

(3) $$R \in \text{funct}.$$

Similarly, we can associate with any element y of set B^0 exactly one element of set A^0. Clearly relation R also holds between the elements x and y associated in this way. Hence

(4) $$D_r(R) = B,$$

(5) $$R^{-1} \in \text{funct}.$$

It follows from formulae (3) and (5) that $R \in 1-1$. It is left to the reader to check that the third formula given in the proof of T5.1 is also satisfied.

Hence sets A^0 and B^0 are similar.

The theorem proved above makes it possible to adopt the following definition:

D5.3. $\lambda = \overline{A^0} \equiv$ set A^0 satisfies conditions a''–c''.

It can easily be seen that the set $\langle \mathbf{R}, < \rangle$ satisfies conditions a''–c'', while the set W, specified by condition c'', is the set of all rational numbers $\mathbf{M}$.

Hence

CONCLUSION 5.4.

$$\lambda = \overline{\langle \mathbf{R}, < \rangle}.$$

This formula may be used as a definition of the order type λ.

Among the cuts of a given ordered set A^0 which are neither jumps nor gaps there occur pairs of cuts which differ only in that an element a of set A^0 is the last element in the lower class of one cut, and the first element of the upper class in the other. For example, one can construct cuts of the set $\langle \mathbf{R}, < \rangle$ such that in one of them the upper class is the set of all positive numbers, and in the other the lower class is the set of all negative numbers.

If we now remove from among all the cuts of set A^0 those belonging to the pairs defined above, in which the lower class has a last element, we obtain a set which will be called the set of all *proper* cuts of set A^0.

This definition and D4.4 lead to

CONCLUSION 5.5. *No cut belonging to the set of proper cuts of a dense set is such that its lower class has a last element.*

It is clear that the relation R defined by the formula

II. $\langle A_1^0, B_1^0 \rangle \ R \ \langle A_2^0, B_2^0 \rangle \equiv A_1 \underset{\neq}{\subseteq} A_2$

orders the set of all proper cuts of set A^0.

L5.2. *If an ordered set A^0 is dense, then the set $\mathfrak{A}^0$ of all its proper cuts, ordered by the relation R defined by formula II, is also dense.*

Proof. It has to be shown that no cut $\langle \mathfrak{A}_1^0, \mathfrak{B}_1^0 \rangle$ of the set $\mathfrak{A}^0$ is a jump. We assume, contrary to what we intend to prove that the cut $\langle A_1^0, B_1^0 \rangle$ of set A^0 is the last element of the class $\mathfrak{A}_1^0$, and the cut $\langle A_2^0, B_2^0 \rangle$ of set A^0 is the first element of class $\mathfrak{B}_1^0$. It is evident that

$$\langle A_1^0, B_1^0 \rangle \prec_\mathfrak{A} \langle A_2^0, B_2^0 \rangle.$$

Hence from formula II $A_1 \underset{\neq}{\subseteq} A_2$. Hence there exists an element a of set A such that

(1) $a \notin A_1 \wedge a \in A_2$.

It follows from Conclusion 5.5 that there exists an element a_1 which satisfies the condition

(2) $a \prec_A a_1 \wedge a_1 \in A_2$.

The following formulae determine subsets of set A^0:

(3) $x \in A_3^0 \equiv x \prec_A a_1, \quad x \in B_3^0 \equiv a_1 \prec_A x \vee x = a_1$.

It is clear that the ordered pair $\langle A_3^0, B_3^0 \rangle$ is an element of set $\mathfrak{A}$. But it follows from formulae (1)–(3) and II that

$$\langle A_1^0, B_1^0 \rangle \prec_\mathfrak{A} \langle A_3^0, B_3^0 \rangle \prec_\mathfrak{A} \langle A_2^0, B_2^0 \rangle.$$

Thus the cut $\langle A_3^0, B_3^0 \rangle$ is neither an element of class $\mathfrak{A}_1^0$ nor an element of class $\mathfrak{B}_1^0$. This conclusion contradicts the assumption that the ordered pair $\langle \mathfrak{A}_1^0, \mathfrak{B}_1^0 \rangle$ is a cut of set $\mathfrak{A}^0$. Hence $\mathfrak{A}^0$ is dense.

T5.6. *If an ordered set A^0 is dense, then the set $\mathfrak{A}^0$ of all its proper cuts, ordered by the relation R defined by formula II, is continuous.*

Proof. Let an ordered pair $\langle \mathfrak{A}_1^0, \mathfrak{B}_1^0 \rangle$ be a cut of set $\mathfrak{A}^0$. We assume that the cut is a gap. The following equivalences determine the subsets A_1^0 and B_1^0 of A^0:

(1) $$x \in A_1^0 \equiv \sum_X \sum_Y (\langle X^0, Y^0 \rangle \in \mathfrak{A}_1^0 \wedge x \in X);$$

$$y \in B_1^0 \equiv \sum_X \sum_Y (\langle X^0, Y^0 \rangle \in \mathfrak{B}_1^0 \wedge y \in Y).^{(1)}$$

It is evident that the ordered pair $\langle A_1^0, B_1^0 \rangle$ is a proper cut of set A^0.[2] We assume that $\langle A_1^0, B_1^0 \rangle \in \mathfrak{A}_1^0$. From the assumption that the cut $\langle \mathfrak{A}_1^0, \mathfrak{B}_1^0 \rangle$ is a gap it follows that there exists a cut $\langle A_2^0, B_2^0 \rangle$ of a set A^0 such that

$$\langle A_2^0, B_2^0 \rangle \in \mathfrak{A}_1^0 \wedge \langle A_1^0, B_1^0 \rangle \prec_{\mathfrak{A}} \langle A_2^0, B_2^0 \rangle.$$

Hence by formula II $A_1 \subsetneq A_2$. But this conclusion contradicts formula (1). Similarly we can demonstrate that the cut is not an element of class $\mathfrak{B}_1^0$. Hence no cut of set $\mathfrak{A}^0$ is a gap.

It also follows from L5.2 that no cut of set $\mathfrak{A}^0$ is a jump. Hence the set is continuous.

In Dedekind's theory of real numbers, mentioned at the beginning of the preceding Section, real numbers are defined as proper cuts of the set of rational numbers, ordered by magnitude. Formula II defines the relation "less than" for real numbers so interpreted. It follows from T5.6 that the set of all real numbers ordered by magnitude is a continuous set.

D5.4. $$\beta = \alpha^* \equiv \sum_X \sum_R (\alpha = \overline{\langle X, R \rangle} \wedge \beta = \overline{\langle X, R^{-1} \rangle}).$$

This definition can be written in a shorter form:

$$\alpha = \overline{\langle X, R \rangle} \rightarrow \alpha^* = \overline{\langle X, R^{-1} \rangle},$$

---

(1) The set A_1 is thus the sum of all lower classes of cuts belonging to the class $\mathfrak{A}_1^0$, and the set B_1 is the sum of all upper classes of cuts belonging to class $\mathfrak{B}_1^0$.

(2) Cf. Exercise 5 in Sec. 4.

or in the form

$$(\overline{\langle X, R \rangle})^* = \overline{\langle X, R^{-1} \rangle}.$$

The order type α^* is called the *type converse* to the order type α. If α is the order type of a set X ordered by a relation R, then α^* is the order type of the set X ordered by the relation converse to R. It follows from D5.4 and Conclusion 5.2 that $\omega^* = \overline{\langle \mathbf{N}, > \rangle}$. Thus the order type ω^* is the order type of those ordered sets which have no first element, have a last element, and in which every cut is a jump. It is obvious that this is the order type of the set of negative integers, ordered by the relation $<$.

This example shows that the type α^* may be different from the type α. In certain cases, however, the type converse to the order type α is identical with α. Thus, for instance, it follows from T5.1 that $n^* = n$ (where n is the order type of a finite set); from T5.4 and T5.5 it follows that $\eta^* = \eta$ and $\lambda^* = \lambda$. It can also be shown that $(\omega^* + \omega)^* = \omega^* + \omega$.

The order type $\omega^* + \omega$ is the order type of the set of integers ordered by the relation $<$. Ordered sets of this type have neither a first nor a last element, and all their cuts are jumps.

Exercises

1. Give examples of ordered sets which satisfy two of conditions a–c, specified in T5.3, but do not satisfy the third. Do the same for conditions a′–c′, specified in T5.4, and conditions a″–c″, specified in T5.5.

2. Show that the order type α of any enumerable dense set satisfies one of the four conditions:

$$\alpha = \eta, \quad \alpha = 1 + \eta, \quad \alpha = \eta + 1, \quad \alpha = 1 + \eta + 1.$$

3. For any order type α, prove the formula

$$\alpha . 2 = \alpha + \alpha.$$

4. Prove the formulae
   (a) $\omega = 1 + \omega = 2 + \omega = \ldots = n + \omega$
   (b) $\omega = 2\omega = 3\omega = \ldots = n . \omega$
   (c) $\eta = \eta + \eta = \eta + 1 + \eta = \eta . \eta$
   (d) $\lambda = \lambda + 1 + \lambda$
   (e) $(\omega + 1) + \omega = \omega + \omega$.

5. Show that the following formulae are false.
   (a) $\omega = \omega + 1$
   (b) $2\omega = \omega + \omega$
   (c) $\eta = \eta + 2 + \eta$

(d) $\lambda = \lambda + 2 + \lambda$
(e) $\lambda = \lambda + \lambda$
(f) $2\eta = \eta$
(g) $\lambda = \lambda \cdot \lambda$.

6. Prove the formulae
   (a) $(\alpha^*)^* = \alpha$
   (b) $(\alpha + \beta)^* = \beta^* + \alpha^*$
   (c) $(\alpha \cdot \beta)^* = \alpha^* \cdot \beta^*$
   (d) $\omega^* = \omega^* + 1$.

7. Show that the following formulae are false.
   (a) $\omega + \omega^* = \omega^* + \omega$
   (b) $(\alpha \cdot \beta)^* = \beta^* \cdot \alpha^*$.

8. Show that two ordered sets each of which is similar to a subset of the other need not be similar.

6. Well-ordered Sets. Ordinal Numbers

D6.1.   An ordered set A^0 is a *well-ordered set*

$$\equiv \prod_{\varnothing \neq B^0 \subset A^0} \sum_b (b \text{ is the first element of the set } B^0).$$

Thus a set is well-ordered if and only if each of its non-empty subsets has a first element.

It follows immediately from D6.1 and L5.1 that every finite ordered set is well-ordered. The set $\langle \mathbf{N}, < \rangle$ is also well-ordered, whereas the sets $\langle \mathbf{M}, < \rangle$ and $\langle \mathbf{R}, < \rangle$ are not well-ordered.

T4.2a leads to

CONCLUSION 6.1. *A set which is similar to a well-ordered set is also well-ordered.*

D6.2. $\alpha \in \mathrm{On} \equiv \sum_{A^0} (A^0 \text{ is a well-ordered set} \wedge \alpha = \overline{A^0})$.

The formula $\alpha \in \mathrm{On}$ is read: α is an ordinal number. Hence, ordinal numbers are order types of well-ordered sets.

It follows from the previous remarks and Conclusions 5.2–5.4 that of the order types ω, η and λ only ω is an ordinal number. Note further that if natural numbers are treated as order types of finite sets (see remarks in the preceding Section), then natural numbers are special cases of ordinal numbers. By D6.1 and D6.2, the number zero, interpreted as the order type of the null set (cf. Exercise 8 in Sec. 2) is an ordinal number.

T.6.1. *Every subset of a well-ordered set is well-ordered.*

Proof. Let B^0 be any subset of a well-ordered set A^0, and C^0 any non-empty subset of B^0. It is clear that C^0 is a subset of A^0. Hence C^0 has a first element. We have thus shown that every non-empty subset of set B^0 has a first element. It then follows from D6.1 that the set is well-ordered.

T6.2. *The sum and product of well-ordered sets are well-ordered sets.*

Proof. Let A^0 and B^0 be any well-ordered sets and let C^0 be any non-empty subset of the set $A^0 \cup B^0$. If $A \cap C \neq \emptyset$, then we assume that the elements of the set D^0, whose range is the set $A \cap C$, are ordered as in the set A^0. The set D^0 is, of course, a non-empty subset of the well-ordered set A^0, and hence has a first element. It can easily be seen that this element is also the first element of the subset C^0. And if $A \cap C = \emptyset$, then set C^0 is a subset of the well-ordered set B^0, and hence has a first element. We have thus demonstrated that in both cases any non-empty subset of the set $A^0 \cup B^0$ has a first element. Hence the set $A^0 \cup B^0$ is well-ordered. Now let C^0 be any non-empty subset of the set $A^0 \cap B^0$. Denote by A_1^0 the subset of set A^0 which includes those and only those elements of set A which are the first elements of the ordered pairs belonging to the set C. In an analogous way we define a subset B_1^0 of set B^0. It is clear that sets A_1^0 and B_1^0 are non-empty. Let a_0 and b_0 be, respectively, the first elements of these sets. It is obvious that the pair $\langle a_0, b_0 \rangle$ is the first element of the set C^0. Thus the set $A^0 \cap B^0$ is also well-ordered.

From the theorem proved above it follows that operations on ordinal numbers may be defined as special cases of analogous operations on order types. This leads immediately to the conclusion that the theorems of the arithmetic of order types, proved in Sec. 3, apply also to ordinal numbers.

Note also that the commutative law for addition and multiplication fail to hold, not only for order types, but also for ordinal numbers. This is shown by the following inequalities:

$$\omega + 1 \neq 1 + \omega,$$
$$\omega . 2 \neq 2 . \omega,$$

the proofs of which are left to the reader.

A well-ordered set whose range is the set A will be denoted by A^+.

Exercises

1. Show that the following definition is equivalent to D6.1:

An ordered set A^0 is well ordered

$$\equiv \prod_{\emptyset \neq B^0 \subseteq A^0} \sum_b (b \text{ is the first element of the set } B^0).$$

2. Show that no cut of a well-ordered set is a gap.
3. Prove that no well-ordered set contains a subset of the type ω^*.

7. Segments of a Well-ordered Set

D7.1. $X^+ = A(a) \equiv X^+ \subset A^+ \wedge \prod_x (x \in X \equiv x \prec_A a).$

The set $A(a)$ is called a *segment* of the well-ordered set A^+, determined by the element a of that set. Thus the segment $A(a)$ is the set of all elements of set A^+ which precede a; further, the elements of the segment $A(a)$ have the same order as they do in the set A^+.

D7.1 and T6.1 lead to

CONCLUSION 7.1. *Any segment of a well-ordered set is a well-ordered set.*

Note also that the segment determined by the first element of the set A^+ is an empty set.

D7.2. $a_1, a_2 \in A^+ \rightarrow [A(a_1) < A(a_2) \equiv a_1 \prec_A a_2].$

The relation thus defined is called the relation "less than". The field of this relation is the set of segments of a well-ordered set.

T7.1. *The relation which associates every element a of a set A^+ with a segment $A(a)$ establishes similarity between set A^+ and the set of all its segments ordered by the relation "less than".*

The straightforward proof of this theorem is left to the reader.

T7.1 and Conclusion 6.1 give

CONCLUSION 7.2. *The set of all segments of a set A^+, ordered by the relation "less than", is a well-ordered set.*

L7.1. $A^+ \simeq B^+ \rightarrow \prod_{a \in A} \sum_{b \in B} [A(a) \simeq B(b)].$

Proof. Assume that a relation R establishes similarity between the sets A^+ and B^+, and that a is any element of set A. Assume also that $b = R(a)$. It is clear that $A(a) \simeq_R B(b)$. Thus the lemma is true.

The definition to be given below includes the symbol Y^X, defined by D8.3 in the previous Chapter.

D7.3.　　$R \in \mathrm{if}(A^+) \equiv R \in A^A \wedge \prod_x \prod_y [x \prec_A y \rightarrow R(x) \prec_A R(y)].$

The formula $R \in \mathrm{if}(A^+)$ is read: R is an increasing function defined in the set A^+.

L7.2.　　　　　$R \in \mathrm{if}(A^+) \rightarrow \neg[R(a) \prec_A a].$

An increasing function defined in a set cannot take a value which precedes its argument within the set.

Proof. We assume, contrary to what we intend to prove that for some $a \in A^+$

(1)　　　　　　　　$R(a) \prec_A a.$

Denote by A_1^+ the subset of the set A^+ whose elements are those and only those elements of the set A^+ which satisfy formula (1). Hence the set A_1^+ is non-empty. Let a_0 be its first element. Hence

(2)　　　　　　　　$R(a_0) \prec_A a_0,$

and for every x:

(3)　　　　　　　$x \prec_A a_0 \rightarrow \neg[R(x) \prec_A x].$

By putting $x = R(a_0)$ we obtain, from (2) and (3), the formula:

(4)　　　　　　　　$\neg[R(R(a_0)) \prec_A R(a_0)].$

On the other hand, from formula (2) and the assumption that function R is increasing we obtain

$$R(R(a_0)) \prec_A R(a_0).$$

This contradicts formula (4). Hence assumption (1) leads to a contradiction.

L7.3.　　　　　　　$\neg[A^+ \simeq A(a)].$

Thus a well-ordered set is not similar to any of its segments.

Proof. We assume, contrary to what we intend to prove that for some $a_1 \in A$

(1)　　　　　　　　$A^+ \simeq A(a_1).$

We denote by R the relation which establishes similarity between these sets. D2.4 then gives

(2)　　　　　　$x \prec_A y \rightarrow R(x) \prec_{A(a_1)} R(y).$

Thus the relation R is an increasing function defined in set A^+. Hence by L7.2

(3) $$\daleth[R(a_1) \prec_A a_1].$$

But since $A^+ \simeq_R A(a_1)$ and $a_1 \in A$, it follows that $R(a_1) \prec_A a_1$. This conclusion contradicts formula (3). Thus assumption (1) leads to a contradiction.

L7.4. $$a_1, a_2 \in A^+ \wedge A(a_1) \simeq A(a_2) \to a_1 = a_2.$$

Proof. Assume that $a_1 \neq a_2$. Then by L7.3, and the fact that of two different segments of the same well-ordered set one is always a segment of the other, it follows that

$$\daleth[A(a_1) \simeq A(a_2)].$$

Thus, the lemma is true.

L7.5. $$\prod_{a \in A^+} \sum_{b \in B^+} [A(a) \simeq B(b)] \wedge \prod_{b \in B^+} \sum_{a \in A^+} [A(a) \simeq B(b)] \to A^+ \simeq B^+.$$

Thus, if for every segment of set A^+ there exists a segment of set B^+ similar to it, and conversely, then sets A^+ and B^+ are similar.

Proof. We shall show that the similarity of sets A^+ and B^+ is established by relation R, defined by the equivalence

(1) $$aRb \equiv A(a) \simeq B(b).$$

From the assumptions of the lemma and formula (1) it follows that

(2) $$D_l(R) = A, \quad D_r(R) = B.$$

Assume that

$$aRb_1 \quad \text{and} \quad aRb_2.$$

Hence from formula (1)

$$A(a) \simeq B(b_1), \quad A(a) \simeq B(b_2).$$

From T2.1b, c it follows that $B(b_1) \simeq B(b_2)$. By L7.4 we conclude that $b_1 = b_2$. Similarly we can show that from the assumptions a_1Rb and a_2Rb it follows that $a_1 = a_2$. Hence relation R is perfect. This and formulae (2) give

(3) $$A \sim_R B.$$

We now assume that

(4) $$a_1 \prec_A a_2,$$

and that

(5) $a_1 R b_1$ and $a_2 R b_2$.

It follows from formulae (1) and (5) that:

(6) $A(a_1) \simeq B(b_1)$,

(7) $A(a_2) \simeq B(b_2)$.

It follows directly from formula (4) that $a_1 \in A(a_2)$. Hence by L7.1 and formula (7) there exists an element b' such that

(8) $b' \in B(b_2)$,

(9) $A(a_1) \simeq B(b')$.

We conclude from formulae (6) and (9) that $B(b_1) \simeq B(b')$. Hence, by L7.4, $b_1 = b'$. This and formula (8) give $b_1 \in B(b_2)$, and hence

(10) $b_1 \prec_B b_2$.

The lemma then follows from formula (3), and from the fact that formula (10) is deduced from formulae (4) and (5).

L.7.6. *If b_0 is the first element of the set B^+ which satisfies the condition*

$$\prod_{a \in A^+} \daleth [A(a) \simeq B(b)],$$

and if $a_1 \in A, b_1 \in B$, and

(1) $A(a_1) \simeq B(b_1)$,

then

$$b_1 \in B(b_0).$$

Proof. Assume, contrary to what we intend to prove, that

$$b_1 \notin B(b_0).$$

This easily leads to the conclusion that either $b_0 = b_1$ or $b_0 \prec_B b_1$. In the former case, formula (1) becomes

$$A(a_1) \simeq B(b_0).$$

However, this formula contradicts the assumption of the lemma concerning the element b_0. Let us assume, then, that $b_0 \prec_B b_1$. Hence from L7.1, and assumption (1) there exists an element a' of the set $A(a_1)$ such that

$$A(a') \simeq B(b_0).$$

This conclusion also contradicts the original assumption concerning the element b_0. Thus the proof of the lemma is complete.

L7.7. $$\prod_{a \in A^+} \sum_{b \in B^+} [A(a) \simeq B(b)] \lor \prod_{b \in B^+} \sum_{a \in A^+} [A(a) \simeq B(b)].$$

Proof. We assume, contrary to what we intend to prove, that there exists an element $a \in A^+$ such that

(1) $$\prod_{b \in B^+} \neg [A(a) \simeq B(b)],$$

and that there exists an element $b \in B^+$ such that

(2) $$\prod_{a \in A^+} \neg [A(a) \simeq B(b)].$$

We denote by a_0 the first element of the set A^+ satisfying condition (1), and by b_0 the first element of the set B^+ satisfying condition (2). Let us assume that $a_1 \in A(a_0)$. Hence $a_1 \prec a_0$. It follows from the definition of the element a_0 that there exists an element b_1 of the set B^+ such that

$$A(a_1) \simeq B(b_1).$$

From L7.6 we conclude that $b_1 \in B(b_0)$.

We have thus shown that

(3) $$\prod_{a \in A(a_0)} \sum_{b \in B(b_0)} [A(a) \simeq B(b)].$$

Similarly we can demonstrate that

(4) $$\prod_{b \in B(b_0)} \sum_{a \in A(a_0)} [A(a) \simeq B(b)].$$

It follows from formulae (3) and (4) and L7.5 that $A(a_0) \simeq B(b_0)$. This conclusion, however, contradicts the definition of the element a_0 (and also the definition of b_0). The indirect proof of the lemma is thus complete.

T7.2. $$A^+ \simeq B^+ \lor \sum_{b \in B} [A^+ \simeq B(b)] \lor \sum_{a \in A} [B^+ \simeq A(a)].$$

Thus, every two well-ordered sets are either similar or one of them is similar to a segment of the other.

Proof. It follows from L7.7 that at least one of the following two conditions is satisfied:

(1) $$\prod_{a \in A^+} \sum_{b \in B^+} [A(a) \simeq B(b)],$$

(2) $$\prod_{b \in B^+} \sum_{a \in A^+} [A(a) \simeq B(b)].$$

If both conditions are satisfied, then by L7.5

(3) $$A^+ \simeq B^+.$$

Let us assume now that condition (1) is satisfied, while condition (2) is not satisfied; hence there exists an element b of the set B^+ which satisfies the condition

$$\prod_{a \in A^+} \neg[A(a) \simeq B(b)].$$

Let us denote by b_0 the first element of the set B^+ which satisfies this formula. Assume also that

(4) $$C^+ = B(b_0).$$

It follows from the definition of the element b_0 and formula (4) that

(5) $$\prod_{c \subseteq C^+} \sum_{a \in A^+} [A(a) \simeq C(c)].$$

Let us assume that $a_1 \in A^+$. From the assumption that formula (1) is true it follows that there exists an element b_1 of the set B^+ such that

$$A(a_1) \simeq B(b_1).$$

Hence from L7.6 it follows that $b_1 \in B(b_0)$. From formula (4) it follows that $b_1 \in C^+$ and that the set $B(b_1)$ is a segment of set C^+. We have thus shown that

(6) $$\prod_{a \subseteq A^+} \sum_{c \subseteq C^+} [A(a) \simeq C(c)].$$

It follows from L7.5 and formulae (5) and (6) that $A^+ \simeq C^+$. Hence from formula (4) we obtain

(7) $$A^+ \simeq B(b_0).$$

Similarly we can show that for some $a_0 \in A^+$

(8) $$B^+ \simeq A(a_0).$$

Hence at least one of formulae (3), (7), and (8) is true. T7.2 is thus proved.

Finally, we shall prove that

L7.8. $A^+ \simeq A_1^+ \wedge B^+ \simeq B_1^+ \wedge \sum_{b \in B} [A^+ \simeq B(b)] \rightarrow \sum_{b \in B_1} [A_1^+ \simeq B_1(b)]$.

Proof. Let a relation R establish similarity between the sets B^+ and B_1^+. From the third assumption of the lemma it follows that there exists an element b' of set B such that

(1) $$A^+ \simeq B(b').$$

Let b_1 be an element of set B_1 which satisfies the formula

$$b_1 = R(b').$$

From this and the fact that relation R establishes similarity between the sets B^+ and B_1^+ it follows that

(2) $$B(b') \simeq_R B_1(b_1).$$

From the first assumption of the lemma, formulae (1) and (2), and the fact that the relation $\simeq$ is transitive and symmetric it follows that $A_1^+ \simeq B_1(b_1)$. This establishes the lemma.

The theorems and lemmas proved in this Section will be used in the next Section.

8. Inequalities. Limit Numbers

Many of the results in this Section will be analogous to those in Sec. 9 of the preceding Chapter.

D8.1. $$\overline{A^+} < \overline{B^+} \equiv \sum_b A^+ \simeq B(b).$$

Thus the ordinal number of a set A^+ is less than the ordinal number of a set B^+ if and only if set A^+ is similar to a segment of set B^+. Note that it follows from D8.1 and L7.8 that if $\overline{A^+} = \overline{A_1^+}$, $\overline{B^+} = \overline{B_1^+}$, and $\overline{A^+} < \overline{B^+}$, then $\overline{A_1^+} < \overline{B_1^+}$.

D8.1 and T2.1a lead to

CONCLUSION 8.1.

$$\overline{A(a)} < \overline{A^+}.$$

To simplify subsequent proofs we introduce the following notation:

$$\alpha = \overline{A^+}, \quad \beta = \overline{B^+}, \quad \gamma = \overline{C^+}.$$

Thus the letters α, β, γ stand for ordinal numbers.

It follows from D8.1 and L7.3 that

L8.1. $$\alpha < \beta \rightarrow \alpha \neq \beta,$$

L8.2. $$\alpha \neq \beta \rightarrow \alpha < \beta \vee \beta < \alpha.$$

This lemma is derived from T7.2 and D8.1.

L8.3. $\qquad\qquad \alpha < \beta \wedge \beta < \gamma \to \alpha < \gamma.$

Proof. It follows from D8.1 and the assumptions of the lemma that the following formulae are satisfied for a certain element b of the set B^+ and for a certain element c of the set C^+:

(1) $\qquad\qquad\qquad A^+ \simeq B(b),$

(2) $\qquad\qquad\qquad B^+ \simeq C(c).$

From L7.1 and (2) it follows that there exists an element c_1 of set C^+ such that $B(b) \simeq C(c_1)$. Hence from (1) it follows that

$$A^+ \simeq C(c_1).$$

Hence by D8.1 $\alpha < \gamma$.

L8.4. $\qquad\qquad\qquad \alpha < \beta \to \neg(\beta < \alpha).$

Proof. Assume, contrary to what we intend to prove, that $\beta < \alpha$. From L8.3 it follows that $\alpha < \alpha$. This conclusion contradicts L8.1.

Lemmas L8.2, L8.3 and L8.4 lead to

T8.1. *The relation "less than" orders the set of ordinal numbers.*

T8.2. *If the set* $\underset{\beta}{E}\,(\beta < \alpha)$ *is ordered by the relation "less than", then*

$$\alpha = \overline{\underset{\beta}{E}\,(\beta < \alpha)}.^{(1)}$$

Proof. Let $\alpha = \overline{A^+}$. To prove the theorem it suffices to show that

$$A^+ \simeq \underset{\beta}{E}\,(\beta < \alpha).$$

The relation which establishes similarity between the sets is the relation R defined by the equivalence

$$\alpha R \beta \equiv \beta = \overline{A(a)}.$$

The details of the proof are left to the reader.

T8.2 and Conclusion 6.1 lead to

CONCLUSION 8.1. *The set* $\underset{\beta}{E}\,(\beta < \alpha)$ *is a well-ordered set.*

T8.3. *Any non-empty set Ψ of ordinal numbers, ordered by the relation "less than", has a first element.*[2]

---

[1] The variables α and β stand for ordinal numbers.

[2] This theorem may also be formulated thus: Every non-empty set of ordinal numbers has a least number.

Proof. We assume, contrary to what we intend to prove, that the set Ψ has no first element, and that α is any element of that set. Hence it follows that the set $\underset{\beta}{E}(\beta \in \Psi \wedge \beta < \alpha)$ is non-empty. Assume that it is ordered by the relation "less than". Hence it is a non-empty subset of the set $\underset{\beta}{E}(\beta < \alpha)$ and thus, by D6.1 and Conclusion 8.1, has a first element. It can easily be seen that this element is also the first element of Ψ. Thus the assumption that this set has no first element leads to a contradiction.

T8.4. *Any set of ordinal numbers ordered by the relation "less than" is well-ordered.*

Proof. The theorem results immediately from D6.1 and T8.3.

We adopt the convention that the set of all ordinal numbers, ordered by the relation "less than", will be denoted by On^+, and a segment of that set determined by an element α will be represented by $On(\alpha)$. Thus, T8.2 may be written

$$\alpha \in On^+ \to \overline{On(\alpha)} = \alpha.$$

D8.2. $\alpha \in Ln \equiv \alpha \in On \wedge \alpha \neq 0 \wedge \prod_{\beta \in On} \alpha \neq \beta+1.$

The formula "$\alpha \in Ln$" is read: α is a limit number. For instance, the numbers $\omega, \omega+\omega, \omega.\omega$ are limit numbers; on the other hand, no natural number and no number of the form $\omega+n$ (where $n \in N$) is a limit number.

T8.5. $\alpha \neq 0 \wedge \alpha \notin Ln \wedge \alpha \notin N \to \sum_{\beta \in Ln} \sum_{n \in N} \alpha = \beta+n.$

Proof. We assume that

(1) $\alpha \neq 0 \wedge \alpha \notin Ln \wedge \alpha \notin N$

and, contrary to what we intend to prove, that

(2) $\prod_{\beta \in Ln} \prod_{n \in N} \alpha \neq \beta+n.$

We denote by α_0 the least ordinal number satisfying (1) and (2). The existence of such a number is guaranteed by T8.3 and the assumptions made above. Hence there also exists an ordinal number β_0 such that

(3) $\alpha_0 = \beta_0+1.$

If $\beta_0 \in \mathrm{Ln}$, then the number α_0 would not satisfy formula (2). Hence the formula

(4) $$\beta_0 \notin \mathrm{Ln}$$

must be satisfied.

It follows directly from (1) and (3) that

(5) $$\beta_0 \neq 0 \wedge \beta_0 \notin \mathbf{N}.$$

It follows from formula (3) and D8.1 that $\beta_0 < \alpha_0$. This formula, formulae (4) and (5), and the assumption that α_0 is the least ordinal number satisfying both (1) and (2), lead to the conclusion that there exist a limit number β_1 and a natural number n_1 such that

$$\beta_0 = \beta_1 + n_1.$$

It then follows from formula (3) and T3.1a that

$$\alpha_0 = \beta_1 + (n_1 + 1).$$

The formula thus obtained contradicts the assumption that α_0 satisfies formula (2). The indirect proof of the theorem is thus complete.

Exercises

1. Show that the following definition of the relation "less than" between ordinal numbers is equivalent to D8.1:
$$\alpha < \beta \equiv \sum_{\gamma \neq 0} \beta = \alpha + \gamma.$$
   Use this definition to prove L8.3.
2. Prove the formulae
   (a) $\alpha < \beta \rightarrow \gamma + \alpha < \gamma + \beta$,
   (b) $\alpha < \beta \rightarrow \alpha.\gamma < \beta.\gamma$.
3. Show that the following formulae are false.
   (a) $\alpha < \beta \rightarrow \alpha + \gamma < \beta + \gamma$,
   (b) $\alpha < \beta \rightarrow \gamma.\alpha < \gamma.\beta$.
4. Prove the formulae
   (a) $\beta \in \mathrm{Ln} \rightarrow \alpha + \beta \in \mathrm{Ln}$,
   (b) $\beta \in \mathrm{Ln} \rightarrow \alpha.\beta \in \mathrm{Ln}$.

9. The Principle of Induction

In the present Section we shall generalize the principle of induction, which has so far been used only with natural numbers.[1] The prin-

---

[1] Detailed information on most of the theorems of the arithmetic of natural numbers referred to here, can be found in W. Sierpiński, *Arytmetyka teoretyczna* (Theoretical Arithmetic).

ciple of induction is usually formulated either as an arithmetical theorem or as a rule for proving theorems. In the former case it can be stated[1]

(1) $$0 \in Z \wedge \prod_{k \in Z} (k+1 \in Z) \to \mathbf{N} \subset Z.$$

When quoting the principle of induction as a rule for proving theorems we use the symbol $W(n)$ to stand for any formula (or, more generally, any theorem) written with the variable n ranging over the set of natural numbers $\mathbf{N}$. The rule for induction will be written in the notation of Part I as follows

I. $$W(0)$$
$$\prod_{k \in \mathbf{N}} [W(k) \to W(k+1)]$$
$$\overline{\prod_{n \in \mathbf{N}} W(n).}$$

By using Rule I we can prove Theorem (1) and conversely, we can show that Rule I can be derived from (1) using the logical rules.

In proofs of mathematical theorems we often make use of a rule which differs from Rule I, but is also called the *rule of induction*. Its schema has the form:

II. $$\prod_{m \in \mathbf{N}} [\prod_{k \in \mathbf{N}} [k < m \to W(k)] \to W(m)]$$
$$\overline{\prod_{n \in \mathbf{N}} W(n).}$$

The arithmetical theorem which is the analogue of Rule II in the same sense as Theorem (1) is the analogue of Rule I, is the theorem

(2) $$\prod_{m \in Z} (\prod_{k < m} k \in Z \to m \in Z) \to \mathbf{N} \subset Z.$$

In the arithmetic of natural numbers Rules I and II are equivalent, but only Rule II can be generalized in the arithmetic of ordinal numbers. Rule I cannot be generalized to cover all ordinal numbers, since not every ordinal number can be obtained from zero by adding units, though every natural number can of course be obtained in this way. Examples of ordinal numbers which cannot be thus obtained are ω, $\omega+1$, $\omega+\omega$, $\omega.\omega$.

---

[1] In this Section it proves convenient to include zero among the natural numbers.

Rule II, when generalized, has the form[1]

III.
$$\frac{\prod_\alpha [\prod_{\beta < \alpha} W(\beta) \to W(\alpha)]}{\prod_\varphi W(\varphi).}$$

Thus to demonstrate that the formula $W(\varphi)$ is true for every ordinal number φ it suffices to prove that it is true for any number α, provided it is true for every number $\beta < \alpha$.

T9.1. *The rule of induction in schema* III *derives from the logical rules.*

Proof. Let $W(\varphi)$ be any formula in which the variable φ occurs ranging over the set of all ordinal numbers. It is assumed that

(1)
$$\prod_\alpha [\prod_{\beta < \alpha} W(\beta) \to W(\alpha)]$$

and that the formula $W(\varphi)$, contrary to what we intend to prove, is false for some ordinal number φ_1, so that the statement

(2)
$$\neg W(\varphi_1)$$

is true.

Denote by Φ the set of all ordinal numbers for which the formula $W(\varphi)$ is false. It follows from (2) that this set is non-empty. Hence by T8.3 it has a least member φ_0. Therefore, the following formulae are satisfied:

(3)
$$\neg W(\varphi_0),$$

(4)
$$\prod_{\beta < \varphi_0} W(\beta).$$

$W(\varphi_0)$ follows from formulae (1) and (4). But this conclusion contradicts (3). Thus the indirect proof of the theorem is complete. The proof given above is analogous to the proof of Rule II by virtue of the so-called *minimum principle*:

Any non-empty set of natural numbers has a least element.

T8.3 is a generalization of this principle from natural numbers to ordinal numbers.

The theorems given in this Section are generalizations of certain arithmetical results. In a similar way we can generalize the method

---

[1] In this Section we maintain the convention adopted previously that the variables $\alpha, \beta, \gamma, ..., \varphi$ stand for ordinal numbers.

of definition which is known as the inductive method and is often used in arithmetic and other branches of mathematics.

We shall confine ourselves here to the following informal description of this method of definition:

To define a property which is an attribute of every element of a given well-ordered set we define it for the first element of the set and then, assuming that it has already been defined for all elements preceding a given element a of that set, we define it for the element a.

As an example we give the definition of raising an ordinal number to a power:

D9.1a. $$\alpha^0 = 1,$$

b. $$\alpha^{\beta+1} = \alpha^\beta \cdot \alpha,$$

c. if $\gamma \in$ Ln and $\alpha = 0$, then $\alpha^\gamma = 0$; if $\gamma \in$ Ln and $\alpha \neq 0$, then α^γ is defined as the least ordinal number equal to or greater than any number α^β, where β is an ordinal number less than γ.

We can also generalize the concept of an infinite sequence as follows:

D9.2. *A transfinite sequence of type α is a function whose left domain is the set of all ordinal numbers less than α, where $\alpha \geqslant \omega$.*

When $\alpha = \omega$ we obtain an ordinary infinite sequence.

Indices of the terms of a transfinite sequence of type α are ordinal numbers less than α. A sequence of the type α is written

$$a_0, a_1, a_2, \ldots, a_\xi, \ldots, \qquad \xi < \alpha,$$

or more briefly

$$\{a_\xi\}_{\xi < \alpha}.$$

If in D9.2 we omit the condition $\alpha \geqslant \omega$, we obtain the most general type of sequence, including finite sequences as special case.

Transfinite sequences are often defined by induction.

Exercises

1. Prove the formula

$$\prod_\alpha \left[\prod_{\beta < \alpha} \beta \in \Phi \to \alpha \in \Phi \right] \to \text{On} \subset \Phi.^{(1)}$$

---

(1) This formula is an analogue of Rule III in the sense in which formula (2) is an analogue of Rule II.

2. Using formula (1), given at the beginning of the present Section, show that the relation "less than" well-orders the set of all natural numbers.

3. Prove the formula $1^\alpha = 1$.

10. Zermelo's Theorem. Alephs. The Continuum Hypothesis

The following important result, which has applications in many branches of mathematics, is called *Zermelo's theorem*:

T10.1. *For every set A there exists a well-ordered set whose range is set A.*

This theorem can also be formulated as follows:

For every set A there exists a relation R such that the ordered pair $\langle A, R \rangle$ is a well-ordered set.

This theorem is often stated briefly, but imprecisely

Every set can be well-ordered.

The proof of T10.1 requires a number of lemmas and definitions.

Let Z be an arbitrary but definite set, and let R be a function that for every non-empty subset of X of the set Z satisfies the condition:

(I) $R(X) \in X.$ [1]

The existence of such a function follows from T14.2, Chap. I.

Let W stand for the family of all those subsets W^+ of the set Z which satisfy the conditions:

(i) W^+ *is a well-ordered set,*

(ii) *if* $w \in W^+$, *then* $w = R[Z - W(w)]$.

Thus every element of the set W^+ is associated by the function R with the set of those elements of the set Z which in the set W^+ are not earlier than w.

L10.1a. $U^+, V^+ \in W \land v \in V^+ \land U^+ \simeq V(v) \to U^+ = V(v)$.

Thus every set belonging to the family W is identical with a segment of the set $V^+ \in W$, whenever it is similar to that segment.

Proof. (1) $U^+, V^+ \in W$ ⎫
(2) $v \in V^+$ ⎬ $\{a.\}$
(3) $U^+ \simeq V(v)$ ⎭

---

[1] Cf. footnote on p. 237.

(4) $U^+ \neq V(v)$ {a.i.p.}
(5) $U^+ \simeq_{S_1} V(v)$ {D2.5;3}
(6) $U^+ \sim_{S_1} V(v)$ {D2.4; 5}
(7) $S_1 \in 1-1$ {D7.1. Chap. I; 6}
(8) $V(v) = S_1(U^+)$ {L7.2. Chap. I; 6}
(9)   Let us denote by u_1 the first element of the set U^+, such that

$$u_1 \neq S_1(u_1)$$

The existence of such a set follows simply from (4), (7), and (8).

(10) $U(u_1) = V\big(S_1(u_1)\big)$ {8; 9}
(11) $u_1 = R[Z-U(u_1)] \wedge S_1(u_1) = R[Z-V\big(S_1(u_1)\big)]$
                                                        {ii; 1}
(12) $u_1 = S_1(u_1)$ {10; 11}
       contr.   {9; 12}

L10.1b $U^+, V^+ \in W \wedge U^+ \simeq V^+ \to U^+ = V^+.$

The proof of this lemma is analogous to the proof of the preceding lemma.

L10.1c. $U^+, V^+ \in W \to U^+$

$$= V^+ \vee \sum_{v \in V} [U^+ = V(v)] \vee \sum_{u \in U} [V^+ = U(u)]$$

{T7.2; L10.1a; L10.1b}

L10.2a. $u, v \in \bigcup_{W^+ \in W} W^+ \to \sum_{W^+ \in W} (u, v \in W^+).$

Proof. (1) $u, v \in \bigcup_{W^+ \in W} W^+$ {a.}

(2) $u \in U_1^+ \wedge U_1^+ \in W$ ⎫
(3) $v \in U_1^+ \wedge U_1^+ \in W$ ⎬   {D4.3. Chap. I; 1}

$\sum_{W^+ \in W} (u, v \in W^+)$ {L10.1c; 2; 3}

L10.2b. $u, v, w \in \bigcup_{W^+ \in W} W^+ \to \sum_{W^+ \in W} (u, v, w \in W^+).$

This lemma follows simply from the preceding one.

L10.3. $U^+, V^+ \in W \wedge u \prec_U v \wedge u, v \in V^+ \to u \prec_V v.$

Proof. (1) $U^+, V^+ \in W$ ⎫
(2) $u \prec_U v$ ⎬
(3) $u, v \in V^+$ ⎭   {a.}
(1.1) $U^+ = V^+$ {ad.a.}

(1.2)	$u \prec_V v$	$\{1.1; 2\}$
(2.1)	$U^+ = V(v_1)$	$\{ad.a.\}$
(2.2)	$u \prec_{V(v_1)} v$	$\{2.1; 2\}$
(2.3)	$u \prec_V v$	$\{2.2; D7.1; D4.5\}$
(3.1)	$V^+ = U(u_1)$	$\{ad.a.\}$
(3.2)	$u, v \in U(u_1)$	$\{3.1; 3\}$
(3.3)	$u \prec_{U(u_1)} v$	$\{2; 3.2; D7.1; D4.5\}$
(3.4)	$u \prec_V v$	$\{3.1; 3.3\}$
	$u \prec_V v$	$\{L10.1c;\ 1.1 \to 1.2;$
		$2.1 \to 2.3;\ 3.1 \to 3.4\}$

We introduce one more definition:

(II) $$u T v \equiv \sum_{W^+ \in W} (u \prec_W v).$$

L10.4a. $$C(T) = \bigcup_{W^+ \in W} W^+.$$

This lemma follows simply from (II) and D1.1.

L10.4b. $$T \in \mathrm{con}\,(\bigcup_{W^+ \in W} W^+).$$

Proof. (1.1) $u, v \in \bigcup_{W^+ \in W} W^+ \wedge u \neq v$ $\quad \{a.\}$

(1.2) $W_1^+ \in W \wedge u, v \in W_1^+$ $\quad \{1.1; L10.2a\}$

(1.3) $u \prec_{W_1} v \vee v \prec_{W_1} u$ $\quad \{\text{formula (a), p. 248; 1.2}\}$

(1.4) $u T v \vee v T u$ $\quad \{\text{II; 1.2; 1.3}\}$

$T \in \mathrm{con}\,(\bigcup_{W^+ \in W} W^+)$ $\quad \{D2.1a; 1.1 \to 1.4\}$

L10.4c. $$T \in \mathrm{trans}\,(\bigcup_{W^+ \in W} W^+).$$

Proof. (1.1) $u T v \wedge v T w$ $\quad \{ad.a.\}$

(1.2) $u, v, w \in \bigcup_{W^+ \in W} W^+$ $\quad \{L10.4a; D1.1; 1.1\}$

(1.3) $W_1^+ \in W \wedge u, v, w \in W_1^+$ $\quad \{L10.2b; 1.2\}$

(1.4) $W_2^+, W_3^+ \in W \wedge u \prec_{W_2} v \wedge v \prec_{W_3} u$
$\quad \{\text{II; 1.1}\}$

(1.5) $u \prec_{W_1} v \wedge v \prec_{W_1} w$ $\quad \{L10.3; 1.3; 1.4\}$

(1.6) $u \prec_{W_1} w$ $\quad \{\text{formula (b), p. 248; 1.5}\}$

(1.7) $u T w$ $\quad \{\text{II; 1.6}\}$

$T \in \mathrm{trans}\,(\bigcup_{W^+ \in W} W^+)$ $\quad \{D2.1b; 1.1 \to 1.7\}$

L10.4d. $$T \in \mathrm{asym}\,(\bigcup_{W^+ \in W} W^+).$$

Proof. (1.1) $uTv \wedge vTu$ {ad.a.}

    (1.2) $U_1^+, V_1^+ \in W \wedge u \prec_{U_1} v \wedge v \prec_{V_1} u$ {II; 1.1}

    (1.3) $u \prec_{V_1} v$ {L10.3; 1.2}

    (1.4) $\neg (v \prec_{V_1} u)$ {formula (c), p. 248; 1.3}

    (1.5) contr.   {1.2; 1.4}

    (1) $uTv \rightarrow \neg(vTu)$ {1.1 → 1.5}

        $T \in \text{asym} (\bigcup_{W^+ \in W} W^+)$ {D2.1c; 1}

L10.4e. $T \in \text{ord} (\bigcup_{W^+ \in W} W^+)$ {D2.2; L10.4a–L10.4d}

It follows from this lemma and from D4.5 that every subset of the set $\bigcup_{W^+ \in W} W^+$ is ordered. We make use of this when recording the next lemma.

L10.5. *If* $U^0 \subset \bigcup_{W^+ \in W} W^+$ *and* $U^0 \neq \varnothing$, *then the set* U^0 *has a first element.*

Proof. (1) $U^0 \subset \bigcup_{W^+ \in W} W^+$ ⎫

    (2) $U^0 \neq \varnothing$ ⎬ {a.}

    (3) $u_1 \in U^0$ {2}

    (4) $u_1 \in \bigcup_{W^+ \in W} W^+$ {1; 3}

    (5) $W_1^+ \in W \wedge u_1 \in W_1^+$ {D4.3, Chap. I; 4}

    (6) $U^0 \cap W_1^+ \subset W_1^+ \wedge U^0 \cap W_1^+ \neq \varnothing$ {3; 5}

    (7) u_2 is the first element of the set $U^0 \cap W_1^+$

                       {def. D6.1; 6}

We shall demonstrate now that u_2 is the first element of the set U^0.

    (1.1) $u \in U^0$ {ad.a.}

    (1.1.1) $u \in W_1^+$ {ad.a.}

    (1.1.2) $u \in U^0 \cap W_1^+$ {1.1; 1.1.1}

    (1.1.3) $u_2 = u \vee u_2 \prec_U u$ {7; 1.1.2}

    (1.2.1) $u \notin W_1^+$ {ad.a.}

    (1.2.2) $W_2^+ \in W \wedge u, u_2 \in W_2^+$ {L10.2a}

    (1.2.3) $W_1^+ = W_2(w_1)$ {L10.1c; 5; 1.2.1; 1.2.2}

    (1.2.4) $w_1 = u \vee w_1 \prec_{W_2} u$ {1.2.1–1.2.3}

    (1.2.5) $u_2 \prec_{W_2} w_1$ {7; 1.2.3}

    (1.2.6) $u_2 \prec_{W_2} u$ {formula (b), p.248; 1.2.4; 1.2.5}

(1.2.7)	$u_2, u \in U^0$	$\{7; 1.1\}$
(1.2.8)	$u_2 \prec_U u$	$\{1.2.6; 1.2.7\}$
(1.3)	$u_2 = u \lor u_2 \prec_U u$	$\{1.1.1 \to 1.1.3, 1.2.1 \to 1.2.8\}$

u_2 is the first element of the set U^0.

$$\{1.1 \to 1.3\}$$

L10.6. *The set* $\bigcup_{W^+ \in W} W^+$ *is well ordered.* $\qquad \{L10.4e; L10.5\}$

L10.7a. *If* $W^+ \in W$ *and* $w \in W^+$, *then* $W(w)$ *is a segment of the set* $\bigcup_{W^+ \in W} W^+$ *determined by the element* w.

The easy proof of this lemma is left to the reader.

L10.7b. *If* $w \in \bigcup_{W^+ \in W} W^+$ *and* V^+ *is a segment of the set* $\bigcup_{W^+ \in W} W^+$ *determined by the element* w, *then*

$$w = R(Z - V^+).$$

Proof.

(1)	$w \in \bigcup_{W^+ \in W} W^+$	$\{a.\}$
(2)	V^+ is a segment of the set $\bigcup_{W^+ \in W} W^+$ determined by the element w.	$\{a.\}$
(3)	$W_1^+ \in W \land w \in W_1^+$	$\{D4.3, \text{Chap. I}; 1\}$
(4)	$w = R[Z - W_1(w)]$	$\{ii, 3\}$
(5)	$W_1(w) = V^+$	$\{L10.7a; 2\}$
	$w = R(Z - V^+)$	$\{4; 5\}$

We shall now try to demonstrate that $\bigcup_{W^+ \in W} W^+ = Z$. In view of L10.6 this conclusion will terminate the proof of T10.1. Now, T4.2a Chap. I, yields the formula

$$\text{(III)} \qquad\qquad \bigcup_{W^+ \in W} W^+ \subset Z.$$

Assume temporarily that

$$\text{(IV)} \qquad\qquad \bigcup_{W^+ \in W} W^+ \underset{\neq}{\subset} Z.$$

Hence

$$Z - \bigcup_{W^+ \in W} W^+ \neq \varnothing.$$

We may thus introduce the following definition:

$$\text{(V)} \qquad\qquad a = R(Z - \bigcup_{W^+ \in W} W^+).$$

It follows from (I) that $a \in Z - \bigcup\limits_{W^+ \in W} W^+$. Hence

(VI) $$a \notin \bigcup_{W^+ \in W} W^+.$$

We introduce one more definition:

$$u T_1 v \equiv u T v \vee u \in \bigcup_{W^+ \in W} W^+ \wedge v = a.$$

L10.8a. *The set $\bigcup\limits_{W^+ \in W} W^+ \cup \{a\}$, which is a subset of the set Z, is well ordered, the ordering relation being T_1, and a is the last element of the set $\bigcup\limits_{W^+ \in W} W^+ \cup \{a\}$.*

The easy proof of this lemma is left to the reader.

L10.8b. *If $w \in \bigcup\limits_{W^+ \in W} W^+ \cup \{a\}$ and V^+ is a segment of the set $\bigcup\limits_{W^+ \in W} W^+ \cup \{a\}$, determined by the element w, then*

(1) $$w = R(Z - V^+).$$

Proof. If $w \in \bigcup\limits_{W^+ \in W} W^+$, then formula (1) follows simply from L10.7b. Assume then that $w = a$. The set $\bigcup\limits_{W^+ \in W} W^+$ is a segment of the set $\bigcup\limits_{W^+ \in W} W^+ \cup \{a\}$ determined by the element a. It follows from this fact and from (V) that the element a also satisfies formula (1).

It follows from L10.8a and L10.8b that the set $\bigcup\limits_{W^+ \in W} W^+ \cup \{a\}$ satisfies the conditions (i) and (ii). Hence that set is an element of the family of sets W. Consequently, $a \in \bigcup\limits_{W^+ \in W} W^+$.

This conclusion contradicts formula (VI), obtained from assumption (IV). Thus, that assumption is false. From this fact, and from (III), it follows that $Z = \bigcup\limits_{W^+ \in W} W^+$. This, and L10.6, leads to the conclusion that the set Z is well ordered. This conclusion completes the proof of T10.1.

The proof of Zermelo's theorem has been based on formula (I), the correctness of which results from T14.2, Chap. I. But conversely, the former is a consequence of the latter.

Let Z be any family of sets satisfying the assumptions of the axiom of choice. By Zermelo's theorem, we may assume that the set $\bigcup\limits_{Z \in Z} Z$ is well ordered. It can easily be seen that every element of the family Z is a non-empty subset of the set $\bigcup\limits_{Z \in Z} Z$, and hence has a first element. Let us form a set Y_1, whose elements are those and only those objects which are the first elements of the sets belonging to the family Z. It can easily be seen that the set Y_1 satisfies the thesis of Axiom A14.1.

D10.1. $$\bar{\alpha} = \overline{\overline{A^+}} \equiv \alpha = \overline{A^+}.^{(1)}$$

This definition is correct since, by Conclusion 2.2, sets whose order types are equal have the same power.

Thus the symbol $\bar{\alpha}$ stands for the power of a well-ordered set whose ordinal number is α. This is expressed by saying that $\bar{\alpha}$ is the cardinal number of the ordinal number α.

D10.1 and Conclusion 5.2 lead to

CONCLUSION 10.1.
$$\bar{\omega} = \overline{\overline{\langle \mathbf{N}, < \rangle}} = \aleph_0.$$

D10.2. $$\alpha \in Z(\mathfrak{m}) \equiv \bar{\alpha} = \mathfrak{m}.$$

Thus the set $Z(\mathfrak{m})$ is the set of all ordinal numbers whose cardinal number is $\mathfrak{m}$.

T10.2. $$\mathfrak{m} \in \mathrm{Cn} \to Z(\mathfrak{m}) \neq \emptyset.$$

Proof. By T7.2c in Chap. I, there exists a set A which satisfies the formula

(1) $$\overline{\overline{A}} = \mathfrak{m}.$$

By T10.1 there exists a well-ordered set A^+ whose range is the set A. By T2.2b, there exists an ordinal number α such that $\alpha = \overline{A^+}$. Hence from D10.1 and formula (1) we conclude that $\bar{\alpha} = \mathfrak{m}$. Hence by D10.2 $\alpha \in Z(\mathfrak{m})$. Thus the set $Z(\mathfrak{m})$ is non-empty.

D10.3a. α is the *initial number* of the set $Z(\mathfrak{m})$ if and only if $\mathfrak{m} \geqslant \aleph_0$ and α is the least ordinal number of the set $Z(\mathfrak{m})$;

b. α is an initial number if and only if there exists an $\mathfrak{m}$ such that α is the initial number of the set $Z(\mathfrak{m})$.

T10.2 and T8.3 lead to

---

(1) The convention is maintained that lower-case Greek letters are variables which stand for ordinal numbers.

CONCLUSION 10.2. *If* $\mathfrak{m} \geqslant \aleph_0$, *then there exists an initial number of the set* $Z(\mathfrak{m})$.

D10.4. $\alpha = \omega_\xi$ if and only if α is an initial number and the set of all initial numbers $< \alpha$, ordered by the relation "less than", is of type ξ.

It is clear that ω is the initial number of the set $Z(\aleph_0)$. It is thus the least initial number. Hence by D10.4, and since zero is the order type of the null set,

(1) $$\omega = \omega_0.$$

Note also that the initial number ω_1 is usually denoted by Ω.

D10.5. $\mathfrak{m} = \aleph_\xi$ if and only if ω_ξ is the initial number of the set $Z(\mathfrak{m})$.

Cardinal numbers which can be represented by the letter $\aleph$ with an index are called *alephs*.

We shall now examine whether the meaning of the symbol $\aleph_0$, as defined by D10.5, is in line with the interpretation given earlier. By D10.5, the number $\omega_0 = \omega$ is the initial number of the set $Z(\aleph_0)$. Hence, by D10.3a and D10.2, $\overline{\omega} = \aleph_0$. Hence by Conclusion 10.1 it follows that the meaning given to $\aleph_0$ by D10.5 agrees with the previous interpretation.

T10.3. *For every cardinal number* $\mathfrak{m} \geqslant \aleph_0$ *there exists an ordinal number* ξ *such that*
$$\mathfrak{m} = \aleph_\xi.$$

This theorem can also be stated as follows:

Every transfinite cardinal number is an aleph.

Proof. Let α be the initial number of the set $Z(\mathfrak{m})$. The existence of such a number is guaranteed by Conclusion 10.2 and the assumption that $\mathfrak{m} \geqslant \aleph_0$. Denote by ξ the ordinal number of the set of all initial numbers $< \alpha$.

By D10.4 $\alpha = \omega_\xi$. Hence by D10.5 $\mathfrak{m} = \aleph_\xi$. Thus the theorem holds.

D10.2, D10.3a, D10.4 and D10.5 lead to

CONCLUSION 10.3.
$$\aleph_\xi = \overline{\omega}_\xi.$$

T10.4. $$\aleph_\xi = \overline{\overline{E(\alpha < \omega_\xi)}}.$$

Thus the cardinal number $\aleph_\xi$ may be defined as the power of the set of all ordinal numbers $< \omega_\xi$.

Proof. We assume that the set $\underset{\alpha}{E}(\alpha < \omega_\xi)$ is ordered by the relation "less than". It follows from T8.2 that

$$\omega_\xi = \overline{\underset{\alpha}{E}\,(\alpha < \omega_\xi)}.$$

Hence by D10.1 we obtain

$$\overline{\omega}_\xi = \overline{\overline{\underset{\alpha}{E}\,(\alpha < \omega_\xi)}}.$$

Hence from Conclusion 10.3 the theorem follows.

T10.5. $\qquad\qquad \aleph_\xi < \aleph_{\xi+1}.$

Proof. It follows from D10.4 that $\omega_\xi \in \underset{\alpha}{E}(\alpha < \omega_{\xi+1})$. Hence

(1) $\qquad\qquad \underset{\alpha}{E}(\alpha < \omega_\xi) \subset \underset{\alpha}{E}(\alpha < \omega_{\xi+1}).$

Formula (1) and Conclusion 9.1 of Chap. I give

$$\overline{\underset{\alpha}{E}(\alpha < \omega_\xi)} \leqslant \overline{\underset{\alpha}{E}(\alpha < \omega_{\xi+1})}.$$

Hence by T10.4

$$\aleph_\xi \leqslant \aleph_{\xi+1}.$$

If $\aleph_\xi = \aleph_{\xi+1}$, then by D10.5 ξ would equal $\xi+1$, which, of course, cannot be the case. Hence the inequality

$$\aleph_\xi < \aleph_{\xi+1}$$

holds.

T10.6. $\qquad\qquad \aleph_1 \leqslant c.$

The proof of this theorem is omitted.

The *continuum problem* is the question whether the numbers $\aleph_1$ and c are equal, or whether the first is less than the second. The *continuum hypothesis* is the supposition that the former is true. It is clear that this hypothesis results from the assumption that there exists a relation R which well orders the set $\mathbf{R}$ of all real numbers,[1] and such that

$$\langle \overline{\mathbf{R}, R} \rangle = \Omega = \omega_1.$$

---

[1] The relation R must, of course, be different from the relation "less than".

Note also that the continuum problem is equivalent to the question whether there exists a cardinal number $\mathfrak{m}$ satisfying the inequality

$$\aleph_0 < \mathfrak{m} < \mathfrak{c},$$

that is, whether there exists an infinite and non-enumerable set of real numbers which is not equinumerous with the set R.

The continuum problem was formulated by Georg Cantor, the founder of set theory. The problem remains unsolved, notwithstanding the fact that it has been investigated by many eminent mathematicians, including Wacław Sierpiński. It is known, however, (Kurt Gödel, 1940) that the continuum hypothesis cannot lead to a contradiction in set theory.

SUPPLEMENT

1. Antinomies of Set Theory

The assumptions on which Cantor based his set theory seem self-evident, and yet when expanding that theory mathematicians soon encountered contradictions whose source they were at first unable to find. An argument which seems intuitively correct but leads to a contradiction is called an *antinomy*. The discovery of antinomies in set theory led to intensive work on the foundations of mathematics. These investigations showed that certain mathematical concepts, apparently quite simple and clear, have to be subjected to a detailed and thorough analysis. Fundamental difficulties are caused even by such basic concepts as those of a set and the relation of element-hood. That an imprecise use of these concepts may lead to contradictions is shown by the following argument:

Let R be the family of those and only those sets X which do not belong to themselves, and hence satisfy the condition $X \notin X$. This gives the equivalence

$$X \in R \equiv X \notin X.$$

By substituting the symbol R for the variable X we obtain

$$R \in R \equiv R \notin R.$$

But this formula immediately reveals a contradiction.

The result described above is called *Russell's antinomy*.

We shall now describe the *antinomy of the set of all sets*, which was known even to Cantor. We assume that every set is an element of the family of sets Z. Hence

$$2^Z \subset Z,$$

since every subset of the set Z is, of course, a set. Hence, by Conclusion 9.1 and T10.1 in Chap. I,

$$\overline{\overline{(2^Z)}} \leqslant \overline{\overline{Z}}, \quad \overline{\overline{Z}} < \overline{\overline{(2^Z)}}.$$

But it follows from the theorems proved in Sec. 9 of Chap. I that these formulae cannot be simultaneously true.

The concept of the set of all ordinal numbers also leads to a contradiction. In Chap. II we represented that set as On and showed (T8.4) that it is well-ordered by the relation "less than". Denote its ordinal number by φ, so that $\varphi \in$ On. By T8.2 the number φ is equal to the ordinal number of the segment of the set On determined by the element φ. Hence, the set On is similar to one of its segments. But this contradicts L7.3. This antinomy was chronologically the first antinomy of set theory. Published in 1897 by the Italian mathematician Burali-Forti, it was known to Cantor two years earlier. The concept of the set of all cardinal numbers also leads to a contradiction, but the derivation requires certain theorems on cardinal numbers which have not been given in the main text.

There are several ways of constructing a set theory that is free of contradictions. Two of these have particular significance in logic and mathematics. The first is called the theory of logical types and is due to Bertrand Russell, whose antinomy has been described in this Section. The other was developed by Ernst Zermelo.

Exercise

Show that in no town can there be a barber who shaves all and only those inhabitants of the town who do not shave themselves. Compare this reasoning with Russell's antinomy.

2. Semantic Categories

Before discussing the theory of logical types we shall outline the theory of division of logical expressions into what are called semantic categories, since these two theories are closely interrelated.[1]

The basic semantic categories are the category of sentences and the category of terms. Sentences in logic are those expressions which are either true or false. Hence the formulae "London lies on the Thames" and "the number 373 is divisible by 3" are sentences because the former is true and the latter false. The semantic category of sentences includes sentential functions, i.e., formulae incorporating

---

[1] This theory originates from the Polish logician Stanisław Leśniewski. The remarks given here are partly a repetition of those made in Chap. 1.

variables which become sentences when constants are substituted for the variables. As examples of terms we may quote "the number π", "London", "the greatest common divisor of the numbers 12 and 18". These are *singular* terms, i.e. terms which denote exactly one object each. We also distinguish *general* terms denoting more than one object, and empty terms which denote no object. As examples of the former we may quote "man", "natural number", and as examples of the latter, "golden mountain", "the greatest natural number". In mathematical logic usually only singular terms occur, and hence the subsequent discussion will be confined to such terms. The category of terms also includes variables for which constant terms may be substituted. In general, if certain constants are included in a given category, then that category also includes those variables for which the constants may be substituted.

In addition to sentences and terms we also distinguish logical expressions called *functors* and *operators*. We have seen examples of functors in Part I. Reference was made there to the functors of the sentential calculus (symbols of negation, implication, etc.) and the functors of the functional calculus (symbols A, B, C, ..., R, S, T, ...). These functors are called *sentence-forming*, since together with their arguments they form expressions of the sentential category. *Term-forming* functors are also common in mathematics. Symbols of arithmetical operations are examples of such functors. For instance, an expression constructed of the symbol "$+$" and the names of two numbers, that is an expression of the form "$3+5$", is the name of a number, whose simplest form is the symbol "8".

Expressions of various types can be arguments of functors. For instance, expressions of the sentential category are arguments of the functors of the sentential calculus; expressions of the term category are arguments of the functors of the functional calculus of the first order. Sentence-forming functors of sentential arguments are included in the same semantic category if and only if they are functors of the same number of arguments. Thus the symbols of implication, conjunction, alternation and equivalence are included in the same semantic category which, however, does not include the negation sign since it is a functor of one argument only. Similarly we divide the functors of arguments which are terms into the various semantic categories.

In more complicated logical systems there are functors whose arguments are neither sentences not terms, but expressions belonging to the category of functors. By way of example we give two definitions of functors of functor arguments:

D1.1. $$O(A) \equiv \prod_x \neg A(x).$$

D1.2. $$\text{Refl}(R) \equiv \prod_x xRx.$$

The sentence-forming functors "O" and "Refl" are placed in different semantic categories, since the former has as its argument a functor of one argument, and the latter a functor of two arguments, so that the arguments of the functors "O" and "Refl" belong to different semantic categories. In general, *two sentence-forming functors belong to the same semantic category if and only if they are functors of the same number of arguments and their respective arguments belong to the same semantic category.* An analogous remark applies to the term-forming functors. It is clear that two functors, one of which is sentence-forming and the other is term-forming, cannot belong to the same semantic category. Hence any given semantic category of functors has a specified number of arguments, and, further, the semantic categories of these arguments are also specified. It is clear that those and only those expressions are meaningful in which every functor has the number of arguments characteristic of its semantic category and these arguments belong to the appropriate semantic categories. Thus, for instance, an expression built of a functor and two terms is meaningful if and only if the functor belongs to the semantic category of functors of two term arguments.

The operators include the quantifiers "$\prod$" and "$\sum$", the description operator "ι", the symbol "E" as used in the context $\underset{\alpha}{E}\phi$, and the symbols of infinite and generalized sums and products of sets.

The common property of the operators is that they bind variables. Without formulating here any precise definition of this property we note that the rule of substitution does not apply to the bound variables.

As in the case of functors, we speak of arguments of operators. The variable under the symbol of the operator is not included in the arguments of that operator. For instance, in the formula "$\prod_{x} A(x)$" the formula "$A(x)$" is the argument of the operator "$\prod$", whereas in the formula "$\prod_{A(x)} B(x)$" the arguments of the operator "$\prod$" are the formulae "$A(x)$" and "$B(x)$".

The operators are divided into semantic categories analogous to functors, that is, with respect to the semantic category of the entire expression (consisting of an operator, the variable under its symbol, and its arguments) and the number and semantic categories of the consecutive arguments, also with consideration of the semantic category of the variable bound by a given operator.[1] Ordinary quantifiers are sentence-forming operators of one sentential argument, while limited-scope quantifiers are sentence-forming operators of two sentential arguments. The description operator introduced in the first-order functional calculus is a term-forming operator of one sentential argument. In the functional calculus of higher orders we may introduce description operators that bind variables standing for functors (of the ith order). The expression $\iota\, \Phi(\varphi)$ then belongs to the same semantic category as the variable φ.

It is evident that expressions belonging to different semantic categories cannot have the same meaning although, as we shall see later, functors of different categories may have clearly analogous properties. Hence symbols belonging to different semantic categories should be kept distinct. It is assumed henceforth that this condition is satisfied.

Consider the expression

(1) $\qquad\qquad\qquad 5+3.$

By replacing the symbol "$+$", which is a term-forming functor of two term arguments, by the symbol "$-$", which is also a term-forming functor of two term arguments, we obtain the expression

(2) $\qquad\qquad\qquad 5-3,$

---

[1] In the case of sentence-forming operators with sentential arguments we may assume that their semantic category is independent of the semantic category of the variables bound by those operators.

i.e., an expression which has a definite meaning. But if we replace "$+$" by a functor of a different semantic category, for instance by the sentence-forming functor "$\vee$", we would obtain an expression which is meaningless. Similarly if we replace the symbol "5" in expression (1) by any expression of the same semantic category, e.g., by expression (1) or (2), we obtain a meaningful expression. If, however, we replace the symbol "5" by an expression of a different semantic category, e.g., the formula "$p \to q$" or "$+$", we obtain a meaningless expression. By generalizing these remarks we arrive at

THEOREM I. *If in a logical expression which has a definite meaning we replace any symbol by a symbol belonging to the same semantic category, then the expression thus obtained continues to be meaningful, but if these two symbols belong to different semantic categories, then the expression thus obtained is meaningless.* [1]

The converse theorem is also true.

THEOREM II. *If two symbols have the property that when we replace one by the other in any meaningful expression we obtain a meaningful expression, then the symbols belong to the same semantic category.*

These theorems follow immediately from the interpretation given to the concepts under discussion.

We now introduce symbols denoting certain semantic categories which have particular importance in the subsequent analysis. Let K^0 stand for the category of terms, K^1 for the category of sentence-forming functors of one term argument, K^2 for the category of sentence-forming functors of one argument belonging to category K^1, and so on. Thus the symbol K^{i+1} ($i = 0, 1, 2, \ldots$) stands for the category of sentence-forming functors of one argument belonging to K^i. Let x be a variable of category K^0, and X^i a variable of category K^i. It is clear that the formula $X^i(x)$ is meaningful if and only if $i = 1$; the formula $X^i(X^j)$ is meaningful if and only if $i = j+1$ ($j \geqslant 1$).

---

[1] This theorem requires an additional restriction, namely that neither in the expression $\prod\limits_{\alpha}$ nor in the expression $\sum\limits_{\alpha}$ can the variable α be replaced by a constant of the same semantic category.

3. The Theory of Logical Types

At the end of the preceding Section we introduced a notation for formulae of the functional calculus which differs from that introduced in Part I: instead of writing $A(x)$, $A(y)$, $B(x)$, ..., we shall now write $X^1(x)$, $X^1(y)$, $Y^1(x)$, ... We shall further modify the symbolism by writing the formulae

(1) $$X^1(x), X^2(X^1), X^3(X^2), ...$$

in the form

(2) $$x \in_1 X^1, \quad X^1 \in_2 X^2, \quad X^2 \in_3 X^3, \quad ... \quad {}^{(1)}$$

These formulae are read: the object x is an element of set X^1; set X^1 is an element of the family of sets X^2; the family of sets X^2 is an element of the set of families X^3; etc. Thus, the letter x stands for any individual, the symbol X^1 for any set of individuals, X^2 for any family of such sets, and so on.

We shall now write two theorems of the functional calculus in the new notation:

(3) $$\prod_x (x \in_1 X^1 \to x \in_1 Y^1) \wedge \prod_x (x \in_1 X^1) \to \prod_x (x \in_1 Y^1),$$

(4) $$\prod_x (x \in_1 X^1) \to \prod_x (x \in_1 X^1 \vee x \in_1 Y^1).$$

Written in the symbolism used in Part I these formulae have the form

(3') $$\prod_x [A(x) \to B(x)] \wedge \prod_x A(x) \to \prod_x B(x),$$

(4') $$\prod_x A(x) \to \prod_x [A(x) \vee B(x)].$$

The definitions and theorems of the algebra of sets may likewise be written either in the symbolism of Part I or in that introduced in the present Section. We give below examples of such theorems and definitions, placing side by side the corresponding theorems and definitions of the functional calculus written in both symbolisms:

$$X = 1 \equiv \prod_x x \in X, \quad X^1 =_1 1^1 \equiv \prod_x x \in_1 X^1,$$
$$A =_1 1^1 \equiv \prod_x A(x),$$

---

(1) The number of indices in the symbolism here introduced may seem excessive, but it will be seen that these indices play an essential role in the theory of logical types.

$$X \subset Y \equiv \prod_x (x \in X \to x \in Y), \quad X^1 \subset_1 Y^1 \equiv \prod_x (x \in_1 X^1 \to x \in_1 Y^1),$$

$$A \subset_1 B \equiv \prod_x [A(x) \to B(x)],$$

$$x \in X \cup Y \equiv x \in X \lor x \in Y, \quad x \in_1 X^1 \cup_1 Y^1 \equiv x \in_1 X^1 \lor x \in_1 Y^1,$$

$$[A \cup_1 B](x) \equiv A(x) \lor B(x).$$

By using these equivalences we can express formulae (3) and (4) as follows:

(3″) $X^1 \subset_1 Y^1 \land X^1 =_1 1^1 \to Y^1 =_1 1^1,$

(4″) $X^1 =_1 1^1 \to X^1 \cup_1 Y^1 =_1 1^1.$

These formulae are analogues of the following theses of the algebra of sets:

$$X \subset Y \land X = 1 \to Y = 1,$$
$$X = 1 \to X \cup Y = 1.$$

It appears that the algebra of sets can be interpreted as a part of the functional calculus including only theses containing functors of not more than one term argument. It is convenient (though not necessary) to assume that in the system of logic of which the algebra of sets is to be a part the symbol of identity should be a primitive term.

The primitive terms of the algebra of sets given in Chapter I were the relation of being an element of a set and the notion of a universal set of individuals. The formula of the algebra of sets $x \in X$ has its analogue in the formula of the functional calculus $x \in_1 X^1$ while the latter formula is equivalent to the formulae $X^1(x)$ or $A(x)$. Hence the theses of the system now under consideration can be written without having to use the symbol $\in$. This symbol has been introduced into the functional calculus in order to bring out the relationship between that system and the algebra of sets. The concept of a universal set of individuals can be defined in logic as follows:

$$x \in_1 1^1 \equiv x =_1 x.$$

This definition immediately gives

$$x \in_1 1^1$$

which is an analogue of Axiom A1, given in Chap. I.

The second axiom used in the construction of the algebra of sets was the axiom of extensionality

$$\prod_x (x \in X \equiv x \in Y) \to X = Y.$$

Its analogue is the thesis of the functional calculus

(5) $$\prod_x (x \in_1 X^1 \equiv x \in_1 Y^1) \to X^1 =_1 Y^1$$

which can also be written in the following form:

$$\prod_x [A(x) \equiv B(x)] \to A = B.$$

Formula (5) is included among the axioms of logic. Without that axiom we could construct, within the sphere of logic, only a tiny fragment of set theory. It is also essential that the logical system now being considered should include definitions. If set theory is constructed as a part of logic having no definitions, we would have to adopt one more axiom. [1] Logical axioms usually include the axiom of infinity, which guarantees the existence of a infinite number of individuals. This axiom can, for example, be written in symbols as follows:

$$\sum_{X^1 \underset{\neq}{\subset} 1^1} X^1 \sim 1^1.$$

This axiom uses the symbol for sets being equinumerous. However, we shall see in the next section that this symbol can be defined in the field of logic. Note also that the proofs of the theses of the algebra of sets and of those theses of the functional calculus which are their analogues, are also quite analogous. This point can be verified by referring to the theses given in Sec. 1 of Chap. I.

We have discussed in some detail the relationship between the algebra of sets and the functional calculus and we have pointed out that other branches of set theory, for instance the arithmetic of cardinal and ordinal numbers, can be interpreted as parts of logic.

We have used the symbol 1^1 to denote the set of all individuals, i.e., objects which are not sets. This set will be called the full set of the first type, and its elements objects of zero type. The constants of

---

[1] The logical systems in which no definitions occur are in certain respects more convenient than systems which include definitions, and for that reason are often used.

the semantic category K^0 are names of definite objects of zero type; the variables of this category are used to describe properties of any objects of zero type. A formula beginning with a universal or existential quantifier binding a variable of category K^0 states that all or some objects of zero type have the property described by the formula that follows the quantifier. We denote by 1^2 the set of all subsets of the set 1^1. These sets are called objects (or sets) of the first type. The constants of category K^1 are names of definite elements (which, of course, are themselves sets) of the set 1^2.

The variables of this category are used to describe the properties of any elements of the set 1^2. Elements of the sets of the first type can only be objects of zero type. In general, the symbol 1^{i+1} ($i = 0, 1, 2, \ldots$) stands for the set of all subsets of the set 1^i. The set 1^{i+1} is called the full set of the $(i+1)$th type, and the elements of that set, objects of the ith type. The constants of category K^i are names of definite elements of the set 1^{i+1} (objects of the ith type); the variables of this category are used to describe the properties of any elements of the set 1^{i+1}. Only objects of the ith type can be elements of sets of the $(i+1)$th type.

We shall discuss in more detail the definition of functors of the semantic category K^{i+1} ($i = 0, 1, 2, \ldots$). The scheme of definition of such a functor has the form

$$X_1^{i+1}(X^i) \equiv \Phi(X^i)$$

where the symbol X_1^{i+1} stands for the constant functor of category K^{i+1} defined by a given definition, the symbol X^i is a variable of category K^i, and the symbol $\Phi(X^i)$ stands for any formula which contains the free variable X^i and no other free variable. We assume that the symbol X_1^{i+1} is the name of a definite set of the $(i+1)$th type, whose elements are those and only those objects of the ith type which satisfy the sentential function $\Phi(X^i)$. We also assume that this set exists if and only if $\Phi(X^i)$ is a meaningful formula.

The division of objects into individuals, sets of individuals, families of sets of individuals, etc., is called the *theory of logical types*. This division has the following properties:

(1) elements of a set of any given type can only be objects of a type one less;

(2) the full set of any type, except for the first type, is the family of all subsets of the full set of type one less.

Sets are often identified with properties in the sense that to say an object is an element of a set is equivalent to saying that the object has a corresponding property. For instance, the statement "x is an element of the set of white objects" is equivalent to the statement "x has the property of whiteness". In this approach individuals are treated as objects which have properties but are not themselves properties; sets of individuals are treated as properties of objects, families of sets of individuals as properties of such properties, etc. That properties can themselves have properties is shown, for example, by the statement that a certain property is not very probable. Thus, in this case a low probability is a property of properties. It is clear that sets so interpreted, when they belong to different logical types, are essentially different objects; no individual can be identical with any set. Thus the interpretation of sets as properties is consistent with the theory of logical types. This interpretation, however, involves certain difficulties which will not be discussed here.

We shall now consider how the theory of logical types prevents antinomies from occurring in set theory. Russell's antinomy has its source in the definition

$$X \in R \equiv X \notin X.$$

This definition, however, is not correct in form, since its right-hand side is not a meaningful formula in the theory of logical types. This is because the variable X on both sides of the symbol $\in$ has the same index "i". On the other hand, the following formula is meaningful in the theory of logical types:

$$X^i \in_{i+1} R^{i+1} \equiv X^i \notin_{i+1} X^{i+1}.$$

But this is not a correct definition, since on its right-hand side it has the free variable X^{i+1}, which does not occur on the left-hand side.[1]

Nor, in the theory of logical types, can we define the set of all sets; we can only define the set of all sets of a given type. That concept, however, does not lead to a contradiction. Thus, in the theory of

---

[1] A correct definition should satisfy the condition of homogeneity (cf. Part I, Chap. 2, Sec. 5).

logical types we can reproduce neither the antinomy of the set of all sets, nor Burali-Forti's antinomy, nor any other known antinomy of set theory.

It is also worth mentioning the theory of division of relations into logical types. We assume that sentence-forming functors of more than one argument are names of certain sets, functors of two arguments being names of sets of ordered pairs, functors of three arguments, names of ordered triples, etc. Sets of ordered n-tuples are called relations. Elements of ordered n-tuples forming relations may be objects of any logical type. The theory of the division of relations into logical types does not differ essentially from the theory of the division of sets which are not relations, but is more complex.

4. Certain Set-Theoretical Concepts as Defined in Logic

As mentioned above all the concepts of set theory can be defined in logic. But for one and the same set-theoretical concept we have in logic not just one analogue, but infinitely many. This is best illustrated by means of the example of the concept of a function, which in Chap. I was defined by the formula

$$\prod_x \prod_y \prod_z (xRy \land xRz \to y = z).$$

The analogues of this formula in logic are the following formulae:

$$\prod_x \prod_y \prod_z (xR^1y \land xR^1z \to y =_1 z),$$

$$\prod_{X^i} \prod_{Y^i} \prod_{Z^i} (X^i R^{i+1} Y^i \land X^i R^{i+1} Z^i \to Y^i =_i Z^i) \quad (i = 1, 2, ...).$$

We thus have an infinite sequence of concepts of functions belonging to different logical types, but defined in analogous ways and having analogous properties. We say that such concepts are "typically ambiguous". The concepts of the right and left domain of a relation and the concepts of a perfect (one–one) relation and of sets being equinumerous are also typically ambiguous. Definitions of these concepts in logic are quite analogous to the corresponding definitions in set theory.

We recall that the only primitive concepts of set theory which go beyond the algebra of sets are the concepts of the power of a set, of the type of an ordered set, and of an ordered pair. These concepts can be defined in logic; they are also typically ambiguous.

We shall confine ourselves to definitions of these concepts in the simplest cases. The definitions of the first two concepts are called *definitions by abstraction*. We shall briefly discuss this kind of definition.

Let R^1 be any equivalence relation (cf. Sec. 6 in Chap. I). Then this relation will satisfy the formulae

I. $$x R^1 x,$$
II. $$x R^1 y \to y R^1 x,$$
III. $$x R^1 y \wedge y R^1 z \to x R^1 z,$$
D4.1. $$y \in_1 [x]_{R^1} \equiv x R^1 y.$$

The set $[x]_{R^1}$ is called the *abstraction class* determined by the element x with respect to the relation R^1.

We shall give some properties of the concept defined above. We assume that R^1 is an equivalence relation.

L4.1. $\qquad\qquad x \in_1 [x]_{R^1}$ $\qquad\qquad\qquad$ {D4.1, I}

L4.2. $\qquad\qquad [x]_{R^1} =_1 [y]_{R^1} \to x R^1 y.$

    Proof. (1) $\quad [x]_{R^1} =_1 [y]_{R^1}$ $\qquad\qquad\qquad$ {a.}
           (2) $\quad y \in_1 [y]_{R^1}$ $\qquad\qquad\qquad\qquad$ {L4.1}
           (3) $\quad y \in_1 [x]_{R^1}$ $\qquad\qquad\qquad\qquad$ {1, 2}
           $\qquad x R^1 y$ $\qquad\qquad\qquad\qquad\qquad$ {D4.1, 3}

L4.3. $\qquad\qquad x R^1 y \to [x]_{R^1} \subset_1 [y]_{R^1}.$

    Proof. (1) $\quad x R^1 y$ $\qquad\qquad\qquad\qquad$ {a.}
          (1.1) $\quad z \in_1 [x]_{R^1}$ $\qquad\qquad\qquad$ {ad.a.}
          (1.2) $\quad x R^1 z$ $\qquad\qquad\qquad\qquad$ {D4.1, 1.1}
          (1.3) $\quad z R^1 y$ $\qquad\qquad\qquad\qquad$ {II, III, 1, 1.2}
          (1.4) $\quad z \in_1 [y]_{R^1}$ $\qquad\qquad\qquad$ {D4.1, II, 1.3}
          $\qquad [x]_{R^1} \subset_1 [y]_{R^1}$ $\qquad\qquad\quad$ {1.1 → 1.4}

L4.4. $\qquad\qquad x R^1 y \to [y]_{R^1} \subset_1 [x]_{R^1}$ $\qquad\qquad$ {L.4.3, II}

T4.1. $\qquad\qquad [x]_{R^1} =_1 [y]_{R^1} \equiv x R^1 y$ $\qquad$ {L4.2, L4.3, L4.4}

In the same way we could prove

T4.1a. $\qquad\qquad [X^1]_{R^2} =_2 [Y^1]_{R^2} \equiv X^1 R^2 Y^1.$

Theorems T4.1 and T4.1a differ only in the semantic categories of the constants and variables which occur in them.

We shall explain the intuitive meaning of D4.1 and T4.1. Every equivalence relation states that two objects are the same with respect to a certain property. For instance the relation of similarity of geometrical figures states that two figures are of the same shape, the congruence relation states that two figures are of the same shape and size. Thus the set $[x]_{R^1}$ is the set of all objects which are the same as x with respect to a certain property. If R^1 is the relation of similarity of geometrical figures, and x is an equilateral triangle, then $[x]_{R^1}$ is the set of all equilateral triangles. If sets are identified with properties, then $[x]_{R^1}$ is a property which is an attribute of x and of all objects which do not differ from x with respect to that property.

Theorem T4.1 states that *the properties $[x]_{R^1}$ and $[y]_{R^1}$ are identical if and only if x and y do not differ from one another with respect to that property.*

We know that the relation of sets being equinumerous is an equivalence relation. Let X^1 be any set of individuals. The power of set X^1 is interpreted as the property which is an attribute of set X^1 and of all sets equinumerous with X^1. Thus the power of set X^1 is an abstraction class determined by X^1 with respect to the relation of being equinumerous. Hence

(1) $$\overline{\overline{X^1}} =_2 [X^1]_\sim.$$

It is clear that the power of a set interpreted in this way is typically ambiguous.

The only axiom on which the theory of equinumerous sets was based in Chap. I, was the axiom

$$\overline{\overline{X^1}} =_2 \overline{\overline{Y^1}} \equiv X^1 \sim Y^1.$$

Using formula (1) we can write this axiom in the following form:

$$[X^1]_\sim =_2 [Y^1]_\sim \equiv X^1 \sim Y^1.$$

The formula thus obtained is a special case of T4.1a, and hence can be derived purely within logic.

The remarks on the power of sets and on equinumerous sets can be transferred without any essential changes to the concepts of type and similarity of ordered sets.

The concept of an ordered pair still remains to be discussed. The pair $\langle x, y \rangle$ is defined in logic as the family of sets whose only elements are the unit set $\{x\}$ and the set of two elements $\{x, y\}$. Hence the following formula is satisfied:

(2) $$\langle x, y \rangle = \{\{x\}, \{x, y\}\}.$$

The only axiom concerning ordered pairs which was adopted in set theory (cf. Sec. 5 in Chap. I) had the form

(3) $$\langle x, y \rangle = \langle z, t \rangle \equiv x = z \wedge y = t.$$

Using formula (2) as a definition of an ordered pair we can establish formula (3).

The equations

(4) $$x = z \quad \text{and} \quad y = t$$

and D1.3a from Chap. I give

(5) $$\{x\} = \{z\} \quad \text{and} \quad \{x, y\} = \{z, t\}.$$

These equations and formula (2) yield the equation

(6) $$\langle x, y \rangle = \langle z, t \rangle.$$

We shall show that conversely equations (4) follow from equation (6). We consider first the case where $x \neq y$. Then formula (6), by formula (2), gives formulae (5), which in turn lead to formulae (4). If $x = y$, then

$$\langle x, y \rangle = \{\{x\}, \{x, y\}\} = \{\{x\}, \{x, x\}\} = \{\{x\}\}.$$

From equation (6) it then follows that $z = t$ and that

$$\langle z, t \rangle = \{\{z\}\} = \{\{x\}\}$$

so that $z = x = y = t$. Thus in both cases formulae (4) follow from formula (6). Formula (3) is therefore true.

Note, however, that the axiom of choice, which was discussed in some detail in Sec. 14, Chap. I, cannot be proved in logic.

5. Axiomatic Set Theory

As previously mentioned, axiomatic set theory was developed by Ernst Zermelo. The axioms he formulated have undergone frequent but rather inessential changes. The set of axioms given below are virtually the same as those given in *Teoria mnogości* (Set Theory) by Kazimierz Kuratowski and Andrzej Mostowski.

The primitive terms used in these axioms are the null set and the relation of being an element.

In writing the axioms of set theory we shall use the same notation as in Part II. In axiomatic set theory there are variables of only one semantic category. They will be written $x, y, z, t, \ldots$. The basic formulae of the system under consideration are of the form

$$x \in y, \quad x = y.$$

In these formulae, the symbol "$\emptyset$", which stands for the null set, may appear instead of variables. Formulae of these forms may be preceded by the negation symbol and by quantifiers, and may be connected by the symbols of implication, conjunction, alternation and equivalence. The same operations can be performed again on the formulae thus obtained, and so on. Any formulae obtained by the process just described will be called *meaningful* (*well-formed*) *formulae* of axiomatic set theory.

The identity symbol may be regarded as a logical symbol. Set theory is then said to be based on the first-order functional calculus with identity (which is often formulated metalogically). But identity can also be defined by means of the symbol $\in$. The following equivalence can serve as such a definition:

$$(I) \qquad x = y \equiv \prod_z (x \in z \equiv y \in z).$$

In axiomatic set theory this equivalence, although analogous with the equivalence TVI (Part. I, Chap. II, Sec. 6), differs from it by the semantic categories of the variables which occur in it. In (I) the variable z belongs to the same semantic category as the variables x and y, whereas the variable A occurring in TVI belongs to a different semantic category from that to which the variables x and y belong.

To give a lucid form to the axioms of set theory we introduce a number of definitions.

D1. $$S(x) \equiv \sum_y (y \in x) \lor x = \emptyset.$$

The formula $S(x)$ is read: x is a set. Hence, an object x is a set if and only if there exists an object which is an element of x, or if it is identical with the null set.

D2. $$F(x) \equiv S(x) \land \prod_{y \in x} S(y).$$

The formula defined above is read: x is a family of sets. Thus, an object x is a family of sets if and only if it is itself a set and each of its elements is also a set.

Note that the formulae $S(x)$ and $F(x)$ are only abbreviations of the formulae occurring in D1 and D2 to the right of the equivalence symbol. Hence the functors "S" and "F" are not to be interpreted as names of sets.

The remaining definitions do not differ in meaning from the corresponding definitions in Chap. I, but are written in a somewhat different notation. Each of the definitions given below is accompanied on the right by the number of its analogue in Chap. I.

D3. $$x \subset y \equiv S(x) \wedge S(y) \wedge \prod_z (z \in x \to z \in y) \qquad \{\text{D1.2}\}$$

D3a. $$x \underset{\neq}{\subseteq} y \equiv x \subset y \wedge x \neq y \qquad \{\text{D1.9}\}$$

D4. $$x = \bigcup_{z \in y} z \equiv \prod_t [t \in x \equiv \sum_{u \in y} t \in u] \qquad \{\text{D4.3}\}$$

D5. $$x = 2^y \equiv \prod_z (z \in x \equiv z \subset y) \qquad \{\text{D10.1}\}$$

D6. $$x = y \cap z \equiv \prod_t (t \in x \equiv t \in y \wedge t \in z) \qquad \{\text{D1.5}\}$$

D7. $$x = \{y, z\} \equiv \prod_t [t \in x \equiv (t = y \vee t = z)] \qquad \{\text{D1.3b}\}$$

Note also that instead of writing $\neg(x \in y)$ we shall write $x \notin y$. We can now state the axioms of set theory.

A1. $$\prod_{S(x)} \prod_{S(y)} [\prod_z (z \in x \equiv z \in y) \to x = y].$$

This axiom is called the *axiom of extensionality*. It does not differ in meaning from A1.2 in Chap. I.

A2. $$\prod_x (x \notin \varnothing).$$

This axiom states that no object is an element of the null set.

A3. $$\prod_x \prod_y \sum_z [z = \{x, y\}].$$

For any objects x and y there exists a set of which these objects are the only elements.

A4.
$$F(y) \rightarrow \sum_{S(x)} (x = \bigcup_{z \in y} z).$$

Thus for every family of sets there exists a set which is the sum of the elements of that family.

A5.
$$S(y) \rightarrow \sum_{F(x)} (x = 2^y).$$

Thus for every set there exists a family of all its subsets.

A6.
$$\sum_{F(x)} (x \neq \emptyset \wedge \prod_{y \in x} \sum_{z \in x} y \subsetneq z).$$

By this axiom there exists at least one non-empty family of sets such that for every set y which is an element of that family there exists a set which is also an element of that family, and of which the set y is a proper part. Note that this condition is satisfied, for example, by the following family of subsets of the set of all natural numbers:

$$(1), (1, 2), (1, 2, 3), \ldots, (1, 2, \ldots, n), \ldots$$

If we consider only an arbitrary finite set, for instance the set of the natural numbers $\leqslant 1000$, and any arbitrary non-empty family of its subsets, then those subsets would always include a subset which is not a proper part of any subset belonging to the family. Thus A6 guarantees the existence of an infinite number of objects. This axiom is called the *axiom of infinity*.

A7. $F(x) \wedge x \neq \emptyset \wedge \emptyset \notin x \wedge \prod_{y \in x} \prod_{z \in x} (y \neq z \rightarrow y \cap z = \emptyset)$

$$\rightarrow \sum_{S(t)} \prod_{u \in x} \sum_{v} {}_1 v \in t \cap u.$$

This is the *axiom of choice*, which was discussed in Sec. 14, Chap. I.

Let x be any set, and let $\Phi(z)$ be any meaningful formula of axiomatic set theory, containing the free variable z but not the free variables x and y.

A8.
$$S(x) \rightarrow \sum_{S(y)} \prod_z [z \in y \equiv z \in x \wedge \Phi(z)].$$

It is clear that the set y, whose existence is stated by A8, is a subset of the set x. Thus by A8 for every set x and for every condition

$\Phi(z)$ there exists a set of objects satisfying that condition, with the further property that it is a subset of the set x. Properly speaking A8 is not a single axiom, but a schema for an infinite number of axioms, which are obtained by substituting in A8 for $\Phi(z)$ specific meaningful formulae, for instance the formula "$z \neq \emptyset$". The formula thus obtained

$$S(x) \rightarrow \sum_{S(y)} \prod_z (z \in y \equiv z \in x \wedge z \neq \emptyset),$$

is included in axiomatic set theory as one of its theses.

Axiom A8 is called the *axiom of subsets*.

As has already been mentioned, set theory can be constructed without the concepts of power and order type, using instead the concepts of equinumerous sets and similarity. But if we want these concepts to be included in axiomatic set theory, we must introduce them as primitive concepts of the theory, because they cannot be defined in axiomatic set theory. This is one of the major differences between axiomatic set theory and the theory of logical types. We also have to add to the axioms of set theory Axioms A7.1 (Chap. I) and A2.1 (Chap. II).

In axiomatic set theory the left sides of the definitions D7.3 (Chap. I), D2.6 (Chap. II) and D6.2 (Chap. II) will be, respectively: Cn($\mathfrak{m}$), Ot(α), and On(α). Some other definitions, introduced in preceding chapters, must also be appropriately reformulated in terms of that theory. For instance, D5.2 (Chap. I) will be replaced by the definition: $x \in \underset{a \supseteq X}{E} \phi(a) \equiv x \in X \wedge \phi(x)$; D5.3 (Chap. I) will be modified in an analogous way, while in D6.1 (Chap. I) we shall introduce, in its left side, the formula "Funct(R)" or shall restrict D6.1 by defining the set of functions belonging to a given set.

We shall see now how the assumptions of axiomatic set theory safeguard it against antinomies. We shall analyse this issue in some detail using the example of Russell's antinomy. As was emphasized earlier the basis of that antinomy is the equivalence

(1) $$X \in R \equiv X \notin X.$$

The following equivalence is related to it in meaning:

(2) $$R(x) \equiv x \notin x,$$

and may be included in axiomatic set theory since the formula "$x \notin x$"
is meaningful unlike the corresponding formula in the theory of
logical types. But the formula $R(x)$ is only an abbreviation of the
formula "$x \notin x$", and the symbol R is not here the name of a set. Hence
it may not be substituted for the variable x. Formula (2) does not,
therefore, lead to a contradiction.

Let us now consider the formula

$$(3) \qquad S(x) \rightarrow \sum_{S(y)} \prod_z (z \in y \equiv z \notin z),$$

whose meaning also comes close to that of formula (1). We shall
show first that formula (3) leads to a contradiction. By substitution
we obtain

$$(3a) \qquad S(\varnothing) \rightarrow \sum_{S(y)} \prod_z (z \in y \equiv z \notin z).$$

By D1, the antecedent of this implication is true, hence the consequent is also true; hence by the rule for omitting the existential quantifier we obtain

$$\prod_z (z \in y_1 \equiv z \notin z),$$

which in turn gives

$$(4) \qquad y_1 \in y_1 \equiv y_1 \notin y_1,$$

which immediately leads to a contradiction.

Note, however, that formula (3) is not a thesis of axiomatic set
theory since A8 permits us to join to the system the formula

$$(5) \qquad S(x) \rightarrow \sum_{S(y)} \prod_z (z \in y \equiv z \in x \wedge z \notin z),$$

This formula has a structure similar to, but not identical with, that
of formula (3). By using a procedure similar to that by which we
deduced formula (4), we obtain from formula (5) the formula

$$y_1 \in y_1 \equiv y_1 \in \varnothing \wedge y_1 \notin y_1.$$

This formula, however, does not lead to a contradiction. Thus we see
that Russell's antinomy cannot be reproduced in axiomatic set theory,

just as it could not be reproduced in the theory of logical types. The same holds for all the known antinomies of set theory. The question arises whether some so far undiscovered antinomy may lead to a contradiction in axiomatic set theory or in the theory of logical types. It has been proved that if the axiom of infinity is omitted then both theories are consistent. The conviction that these theories together with the axiom of infinity are consistent has to be based on the empirical fact that these theories have been very much developed without resulting in a contradiction.

The question also arises as to how it comes about that these theories avoid antinomies. This is explained by the fact that each of them restricts the way in which sets can be constructed. In the theory of logical types we assume that for every sentential function there exists a set of objects that satisfy that function. But the construction of sentential functions is restricted by the conditions which must be satisfied by the semantic categories of the formulae preceding and following the symbol "ϵ", which in that theory is "typically ambiguous". In axiomatic set theory there are no such restrictions, but on the other hand it is not assumed that for every sentential function there is a set of objects satisfying that function. It is only assumed that given a sentential function every set has a subset consisting of objects satisfying that sentential function. Hence in neither of these theories can we construct the set of all sets or the set of all those sets which do not belong to themselves.

Although the intuitive assumptions on which these two theories are based differ essentially, the theorems they contain are almost the same. To put it more precisely: to every thesis of axiomatic set theory there corresponds a group of theorems of the theory of logical types, which differ from one another only in that in each of them reference is made to sets of different logical types. Conversely, to each such group of theorems of the theory of logical types there corresponds a thesis of axiomatic set theory, which has the same or a closely related meaning.

It would be difficult to decide which of these two theories is more satisfactory. Both have defects in formal structure and in intuitive assumptions. That is why in the exposition of set theory given in Part II neither theory was followed exclusively. Nor are they in other textbooks, even in Wacław Sierpiński's. A detailed exposition of axiomatic set theory can be found in the textbook by Kazimierz Kuratowski and Andrzej Mostowski, and an outline of the theory of logical types in *Logika matematyczna* (Mathematical Logic) by Andrzej Mostowski.

6. Set Theory and Arithmetic

In arithmetic, integers are defined by means of natural numbers, and the arithmetic of integers is based on the arithmetic of natural numbers. Much the same relation holds between the arithmetic of rational numbers and of integers; the arithmetic of real numbers and of rational numbers; the arithmetic of complex numbers and of real numbers. Thus only natural numbers remain undefined, and hence in constructing their arithmetic we have to introduce special axioms. The earliest and best known axiom system for the arithmetic of natural numbers was given by the Italian mathematician and logician Giuseppe Peano. The primitive concepts used in this system are: the number zero, the set of all natural numbers, and the function

$$x = \operatorname{seq} y$$

which is read: x is the successor of y. It is assumed that it is equivalent to the formula

$$x = y+1.$$

In writing down Peano's axioms we used the symbolism introduced in Part II:

I. $0 \in \mathbf{N},$[1]

II. $x \in \mathbf{N} \to \operatorname{seq} x \in \mathbf{N},$

III. $x, y \in \mathbf{N} \wedge \operatorname{seq} x = \operatorname{seq} y \to x = y,$

IV. $x \in \mathbf{N} \to \operatorname{seq} x \neq 0,$

V. $0 \in X \wedge \prod_x (x \in X \to \operatorname{seq} x \in X) \to \mathbf{N} \subset X.$

The last axiom is the axiom of induction.

The fact that the primitive concepts of Peano's arithmetic can be defined in terms of the concepts of set theory, and that Axioms

---

[1] Thus in Peano's arithmetic zero is included among the natural numbers.

I–V can be proved in set theory, is of great significance for the foundations of mathematics. Thus Peano's arithmetic is a part of set theory. We know that the construction of other arithmetics, for instance the arithmetic of real numbers, does not require the introduction of any new primitive concepts or the adoption of new axioms. Consequently, the arithmetics of all kinds of numbers, and hence mathematical analysis too, are part of set theory.

The definitions of the terms "0", "N", and "seq" are essentially the same in the theory of logical types and in axiomatic set theory, though they differ in formal details. In order to avoid the unnecessary complications resulting from this difference of approach, we shall give only verbal, and hence necessarily imprecise definitions of the concepts under consideration.

Zero is defined as the power of the null set, and the set of natural numbers as the set of all cardinal numbers which are not transfinite.[1] To define the concept of a successor we introduce an auxiliary relation ρ, which holds between two sets if and only if one of them includes the other, and their difference is a unit set. The successor can then be defined as follows:

x is the *successor* of y if and only if x and y are powers of sets between which the relation ρ holds.

The proofs of Axioms I–V in set theory will not be given here; we confine ourselves to observing that in order to construct the arithmetic of natural numbers, both in the theory of logical types and in axiomatic set theory, it is necessary to include the axiom of infinity.

7. The Philosophical Aspects of the Concept of a Set

If we do not examine it closely, the concept of a set seems to be quite simple and clear. We know, however, that when used without precision, this concept can lead to contradictions. This is not the only difficulty connected with the concept of a set and its related concepts. Many philosophers, logicians and mathematicians are engaged in lively discussion on the philosophical and intuitive foundations of set theory. So far these discussions have not led to any

---

[1] Cf. Sec. 11 in Chap. 1, Part 2.

generally accepted viewpoint concerning the various points of issue. We shall confine ourselves here to presenting the more important standpoints, mostly in the form of quotations from statements by contemporary logicians and mathematicians.

We note first that in ordinary language the word "set" has two distinct meanings, one of which is called collective and the other distributive.[1] In the collective sense, a set of objects is a single entity consisting of the objects which are its elements in the same way as a chain consists of its links, or a heap of sand consists of grains of sand. In that sense, a set of concrete objects perceivable by the senses, is also a concrete perceivable object. When using the term "set" in this sense, we understand the statement "x is an element of the set A" as meaning "x is part of the set A" (the word "part" being interpreted in the same way as when we say that a leg is part of a table). A theory of sets so interpreted has been built up by the Polish logician Stanisław Leśniewski and called by him *mereology*. When we use the term "set" in the distributive sense, we consider the statement "Mars is an element of the set of planets of the solar system" as equivalent to the statement "Mars is a planet of the solar system". The difference between the two meanings is illustrated by the fact that certain statements which are true in one sense of the term "set" are false in the other. For instance, in the collective sense it is true that one-tenth part of Mars is an element of the set of the planets of the solar system, since it is part of that system; in the distributive sense the statement is false, since one-tenth part of Mars is not a planet of the solar system.

It is not difficult to see that in set theory the term "set" is used in the distributive, and not in the collective, sense, and the relation of being an element is not interpreted as being that of a part of a whole. The latter relation differs from the former in its essential properties, for instance it is transitive, whereas transitivity does not hold in the case of "$\in$". This is shown by the fact that the formulae $\varnothing \in \{\varnothing\}$ and $\{\varnothing\} \in \{\{\varnothing\}\}$ are true, whereas the formula $\varnothing \in \{\{\varnothing\}\}$ (equivalent to the formula $\varnothing = \{\varnothing\}$) is false. As noted by Bertrand

---

[1] Cf. T. Kotarbiński, *Elementy teorii poznania, logiki formalnej i metodologii nauk* (Elements of Epistemology, Formal Logic and the Methodology of Sciences), 2nd. ed. Warszawa 1961, pp. 23–4.

Russell, should we treat sets as aggregates or conglomerates (and hence as sets in the collective sense), "we should find it impossible to understand how there can be such a class as the null-class, which has no members at all and cannot be regarded as a "heap"; we should also find it very hard to understand how it comes about that a class which has only one member is not identical with that one member."[1]

Willard Van Orman Quine says: "The fact that classes are universals, or abstract entities, is sometimes obscured by speaking of classes as mere aggregates or collections, thus likening a class of stones, say, to a heap of stones. The heap is indeed a concrete object, as concrete as the stones that make it up; but the class of stones in the heap cannot properly be identified with the heap. For, if it could, then by the same token another class could be identified with the same heap, namely, the class of molecules of stones in the heap. But actually these classes have to be kept distinct; for we want to say that one has just, say, a hundred members, while the other has trillions. Classes, therefore, are abstract entities."[2]

The issue whether and how sets exist in the distributive sense has much in common with the controversy from the history of philosophy about the existence of universals.[3]

Universals are general objects, such as "man in general" or "horse in general". Instead of speaking of man in general and horse in general we may speak of the species of men and the species of horses. Individual men and individual horses can be perceived, whereas man in general and horse in general (or the species of men and the species of horses) are objects that can be grasped only conceptually.

The following standpoints were outlined in the controversy about universals, which was particularly lively in the Middle Ages.

1. Radical realism, formulated by Plato, assumes an independent existence of universals, outside individual objects. Outside individual men there exists the idea of men, different from every man and existing independently of the existence of individual men.

---

(1) B. Russell, *Introduction to Mathematical Philosophy*, p. 183.

(2) W. V. Quine, *From a Logical Point of View*, p. 114.

(3) The remarks on universals that follow are based on K. Ajdukiewicz's remarks in his book *Zagadnienia i kierunki filozofii* (Problems and Trends in Philosophy), pp. 120–4.

2. Moderate realism, formulated by Aristotle, assumes that universals do not exist independently, but as essential properties of individual objects. Thus "man in general" is the property or the system of properties inherent in individual men and characteristic of them; for instance, rationality is such a property.

3. Conceptualism rejects the existence of universals outside the cognitive mind, and accords them conceptual existence only, assuming that only general concepts exist. Thus "man in general", does not exist, whereas the concept of man does exist.

4. Nominalism rejects the existence of both universals and general concepts, assuming only the existence of general terms. Thus there exists neither an idea of man, nor a concept of man, but only the general term "man".

"The old philosophical controversy about the existence of universals," wrote Kazimierz Ajdukiewicz in his *Zagadnienia i kierunki filozofii* (Issues and Trends in Philosophy), "reappears in recent philosophy under somewhat different form. The contemporary form of that issue is expressed in the question whether *a priori* disciplines, i.e., such disciplines as mathematics, investigate a completely real world, though quite different from the world given to us in sensory experience, namely what is called the world of ideal entities such as numbers, mathematical functions, etc., which exists in a way quite independent of our minds, or whether such a world does not exist at all."

As we know, the concepts of number and function, and other mathematical concepts can be reduced to the concept of a set. Hence the issue reduces to the question whether sets exists, and if so, in what sense.

Not only philosophers, but logicians and mathematicians as well have drawn attention to the analogy between the issues of the existence of sets and of the existence of universals.

In their book on the foundations of set theory, Abraham A. Fraenkel, one of the co-founders of axiomatic set theory, and Yehoshua Bar-Hillel write:

"Since sets, as ordinarily understood,[1] are what philosophers call *universals*, our present problem is part of the well-known and

_____
[1] That is, in the distributive sense.

amply discussed classical problem of the *ontological status of universals*. The three main traditional answers to the general problem of universals, stemming from medieval discussions, are known as *realism, nominalism,* and *conceptualism*. We shall not deal here with these lines of thought in their traditional version, but only with their modern counterparts, known as *Platonism, neo-nominalism* and *neo-conceptualism*".[1] These modern standpoints are characterized by them as follows:

"A *Platonist* is convinced that corresponding to each well-defined (monadic) condition there exists, in general, a set, or class, which comprises all and only those entities that fulfil this condition and which is an entity on its own right of an ontological status similar to that of its members. (To avoid antinomies) ... he reluctantly (...) declares himself ready to accept certain restrictions in the use of the axiom-schema of comprehension, temporarily working with a type theory or a set theory of the Zermelian brand, but hoping that sooner or later someone will be able to show that much less radical interventions will do the trick. Of course, some Platonists may convince themselves, or become convinced by others, that the objects of the world they live in are *really* stratified into types and orders and, as a consequence, accept type theory not as an *ad hoc* device but as an expression of hard fact.

"A *neo-nominalist* declares himself unable to understand what other people mean when they are talking about sets, unless he is able to interpret their talk as a *façon de parler*. The only language he professes to understand is a calculus of individuals, constructed as an applied first-order functional calculus. With regard to many locutions used in scientific or ordinary discourse, which *prima facie* involve sets, he has little trouble in translating them adequately into his restricted language. This is the case, for instance, for such a common statement as 'the set of a's is a subset of the set of b's', which he renders as 'for all x, if x is a, x is b'. With regard to other locutions and devices he has greater trouble. (...) The difficulties in rephrasing *all* of classical mathematics in nominalistic terms seem, and probably are, insurmountable. (...) The only serious ways

(1) A. A. Fraenkel and Y. Bar-Hillel, *Foundations of Set Theory*, p. 333.

out of their predicament are either to go on using all the useful parts of mathematics in the hope — admittedly not too well founded — that one day someone will produce an adequate rephrasing in nominalistic terms, or else to declare that all higher mathematics is an uninterpreted calculus (...). How exactly an uninterpreted (and directly uninterpretable) calculus is able to perform its useful function of mediating between interpreted empirical statements is an issue that is still far from being definitely clarified (...)."

Neo-conceptualists "claim to understand what sets are, though the metaphor they prefer is that of *constructing* (or *inventing*) rather than of *singling out* (or *discovering*), which is the one cherished by the Platonists, these metaphors replacing the older antithesis of *existence in the mind* versus *existence in some outside (real or ideal) world*. They are ready to admit that any well-determined and perspicuous condition indeed determines a corresponding set — since they are able in this case to 'construct' this set out of a stock of sets whose existence is either intuitively obvious or which have been constructed previously — but they are not ready to accept axioms or theorems that would force them to admit the existence of sets which are not constructively characterizable." The principal representatives of this approach are the intuitionists, followers of the Dutch mathematician Luitzer Egbertus Jan Brouwer.

In the above review of the modern approaches to the issue of the existence of sets there is no mention about a modern counterpart of moderate realism. That standpoint is taken when predicates of one argument are interpreted as singular names of properties, i.e.,when the formula "$A(x)$" is read: "the property A is an attribute of the object x". This interpretation of formulae of the form "$A(x)$" is often adopted, and frequently the formula "x is an element of the set of objects having the property A" is interpreted as equivalent to the formula "x has the property A". The concept of set then reduces to the concept of property (attribute). But in the functional calculus we have to do with the principle of extensionality: $\prod_x [A(x) \equiv B(x)]$ $\rightarrow A = B$. But properties which are attributes of the same objects need not be identical. For instance, the properties of having equal angles and equal sides are attributes of the same triangles, and yet we do not consider them to be identical properties of triangles. Hence in interpreting the functional calculus as a calculus of properties we must realize that some of its laws hold only for properties which satisfy the principle of extensionality.

If we define as Platonic the first of the standpoints described above concerning the foundations of set theory, we do not settle the question whether Platonic ideas may be identified with sets.[1] But as Plato assumed that to the term "man" corresponds the idea of man under which all individual men are included, here, too, it is assumed that to the term "man" corresponds the set of all men, of which individual men are elements. By assuming that sets exist and that they differ from the singular objects which are their elements, we assume, as Plato did, the existence of objects different from singular objects. This opinion is shared by many eminent contemporary logicians. Kurt Gödel comments on this point as follows: "It seems to me that the assumption of such objects is quite as legitimate as the assumption of physical bodies and there is quite as much reason to believe in their existence."[2] It would not be difficult to adduce reasons which make us adopt this view concerning set theory. The formula "$x \in y$" is read "x is an element of the set y", while, as mentioned before, the term "set" cannot be interpreted here in the collective sense. Axioms and theorems of set theory include sentences stating the existence of sets satisfying definite conditions (e.g., the axiom of the null set, the axiom of infinity, the axiom of choice), and these sentences cannot be reformulated as statements about individuals. In particular, sentences containing quantifiers that bind variables standing for sets cannot be reformulated as statements about individuals.

Both intuitionism and nominalism render mathematics and logic much poorer. In the mathematics of the intuitionists various theorems of classical mathematics do not hold. This is particularly marked in set theory. The nominalists must renounce many results obtained in modern logic and mathematics, in which essential use is made of variables of higher orders and of quantifiers binding

---

[1] *Ibid.*, footnote 2 on p. 333.

[2] K. Gödel, *Russell's Mathematical Logic*, in *The Philosophy of Bertrand Russell*, p. 137.

such variables. The Platonists do not encounter such difficulties, but the Platonic approach leads to philosophical difficulties since it assumes the existence of non-concrete objects.[1]

---

[1] A more comprehensive treatment of the issue under consideration and of the principal modern standpoints can be found in:

A. A. Fraenkel and Y. Bar-Hillel, *Foundations of Set Theory*, Chap. IV, Sec. 8; Philosophical Remarks, pp. 332–47;

P. Bernays, *Sur le platonisme dans les mathématiques*, in *L'Enseignement Mathématique*, XXXIV (1935);

W. V. Quine, *Logic and the Reification of Universals. On What There Is*. In *From a Logical Point of View*, 1953;

N. Goodman and W. V. Quine, *Steps Towards a Constructive Nominalism*, in *The Journal of Symbolic Logic*, V, 12(1947);

A. Grzegorczyk, *O pewnych formalnych konsekwencjach reizmu* (On Certain Formal Consequences of Reism), in *Fragmenty Filozoficzne*, II;

E. W. Beth, *Foundations of Mathematics*.

APPENDIX

1. Definitions

Part I, Chapter 2

D1 $\qquad \vdash \sum_{\alpha} {}_1 \phi(\alpha) \equiv \sum_{\alpha} \phi(\alpha) \land \prod_{\phi(\alpha)} \prod_{\phi(\beta)} \alpha = \beta\,;$

Part II, Chapter 1

D1.1 $\quad X \doteq Y \equiv \prod_{x} (x \in X \equiv x \in Y).$

D1.2 $\quad X \subset Y \equiv \prod_{x} (x \in X \to x \in Y).$

D1.3a $\quad x \in \{y\} \equiv x = y.$

    b $\quad x \in \{y_1, y_2, \ldots, y_n\} \equiv x = y_1 \lor x = y_2 \lor \ldots \lor x = y_n.$

D1.4 $\quad x \in X \cup Y \equiv x \in X \lor x \in Y.$

D1.5 $\quad x \in X \cap Y \equiv x \in X \land x \in Y.$

D1.6 $\quad x \in X - Y \equiv x \in X \land x \notin Y.$

D1.7 $\quad x \in \overline{X} \equiv x \notin X.$

D1.8 $\quad x \in \emptyset \equiv x \in \overline{1}.$

D1.9 $\quad X \underset{\neq}{\subset} Y \equiv X \subset Y \land X \neq Y.$

D4.1 $\quad x \in \bigcup_{i=1}^{\infty} \varphi(X_i) \equiv \sum_{i \in \mathbb{N}} [x \in \varphi(X_i)].$

D4.2 $\quad x \in \bigcap_{i=1}^{\infty} \varphi(X_i) \equiv \prod_{i \in \mathbb{N}} [x \in \varphi(X_i)].$

D4.3 $\quad x \in \bigcup_{X \in \mathbf{X}} \varphi(X) \equiv \sum_{X \in \mathbf{X}} [x \in \varphi(X)].$

D4.4 $\quad x \in \bigcap_{X \in \mathbf{X}} \varphi(X) \equiv \prod_{X \in \mathbf{X}} [x \in \varphi(X)].$

D5.1 $\quad \langle x, y \rangle \in X \times Y \equiv x \in X \land y \in Y.$

D5.2 $\quad x \in \underset{a}{E} \Phi(a) \equiv \Phi(x).$

D5.3 $\quad \alpha \in \underset{a,b}{E} \Phi(a, b) \equiv \sum_{z} \sum_{t} [\alpha = \langle z, t \rangle \land \Phi(z, t)].$

D6.1 $\quad R \in \mathrm{funct} \equiv \prod_{x} \prod_{y} \prod_{z} (x R y \land x R z \to y = z).$

[327]

D6.2a $x \in D_l(R) \equiv \sum_y xRy.$

   b $y \in D_r(R) \equiv \sum_x xRy.$

D6.3 $R \in \text{funct} \wedge x \in D_l(R) \rightarrow [y = R(x) \equiv xRy].$

D6.4 $y \in R(X) \equiv \sum_{x \in X} (xRy).$

D6.5 $xR^{-1}y \equiv yRx.$

D6.6 $R \in 1-1 \equiv R, R^{-1} \in \text{funct}.$

D6.7 $xR_X y \equiv xRy \wedge x \in X.$

D6.8 $xR\,;Sy \equiv \sum_z (xRz \wedge zSy).$

D6.9   An equivalence relation is a relation which is reflexive, symmetric and transitive, where:

   a relation R is reflexive if $\prod_x xRx$;

   a relation R is symmetric if $\prod_x \prod_y (xRy \rightarrow yRx)$;

   a relation R is transitive if $\prod_x \prod_y \prod_z (xRy \wedge yRz \rightarrow xRz).$

D6.10 $R \doteq S \equiv \prod_x \prod_y (xRy \equiv xSy).$

D6.11 $xR \cup Sy \equiv xRy \vee xSy.$

D7.1 $X \sim_R Y \equiv R \in 1-1 \wedge X = D_l(R) \wedge Y = D_r(R).$

D7.2 $X \sim Y \equiv \sum_R X \sim_R Y.$

D7.3 $\mathfrak{m} \in \text{Nc} \equiv \sum_X \mathfrak{m} = \overline{\overline{X}}.$

D7.4   A set A is the range of the sequence $\{a_n\}$
$$\equiv \sum_a [a \in A \equiv \sum_{i \in \mathbf{N}} a = a_i].$$

D8.1 $X \cap Y = \emptyset \rightarrow \overline{\overline{X}} + \overline{\overline{Y}} = \overline{\overline{X \cup Y}}.$

D8.2 $\overline{\overline{X}} . \overline{\overline{Y}} = \overline{\overline{X \times Y}}.$

D8.3 $R \in Y^X \equiv R \in \text{funct} \wedge D_l(R) = X \wedge R(X) \subset Y.$

D8.4 $\overline{\overline{X}}^{\overline{\overline{Y}}} = \overline{\overline{(X^Y)}}.$

D9.1 $\overline{\overline{X}} \leqslant \overline{\overline{Y}} \equiv \sum_{Z \subset Y} (X \sim Z).$

D9.2 $\overline{\overline{X}} < \overline{\overline{Y}} \equiv \overline{\overline{X}} \leqslant \overline{\overline{Y}} \wedge \overline{\overline{X}} \neq \overline{\overline{Y}}.$

D10.1 $X \in 2^Y \equiv X \subset Y.$

D10.2 $R \in \text{funct} \rightarrow [x \in Z(R) \equiv x \in D_l(R) \wedge x \notin R(x)]$.

D10.3 The function $\varphi_A(m)$ is the characteristic function of the set

$$A \subset M \equiv \text{for every } m \in M, \quad \varphi_A(m) = \begin{cases} 1, & \text{when } m \in A, \\ 0, & \text{when } m \notin A. \end{cases}$$

D11.1 X is an infinite set in Dedekind's sense $\equiv \sum\limits_{\substack{Y \subset X \\ \neq}} Y \sim X$.

D11.2 Transfinite cardinal numbers are powers of infinite sets in Dedekind's sense.

Part II. Chapter II

D1.1 $C(R) = D_l(R) \cap D_r(R)$.

D1.2 $\langle X, S \rangle \text{ is}_R \langle Y, T \rangle \equiv R \in 1-1 \wedge D_l(R) = X \wedge D_r(R) = Y$
$\wedge \prod\limits_x \prod\limits_y \prod\limits_z \prod\limits_u [xRz \wedge yRu \rightarrow (xSy \equiv zTu)]$.

D1.3 $\langle X, S \rangle \text{ is } \langle Y, T \rangle \equiv \sum\limits_R (\langle X, S \rangle \text{ is}_R \langle Y, T \rangle)$.

D2.1a $R \in \text{con}(A) \equiv \prod\limits_{x \in A} \prod\limits_{y \in A} [x \neq y \rightarrow (xRy \vee yRx)]$.

b $R \in \text{trans}(A) \equiv \prod\limits_{x \in A} \prod\limits_{y \in A} \prod\limits_{z \in A} (xRy \wedge yRz \rightarrow xRz)$.

c $R \in \text{asym}(A) \equiv \prod\limits_{x \in A} \prod\limits_{y \in A} [xRy \rightarrow \neg (yRx)]$.

D2.2 $R \in \text{ord}(A) \equiv A = C(R) \wedge R \in \text{con}(A) \cap \text{trans}(A)$
$\cap \text{asym}(A)$.

D2.3 $\langle A, R \rangle$ is an ordered set $\equiv R \in \text{ord}(A)$.

D2.4 $A^0 \simeq_R B^0 \equiv A \sim_R B \wedge \prod\limits_x \prod\limits_y [x \prec_A y \equiv R(x) \prec_B R(y)]$.

D2.5 $A^0 \simeq B^0 \equiv \sum\limits_R (A^0 \simeq_R B^0)$.

D2.6 $\alpha \in \text{Ot} \equiv \sum\limits_{A^0} \alpha = \overline{A^0}$.

D3.1 $A \cap B = \emptyset \rightarrow \{A^0 \cup B^0 = C^0 \equiv C = A \cup B$
$\wedge \prod\limits_x \prod\limits_y [x \prec_C y \equiv (x \prec_A y \vee x \prec_B y \vee x \in A \wedge y \in B)]\}$.

D3.2 $A^0 \cap B^0 = C^0 \equiv C = A \times B \wedge \prod\limits_a \prod\limits_b \prod\limits_x \prod\limits_y [\langle a, b \rangle \prec_C \langle x, y \rangle$
$\equiv (b \prec_B y \vee b = y \wedge a \prec_A x)]$.

D3.3 $A \cap B = \emptyset \rightarrow \overline{A^0} \dot{+} \overline{B^0} = \overline{A^0 \cup B^0}$.

D3.4 $\overline{A^0} \cdot \overline{B^0} = \overline{A^0 \cap B^0}$.

D4.1 $\langle A^0, B^0 \rangle$ is a cut of the ordered set C^0
$\equiv A \neq \emptyset \wedge B \neq \emptyset \wedge A \cap B = \emptyset \wedge A^0 \cup B^0 = C^0$.

D4.2a x is the first element of the ordered set A^0
$$\equiv x \in A \,\wedge\, \prod_{y \in A} (x \neq y \to x \prec_A y).$$

   b x is the last element of the ordered set A^0
$$\equiv x \in A \,\wedge\, \prod_{y \in A} (x \neq y \to y \prec_A x).$$

D4.3a   The cut $\langle A^0, B^0 \rangle$ is a jump $\equiv$ the set A^0 has a last element and the set B^0 has a first element.

   b   The cut $\langle A^0, B^0 \rangle$ is a gap $\equiv$ the set A^0 has no last element and the set B^0 has no first element.

D4.4a   An ordered set is dense $\equiv$ none of its cuts is a jump.

   b   An ordered set is continuous $\equiv$ none of its cuts is either a jump or a gap.

D4.5 $A^0 \subset B^0 \equiv A \subset B \,\wedge\, \prod_{x \in A} \prod_{y \in A} (x \prec_A y \equiv x \prec_B y).$

D5.1 $\omega = \overline{A^0} \equiv$ set A^0 satisfies conditions a–c of Theorem T5.3.

D5.2 $\eta = \overline{A^0} \equiv$ set A^0 satisfies conditions a'–c' of Theorem T5.4.

D5.3 $\lambda = \overline{A^0} \equiv$ set A^0 satisfies conditions a''–c'' of Theorem T5.5.

D5.4 $\beta = \alpha^* \equiv \sum_X \sum_R (\alpha = (\overline{X}, R) \,\wedge\, \beta = \langle \overline{X}, R^{-1} \rangle).$

D6.1   The ordered set A^0 is a well-ordered set
$$\equiv \sum_{\varnothing \neq B^0 \subset A^0} \sum_b (b \text{ is the first element of the set } B^0).$$

D6.2 $\alpha \in \mathrm{On} \equiv \sum_{A^0} (A^0 \text{ is a well-ordered set} \,\wedge\, \alpha = \overline{A_0})$

D7.1 $X^+ = A(a) \equiv X^+ \subset A^+ \,\wedge\, \prod_x (x \in X \equiv x \prec_A a).$

D7.2 $a_1, a_2 \in A^+ \to [A(a_1) < A(a_2) \equiv a_1 \prec_A a_2].$

D7.3 $R \in \mathrm{if}\,(A^+) \equiv R \in A^A \,\wedge\, \prod_x \prod_y [x \prec_A y \to R(x) \prec_A R(y)].$

D8.1 $\overline{A^+} < \overline{B^+} \equiv \sum_b A^+ \simeq B(b).$

D8.2 $\alpha \in \mathrm{Ln} \equiv \alpha \in \mathrm{On} \,\wedge\, \alpha \neq 0 \,\wedge\, \prod_{\beta \in \mathrm{On}} \alpha \neq \beta + 1.$

D9.1a $\alpha^0 = 1.$

   b $\alpha^{\beta + 1} = \alpha^{\beta} . \alpha.$

D9.1c   If $\gamma \in$ Ln and $\alpha = 0$, then $\alpha^\gamma = 0$. If $\gamma \in$ Ln and $\alpha \neq 0$, then α^γ is the least ordinal number equal to or greater than any number α^β, where β is an ordinal number less than γ.

D9.2   A transfinite sequence of type α is a function whose left domain is the set of all ordinal numbers less than α, where $\alpha \geqslant \omega$.

D10.1 $\bar{\alpha} = \overline{\overline{A^+}} \equiv \alpha = \overline{A^+}$.

D10.2 $\alpha \in Z(\mathfrak{m}) \equiv \bar{\alpha} = \mathfrak{m}$.

D10.3a α is the initial number of the set $Z(\mathfrak{m}) \equiv \mathfrak{m} \geqslant \aleph_0$ and α is the least ordinal number of the set $Z(\mathfrak{m})$.

     b α is an initial number $\equiv \sum\limits_{\mathfrak{m}}$ [α is the initial number of the set $Z(\mathfrak{m})$].

D10.4 $\alpha = \omega_\xi \equiv \alpha$ is an initial number and the set of all initial numbers $< \alpha$, ordered by the relation "less than", is of the type ξ.

D10.5 $\mathfrak{m} = \aleph_\xi \equiv \omega_\xi$ is the initial number of the set $Z(\mathfrak{m})$.

2. Theses

Part I

THE SENTENTIAL CALCULUS

T1 $(p \rightarrow q) \wedge (q \rightarrow r) \rightarrow (p \rightarrow r)$.

T1a $(p \rightarrow q) \rightarrow [(q \rightarrow r) \rightarrow (p \rightarrow r)]$.

T2b $p \wedge q \rightarrow r \equiv p \rightarrow (q \rightarrow r)$.

T3 $p \rightarrow (q \rightarrow r) \equiv q \rightarrow (p \rightarrow r)$.

T4 $p \vee q \rightarrow (\neg q \rightarrow p)$.

T4b $\neg p \vee q \rightarrow (p \rightarrow q)$.

T5b $\neg \neg p \equiv p$.

T6 $p \rightarrow q \equiv \neg q \rightarrow \neg p$.

T7 $p \wedge q \rightarrow r \equiv p \wedge \neg r \rightarrow \neg q$.

T8 $(p \rightarrow q) \wedge \neg q \rightarrow \neg p$.

T9 $p \rightarrow (\neg p \rightarrow q)$.

T10 $q \rightarrow (p \rightarrow q)$.

T11 $p \rightarrow p$.

T11a $p \equiv p$.

T12 $(p \equiv q) \to (q \equiv p)$.

T13 $(p \equiv q) \land (q \equiv r) \to (p \equiv r)$.

T14 $(p \equiv q) \equiv (p \to q) \land (q \to p)$.

T15 $(p \to \neg p) \to \neg p$.

T15a $(\neg p \to p) \to p$.

T15b $(p \to q \land \neg q) \to \neg p$.

T15c $(\neg p \to q \land \neg q) \to p$.

T15d $(p \to q) \land (p \to \neg q) \to \neg p$.

T16b $p \to q \land r \equiv (p \to q) \land (p \to r)$.

T17 $(p \to r) \land (q \to r) \equiv p \lor q \to r$.

T18 $(p \to r) \land (q \to r) \land (p \lor q) \to r$.

T19 $\neg(p \lor q) \equiv \neg p \land \neg q$.

T20 $\neg(p \land q) \equiv \neg p \lor \neg q$.

T21 $\neg(p \to q) \equiv p \land \neg q$.

T22 $\neg(p \land \neg p)$.

T23 $p \lor \neg p$.

T24 $(p \to q) \to (p \land r \to q \land r)$.

T25 $(p \to q) \land (r \to s) \to (p \land r \to q \land s)$.

T26 $(p \to q) \to (p \lor r \to q \lor r)$.

T27 $(p \to q) \land (r \to s) \to (p \lor r \to q \lor s)$.

T28 $(p \to q) \land (r \to s) \land (p \lor r) \to (q \lor s)$.

T29 $\neg p \lor q \equiv (p \to q)$.

T29a $p \lor q \equiv (\neg p \to q)$.

T30 $p \land (q \lor r) \equiv p \land q \lor p \land r$.

T31 $p \lor q \land r \equiv (p \lor q) \land (p \lor r)$.

T32 $(p \to q) \land (r \to s) \land (p \lor r) \land \neg(q \land s) \to (q \to p)$
$$\land (s \to r).$$

T33a $(p \equiv q) \to (\neg p \equiv \neg q)$.

  b $(p \equiv q) \to (p \land r \equiv q \land r)$.

  c $(p \equiv q) \to (r \land p \equiv r \land q)$.

  d $(p \equiv q) \to (p \lor r \equiv q \lor r)$.

  e $(p \equiv q) \to (r \lor p \equiv r \lor q)$.

  f $(p \equiv q) \to (p \to r \equiv q \to r)$.

  g $(p \equiv q) \to (r \to p \equiv r \to q)$.

  h $(p \equiv q) \to [(p \equiv r) \equiv (q \equiv r)]$.

  i $(p \equiv q) \to [(r \equiv p) \equiv (r \equiv q)]$.

T34 $p \land q \equiv \neg(\neg p \lor \neg q)$.

T35 $(p \lor q) \lor r \equiv p \lor (q \lor r).$

T36 $(p \land q) \land r \equiv p \land (q \land r).$

T37 $p \lor q \equiv q \lor p.$

T38 $p \land q \equiv q \land p.$

T39 $(p \lor q) \land (r \lor s) \equiv p \land r \lor p \land s \lor q \land r \lor q \land s.$

T40 $p \land q \lor r \land s \equiv (p \lor r) \land (p \lor s) \land (q \lor r) \land (q \lor s).$

Part II

THE FUNCTIONAL CALCULUS

T1 $\prod_x A(x) \to A(y).$

T2 $A(y) \to \sum_x A(x).$

T3 $\neg \prod_x A(x) \equiv \sum_x \neg A(x).$

T4 $\neg \sum_x A(x) \equiv \prod_x \neg A(x).$

T5 $\prod_x [A(x) \to B(x)] \to \left[\prod_x A(x) \to \prod_x B(x) \right].$

T6 $\prod_x [A(x) \to B(x)] \to \left[\sum_x A(x) \to \sum_x B(x) \right].$

T7 $\prod_x [A(x) \land B(x)] \equiv \prod_x A(x) \land \prod_x B(x).$

T8 $\prod_x A(x) \lor \prod_x B(x) \to \prod_x [A(x) \lor B(x)].$

T9 $\sum_x [A(x) \land B(x)] \to \sum_x A(x) \land \sum_x B(x).$

T10 $\sum_x [A(x) \lor B(x)] \equiv \sum_x A(x) \lor \sum_x B(x).$

T11 $\prod_x [p \to A(x)] \equiv p \to \prod_x A(x).$

T12 $\sum_x [p \to A(x)] \equiv p \to \sum_x A(x).$

T13 $\prod_x [A(x) \to p] \equiv \sum_x A(x) \to p.$

T14 $\sum_x [A(x) \to p] \equiv \prod_x A(x) \to p.$

T15 $\prod_x [p \lor A(x)] \equiv p \lor \prod_x A(x).$

T16 $\sum_x [p \land A(x)] \equiv p \land \sum_x A(x).$

T17 $\displaystyle\prod_x [p \wedge A(x)] \equiv p \wedge \prod_x A(x).$

T18 $\displaystyle\sum_x [p \vee A(x)] \equiv p \vee \sum_x A(x).$

T19 $\displaystyle\prod_x \prod_y xRy \equiv \prod_y \prod_x xRy.$

T20 $\displaystyle\sum_x \sum_y xRy \equiv \sum_y \sum_x xRy.$

T21 $\displaystyle\sum_x \prod_y xRy \to \prod_y \sum_x xRy.$

T22 $\displaystyle\prod_x [A(x) \equiv B(x)] \to \left[\prod_x A(x) \equiv \prod_x B(x)\right].$

T23 $\displaystyle\prod_x [A(x) \equiv B(x)] \to \left[\sum_x A(x) \equiv \sum_x B(x)\right].$

T24 $\displaystyle\prod_x A(x) \equiv \neg \sum_x \neg A(x).$

T25 $\displaystyle\sum_x A(x) \equiv \neg \prod_x \neg A(x).$

T3* $\displaystyle\neg \prod_{A(x)} B(x) \equiv \sum_{A(x)} \neg B(x).$

T4* $\displaystyle\neg \sum_{A(x)} B(x) \equiv \prod_{A(x)} \neg B(x).$

T13* $\displaystyle\prod_{B(x)} [A(x) \to p] \equiv \sum_{B(x)} A(x) \to p.$

T16* $\displaystyle\sum_{B(x)} [p \wedge A(x)] \equiv p \wedge \sum_{B(x)} A(x).$

T19* $\displaystyle\prod_{A(x)} \prod_{B(y)} xRy \equiv \prod_{B(y)} \prod_{A(x)} xRy.$

T20* $\displaystyle\sum_{A(x)} \sum_{B(y)} xRy \equiv \sum_{B(y)} \sum_{A(x)} xRy.$

T21* $\displaystyle\sum_{A(x)} \prod_{B(y)} xRy \to \prod_{B(y)} \sum_{A(x)} xRy.$

IDENTITY

A1 $x = x.$

T1 $x = y \to y = x.$

T2 $x = y \wedge y = z \to x = z.$

T2a $x = z \wedge y = z \to x = y.$

T3 $x = y \wedge A(x) \to A(y).$

T3a $A(x) \wedge \neg A(y) \to x \neq y.$

T4 $\displaystyle A(x) \equiv \sum_{y=x} A(y).$

T5 $\quad A(x) \equiv \prod\limits_{y=x} A(y).$

T6 $\quad \neg \sum\limits_{x} 1 A(x) \equiv \neg \sum\limits_{x} A(x) \vee \sum\limits_{A(x)} \sum\limits_{A(y)} x \neq y.$

TVI.§6 $\quad x = y \equiv \prod\limits_{A} [A(x) \equiv A(y)].$

GENERAL SET THEORY

A1.1 $\quad x \in 1.$

T1.1 $\quad X \subset 1.$

T1.4a $\quad X \subset X.$

b $\quad X \subset Y \wedge Y \subset Z \to X \subset Z.$

T1.5 $\quad x = y \equiv \prod\limits_{X} (x \in X \equiv y \in Y).$

A1.2 $\quad X \doteq Y \to X = Y.$

T1.7 $\quad X \doteq Y \equiv X = Y.$

T1.9 $\quad x \notin \emptyset.$

A5.1 $\quad \langle x, y \rangle = \langle z, t \rangle \equiv x = z \wedge y = t.$

T5.1a $\quad (X \cup Y) \times Z = X \times Z \cup Y \times Z.$

b $\quad X \cap Y = \emptyset \to (X \times Z) \cap (Y \times Z) = \emptyset.$

c $\quad \{1, 2, \ldots, n\} \times X = \{1\} \times X \cup \{2\} \times X \cup \ldots \cup \{n\} \times X.$

d $\quad X \subset Y \to X \times Z \subset Y \times Z.$

T5.2 $\quad \langle x, y \rangle \in \underset{a,b}{E} \Phi(a, b) \equiv \Phi(x, y).$

T6.1a $\quad X \subset Y \to R(X) \subset R(Y).$

L6.1a $\quad D_l(R) = D_r(R^{-1}).$

b $\quad D_r(R) = D_l(R^{-1}).$

L6.2 $\quad R \in 1-1 \to R^{-1} \in 1-1.$

L6.3b $\quad R \in 1-1 \wedge x, y \in D_l(R) \wedge x \neq y \to R(x) \neq R(y).$

L6.4 $\quad R \in 1-1 \to R_X \in 1-1.$

L6.5a $\quad X \subset D_l(R) \to D_l(R_X) = X.$

b $\quad X \subset D_l(R) \to D_r(R_X) = R(X).$

L6.6 $\quad R, S \in 1-1 \to R; S \in 1-1.$

L6.7a $\quad D_r(R) = D_l(S) \to D_l(R; S) = D_l(R).$

b $\quad D_r(R) = D_l(S) \to D_r(R; S) = D_r(S).$

A6.1 $\quad R \doteq S \to R = S.$

T6.4 $\quad R \doteq S \equiv R = S.$

L6.8 $\quad X = D_l(R) \to R_X = R.$

T7.1a $X \sim X.$

   b $X \sim Y \to Y \sim X.$

   c $X \sim Y \wedge Y \sim Z \to X \sim Z.$

L7.2 $X \sim_R Y \to Y = R(X).$

L7.3 $R \in 1-1 \wedge X \subset D_l(R) \to X \sim_{R_X} R(X).$

A7.1 $\overline{\overline{X}} = \overline{\overline{Y}} \equiv X \sim Y.$

T7.2a $\overline{\overline{X}} \in \text{Nc}.$

   b $\prod_X \sum_{\mathfrak{m} \in \text{Nc}} {}_1\, \mathfrak{m} = \overline{\overline{X}}.$

   c $\prod_{\mathfrak{m} \in \text{Nc}} \sum_X \mathfrak{m} = \overline{\overline{X}}.$

L7.4 $\prod_A \prod_B \sum_{A_1} \sum_{B_1} (A_1 \sim A \wedge B_1 \sim B \wedge A_1 \cap B_1 = \emptyset).$

L7.5a $A \sim B \wedge A_1 \sim B_1 \wedge A \cap A_1 = B \cap B_1 = \emptyset \to A \cup A_1 \sim B \cup B_1.$

   b $A \sim B \wedge A_1 \sim B_1 \to A \times A_1 \sim B \times B_1.$

T7.3 $\overline{\overline{A}} = \aleph_0 \equiv \sum_{\{a_n\}} [\prod_i \prod_j (i \neq j \to a_i \neq a_j)$

$\wedge \; A \text{ is the range of the sequence } \{a_n\}].$

T8.3 $n \cdot \overline{\overline{X}} = \underbrace{\overline{\overline{X}} + \overline{\overline{X}} + \ldots + \overline{\overline{X}}}_{n} \quad (n \in \mathbf{N}).$

T8.4 $(\overline{\overline{X}})^n = \underbrace{\overline{\overline{X}} \cdot \overline{\overline{X}} \cdot \ldots \cdot \overline{\overline{X}}}_{n} \quad (n \in \mathbf{N}).$

T8.5 $\overline{\overline{X^{\overline{\overline{Y}} + \overline{\overline{Z}}}}} = \overline{\overline{X^Y}} \cdot \overline{\overline{X^Z}}.$

T8.6 $\overline{\overline{(X^Y)^{\overline{\overline{Z}}}}} = \overline{\overline{X^{\overline{\overline{Y}} \cdot \overline{\overline{Z}}}}}.$

L9.1 $X \sim X_1 \wedge Y \sim Y_1 \wedge \sum_{Z \subset Y} (X \sim Z) \to \sum_{U \subset Y_1} (X_1 \sim U).$

Concl.9.1 $X \subset Y \to \overline{\overline{X}} \leqslant \overline{\overline{Y}}.$

T9.1 $\mathfrak{m} \leqslant \mathfrak{n} \equiv \mathfrak{m} = \mathfrak{n} \vee \mathfrak{m} < \mathfrak{n}.$

T9.2b $\mathfrak{m} \leqslant \mathfrak{n} \wedge \mathfrak{n} \leqslant \mathfrak{m} \to \mathfrak{m} = \mathfrak{n}.$

T9.4 $\mathfrak{m} \leqslant \mathfrak{n} \vee \mathfrak{n} \leqslant \mathfrak{m}.$

T9.5a $\aleph_0 \leqslant \mathbf{c}.$

   b $n \in \mathbf{N} \to n < \aleph_0.$

L9.3a $\mathfrak{m} < \mathfrak{n} \to \mathfrak{m} \cdot \mathfrak{k} \leqslant \mathfrak{n} \cdot \mathfrak{k}.$

   b $\mathfrak{m} < \mathfrak{n} \to \mathfrak{m}^{\mathfrak{k}} \leqslant \mathfrak{n}^{\mathfrak{k}}.$

T10.1 $\overline{\overline{M}} < \overline{\overline{(2^M)}}.$

T10.2 $\quad 2^{\overline{\overline{M}}} = (\overline{\overline{2^M}})$.

T10.3 $\quad \overline{\overline{M}} < 2^{\overline{\overline{M}}}$.

L11.1 $\quad P \subset X \wedge \overline{\overline{P}} = \aleph_0 \to \sum\limits_{\substack{Y \subset X \\ \neq}} X \sim Y$.

L11.2 $\quad \sum\limits_{\substack{Y \subset X \\ \neq}} (X \sim Y) \to \sum\limits_{P \subset X} (\overline{\overline{P}} = \aleph_0)$.

T11.1 $\quad$ The set X is infinite in Dedekind's sense $\equiv \sum\limits_{P \subset X} \overline{\overline{P}} = \aleph_0$.

L11.3 $\quad \overline{\overline{X}} = \overline{\overline{Y}} = \aleph_0 \to \overline{\overline{X \cup Y}} = \aleph_0$.

L11.4 $\quad \overline{\overline{X}} = \aleph_0 \wedge \overline{\overline{Y}} = n \to \overline{\overline{X \cup Y}} = \aleph_0 \quad (n \in \mathbf{N})$.

T11.2 $\quad$ If X is an infinite set in Dedekind's sense and $\overline{\overline{Y}} = n$ or $\overline{\overline{Y}} = \aleph_0$, then $\overline{\overline{X \cup Y}} = \overline{\overline{X}}$.

T11.3 $\quad$ If $\overline{\overline{Y}} = n$ or $\overline{\overline{Y}} = \aleph_0$ and $X - Y$ is an infinite set in Dedekind's sense, then $\overline{\overline{X - Y}} = \overline{\overline{X}}$.

T12.1 $\quad \aleph_0 \neq \mathbf{c}$.

T12.3 $\quad a < b \to \overline{\overline{E(a < x < b)}} = \mathbf{c}$.
$\qquad\qquad\;\; x$

T12.4a $\quad \aleph_0 + n = \aleph_0 \quad (n \in \mathbf{N})$.

$\quad$ b $\quad \aleph_0 + \aleph_0 = \aleph_0$.

$\quad$ c $\quad \mathbf{c} + n = \mathbf{c} \quad (n \in \mathbf{N})$.

$\quad$ d $\quad \mathbf{c} + \aleph_0 = \mathbf{c}$.

$\quad$ e $\quad \mathbf{c} + \mathbf{c} = \mathbf{c}$.

T12.5 $\quad \prod\limits_{i \in \mathbf{N}} \overline{\overline{Z_i}} = \aleph_0 \wedge \prod\limits_{i,j} (i \neq j \to Z_i \cap Z_j = \varnothing) \to \overline{\overline{\bigcup\limits_{i=1}^{\infty} Z_i}} = \aleph_0$.

T12.6 $\quad \aleph_0 \cdot \aleph_0 = \aleph_0$.

T12.7 $\quad$ If n is a natural number $\geqslant 2$ then $n^{\aleph_0} = \mathbf{c}$.

T12.9a $\quad n \cdot \aleph_0 = \aleph_0 \quad (n \in \mathbf{N})$.

$\quad$ b $\quad \aleph_0^n = \aleph_0 \quad (n \in \mathbf{N})$.

$\quad$ c $\quad n \cdot \mathbf{c} = \mathbf{c} \quad (n \in \mathbf{N})$.

$\quad$ d $\quad \mathbf{c} \cdot \mathbf{c} = \mathbf{c}$.

$\quad$ e $\quad \mathbf{c}^n = \mathbf{c} \quad (n \in \mathbf{N})$.

f $\mathbf{c}^{\aleph_0} = \mathbf{c}$.

g $\aleph_0 \cdot \mathbf{c} = \mathbf{c}$.

h $\aleph_0^{\aleph_0} = \mathbf{c}$.

A14.1 $Z \neq \varnothing \wedge \varnothing \notin Z \wedge \prod\limits_{P \in Z} \prod\limits_{Q \in Z} (P \neq Q \to P \cap Q = \varnothing)$

$$\to \sum\limits_{Y} \prod\limits_{M \in Z} \sum\limits_{x} {}_1 (x \in M \cap Y).$$

T14.1 $X \neq \varnothing \wedge \prod\limits_{n \in \mathbf{N}} (\overline{\overline{X}} \neq n) \to \sum\limits_{Y \subset X} (\overline{\overline{Y}} = \aleph_0)$.

ORDERED SETS

T2.1a $A^0 \simeq A^0$.

    b $A^0 \simeq B^0 \to B^0 \simeq A^0$.

    c $A^0 \simeq B^0 \wedge B^0 \simeq C^0 \to A^0 \simeq C^0$.

A2.1 $\overline{A^0} = \overline{B^0} \equiv A^0 \simeq B^0$.

T2.2a $\overline{A^0} \in \mathrm{Ot}$.

    b $\prod\limits_{A^0} \sum\limits_{\alpha \in \mathrm{Ot}} {}_1 \; \alpha = \overline{A^0}$.

    c $\prod\limits_{\alpha \in \mathrm{Ot}} \sum\limits_{A^0} \alpha = \overline{A^0}$.

L2.1 $\prod\limits_{A^0} \prod\limits_{B^0} \sum\limits_{A_1^0} \sum\limits_{B_1^0} (A_1^0 \simeq A^0 \wedge B_1^0 \simeq B^0 \wedge A_1 \cap B_1 = \varnothing)$.

L3.1a $A^0 \simeq B^0 \wedge A_1^0 \simeq B_1^0 \wedge A \cap A_1 = B \cap B_1 = \varnothing$

$$\to A^0 \cup A_1^0 \simeq B^0 \cup B_1^0.$$

    b $A^0 \simeq B^0 \wedge A_1^0 \simeq B_1^0 \to A^0 \cap A_1^0 \simeq B^0 \cap B_1^0$.

T4.1 $A^0 \subset B^0 \wedge B^0 \subset C^0 \to A^0 \subset C^0$.

T4.2a If $A^0 \simeq B^0$ and the set A^0 has a first element, then the set B^0 also has a first element.

    b   If $A^0 \simeq B^0$ and the set A^0 has a last element, then the set B^0 also has a last element.

    c   If $A^0 \simeq B^0$ and the set A^0 is dense, then the set B^0 is also dense.

    d   If $A^0 \simeq B^0$ and the set A^0 is continuous, then the set B^0 is also continuous.

T4.3   An ordered set A^0 is dense $\equiv \prod\limits_{x} \prod\limits_{y} \sum\limits_{z} (x \prec_A y \to x \prec_A z \prec_A y)$.

T5.1   If the finite ordered sets A^0 and B^0 are equinumerous, then these sets are similar.

T5.3   If the ordered sets A^0 and B^0 satisfy the conditions:
       a. An ordered set X^0 has a first element;
       b. An ordered set X^0 has no last element;
       c. Every cut of a set X^0 is a jump,
       then these sets are similar.

Concl.5.2 $\omega = \langle \overline{\mathbf{N}}, < \rangle.$

T5.4   If the ordered sets A^0 and B^0 satisfy the conditions:
       a'. An ordered set X^0 has neither a first nor a last element;
       b'. An ordered set X^0 is enumerable;
       c'. An ordered set X^0 is dense,
       then these sets are similar.

Concl.5.3. $\eta = \langle \overline{\mathbf{M}}, < \rangle.$

T5.5   If the ordered sets A^0 and B^0 satisfy the conditions:
       a''. An ordered set X^0 has neither a first nor a last element;
       b''. An ordered set X^0 is continuous;
$$c''. \sum_W \{ W^0 \subset X^0 \wedge \overline{\overline{W}} = \aleph_0 \wedge \prod_x \prod_y [x \prec_x y \to \sum_{w \in W} (x \prec_x w \prec_x y)] \},$$
       then these sets are similar.

Concl.5.4 $\lambda = \langle \overline{\mathbf{R}}, < \rangle.$

T5.6   If an ordered set A^0 is dense, then the set $\mathfrak{A}^0$ of all its proper cuts, ordered by a relation R defined by the formula:
$$\langle A_1^0, B_1^0 \rangle \, R \, \langle A_2^0, B_2^0 \rangle \equiv A_1 \underset{\neq}{\subset} A_2,$$
       is continuous.

T6.1   Every subset of a well-ordered set is well-ordered.

T6.2   The sum and the product of well-ordered sets are well-ordered sets.

T7.1   A relation which associates every element a of a well-ordered set A^+ with a segment $A(a)$ establishes similarity between the set A^+ and the set of all its segments ordered by the relation "less than".

L7.1 $A^+ \simeq B^+ \to \prod_{a \in A} \sum_{b \in B} [A(a) \simeq B(b)].$

L7.3 $\neg [A^+ \simeq A(a)].$

L7.4 $a_1, a_2 \in A^+ \wedge A(a_1) \simeq A(a_2) \to a_1 = a_2.$

L7.5 $\prod_{a \in A^+} \sum_{b \in B^+} [A(a) \simeq B(b)] \wedge \prod_{b \in B^+} \sum_{a \in A^+} [A(a) \simeq B(b)] \to A^+ \simeq B^+.$

L7.6   If b_0 is the first element of a set B^+, which satisfies the condition $\prod_{a \in A^+} \neg[A(a) \simeq B(b)]$, and if $a_1 \in A$, $b_1 \in B$, and $A(a_1) \simeq B(b_1)$, then $b_1 \in B(b_0)$.

L7.7 $\prod_{a \in A^+} \sum_{b \in B^+} [A(a) \simeq B(b)] \vee \prod_{b \in B^+} \sum_{a \in A^+} [A(a) \simeq B(b)]$.

T7.2 $A^+ \simeq B^+ \vee \sum_{b \in B} [A^+ \simeq B(b)] \vee \sum_{a \in A} [B^+ \simeq A(a)]$.

L7.8 $A^+ \simeq A_1^+ \wedge B^+ \simeq B_1^+ \wedge \sum_{b \in B} [A^+ \simeq B(b)] \to \sum_{b \in B_1} [A_1^+ \simeq B_1(b)]$.

T8.1   The relation "less than" orders the set of ordinal numbers.

T8.2   If the set $\underset{\beta}{E}(\beta < \alpha)$ is ordered by the relation "less than", then $\alpha = \overline{\underset{\beta}{E}(\beta < \alpha)}$.

T8.4   Any set of ordinal numbers, ordered by the relation "less than", is well-ordered.

T8.5 $\alpha \neq 0 \wedge \alpha \notin \mathrm{Ln} \wedge \alpha \notin \mathbf{N} \to \sum_{\beta \in \mathrm{Ln}} \sum_{n \in \mathbf{N}} \alpha = \beta + n$.

T10.1 For every set A there exists a well-ordered set whose range is the set A.

T10.2 $\mathfrak{m} \in \mathrm{Cn} \to Z(\mathfrak{m}) \neq \emptyset$.

T10.3 For every cardinal number $\mathfrak{m} \geqslant \aleph_0$ there exists an ordinal number ξ such that $\mathfrak{m} = \aleph_\xi$.

T10.4 $\aleph_\xi = \overline{\underset{\alpha}{E}(\alpha < \omega_\xi)}$.

T10.6 $\aleph_1 \leqslant \mathfrak{c}$.

INDEX

OTHER VOLUMES PUBLISHED IN THIS SERIES

Other Titles in the Series in Pure and Applied Mathematics